Haematology and Blood Transfusion

24

Hämatologie und Bluttransfusion

Edited by
H. Heimpel, Ulm · D. Huhn, München
G. Ruhenstroth-Bauer, München

Aplastic Anemia

Pathophysiology and Approaches
to Therapy

Edited by
H. Heimpel, E. C. Gordon-Smith,
W. Heit and B. Kubanek

With 81 Figures and 71 Tables

Springer-Verlag
Berlin Heidelberg New York 1979

Hermann Heimpel, M. D., Department Innere Medizin, Abteilung Innere Medizin und Hämatologie, Steinhövelstraße 9, 7900 Ulm, Federal Republic of Germany

Edward C. Gordon-Smith, M. D., Department of Haematology, Royal Postgraduate Medical School, Hammersmith Hospital, Ducane Road, London W12, Great Britain

Wolfgang Heit, M. D., Department Innere Medizin, Abteilung Innere Medizin und Hämatologie, Steinhövelstraße 9, 7900 Ulm, Federal Republic of Germany

Bernhard Kubanek, M. D., Department Innere Medizin, Abteilung Innere Medizin und Hämatologie, Steinhövelstraße 9, 7900 Ulm, Federal Republic of Germany

---

Supplement to

BLUT – Journal for Blood Research

Organ of the "Deutsche Gesellschaft für Hämatologie und Onkologie der Deutschen Gesellschaft für Bluttransfusion und Immunohämatologie" and of the "Österreichischen Gesellschaft für Hämatologie und Onkologie"
Edited by H. Heimpel, Ulm   D. Huhn, München   G. Ruhenstroth-Bauer, München

---

ISBN 3-540-09772-4 Springer Verlag Berlin Heidelberg New York
ISBN 0-387-09772-4 Springer-Verlag New York Heidelberg Berlin

Library of Congress Cataloging in Publication Data. Main entry under title: International Symposium on Aplastic Anemia–Pathophysiology and Approaches to Therapy, Reisensburg, Ger., 1978. International Symposium on Aplastic Anemia. (Hämatologie und Bluttransfusion; Bd. 24) Organized by the Department of Internal Medicine and Hematology of the University of Ulm and others. Bibliography: p. Includes index.
1. Aplastic anemia–Congresses. I. Heimpel, Hermann. II. Ulm. Universität. Abteilung Innere Medizin und Hämatologie. III. Series.
RC641.7.A6I583 1978 616.1'52 79-23868
ISBN 0-387-09772-4

© Springer-Verlag Berlin Heidelberg 1979
Printed in Germany

Typesetting, printing and binding: Kösel GmbH & Co., Kempten
2329/3321/543210

Contents

Participants of the Meeting

Barrett, A. J., The Royal Marsden Hospital, London & Surrey, Downs Road, Sutton, Surrey SM 2 5 PT, Great Britain

Böttiger, L., Department of Medicine, Karolinska Hospital, 10401 Stockholm, Sweden

Byrne, P. V., Department Innere Medizin, Abteilung Innere Medizin und Hämatologie, Parkstraße 11, 7900 Ulm (Donau), Federal Republic of Germany

Camitta, B. M., Milwaukee Children's Hospital, 1700 West Wisconsin Avenue, P. O. Box 1977, Milwaukee, Wisconsin 53201, USA

Devergie, A., Hôpital Saint Louis, Service d' Hématologie, Centre Hayem, 2, Place du Docteur Fournier, 75010 Paris, France

Dexter, T. M., Christie Hospital & Holt Radium Institute, Manchester Area Health Authority, South District, Manchester M 20 9 BX, Great Britain

Dresch, C., Service Central de Médécine Nucléaire, Hôpital Saint Louis, 2, Place du Docteur Fournier, 75475 Paris Cedex 10, France

Flad, H. D., Zentrum für Klinische Grundlagenforschung, Abteilung Mikrobiologie, Oberer Eselsberg, 7900 Ulm (Donau), Federal Republic of Germany

Fliedner, T. M., Zentrum für Klinische Grundlagenforschung, Abteilung Klinische Physiologie, Oberer Eselsberg, 7900 Ulm (Donau), Federal Republic of Germany

Gale, R. P., UCLA Bone Marrow Transplant Unit, Department of Microbiology and Immunology, UCLA, Los Angeles, Ca. 90024, USA

Goldmann, S. F., Tissue Typing Labor, Abteilung Klinische Physiologie, Steinhövelstraße 9, 7900 Ulm (Donau), Federal Republic of Germany

Gordon, M. Y., Institute of Cancer Research, Royal Cancer Hospital, Royal Marsden Hospital, Clifton Avenue, Sutton, Surrey, SM 2 5 PX, Great Britain

Gordon, S., Haematology Department, Macquarie Street, Sydney Hospital, Sydney, N. S. W. 2000, Australia

Gordon-Smith, E. C., Department of Haematology, Royal Postgraduate Medical School, Hammersmith Hospital, Ducane Road, London W 12, Great Britain

Gross, R., I. Medizinische Universitätsklinik, Joseph-Stelzmann-Str. 9, 5000 Köln 41, Federal Republic of Germany

Haak, H. L., Isolation Pavilion, J. A. Cohen Instituut, Academisch Ziekenhuis, Leiden, The Netherlands

Hansi, W., Department Innere Medizin, Abteilung Innere Medizin und Hämatologie, Steinhövelstraße 9, 7900 Ulm (Donau), Federal Republic of Germany

Harriss, E. B., Zentrum für Klinische Grundlagenforschung, Abteilung Klinische Physiologie, Parkstraße 11, 7900 Ulm (Donau), Federal Republic of Germany

Heimpel, H., Department Innere Medizin, Abteilung Innere Medizin und Hämatologie, Steinhövelstraße 9, 7900 Ulm (Donau), Federal Republic of Germany

Heit, W., Department Innere Medizin, Abteilung Innere Medizin und Hämatologie, Steinhövelstraße 9, 7900 Ulm (Donau) Federal Republic of Germany

Hellriegel, K.-P., I. Medizinische Universitätsklinik, Joseph-Stelzmann-Straße 9, 5000 Köln 41, Federal Republic of Germany

Iscove, N. N., Basel Institute for Immunology, 487 Grenzacher Straße, 4005 Basel-5, Switzerland

Kleihauer, E., Department für Kinderheilkunde, Abteilung Hämatologie, Pritt-
   witzstraße 43, 7900 Ulm (Donau), Federal Republic of Germany
Kohne, E., Department für Kinderheilkunde, Abteilung Hämatologie, Prittwitz-
   straße 43, 7900 Ulm (Donau), Federal Republic of Germany
Kubanek, B., Department für Innere Medizin, Abteilung Innere Medizin und
   Hämatologie, Steinhövelstraße 9, 7900 Ulm (Donau), Federal Republic of
   Germany
Lohrmann, H.-P., Department für Innere Medizin, Abteilung Innere Medizin und
   Hämatologie, Steinhövelstraße 9, 7900 Ulm (Donau), Federal Republic of
   Germany
Lucarelli, G., Ospedale S. Salvatore di Pesaro, Divisione Ematologica, Piazza
   Albani, 28, 61100 Pesaro, Italy
Mangalik, Aarop, University of Colorado, Medical Center, 4200 East Ninth
   Avenue, Denver, Col. 80262, USA
Mangalik, Asha, University of Colorado, Medical Center, 4200 East Ninth
   Avenue, Denver, Col. 80262, USA
Moore, M. A. S., Memorial Sloan Kettering Cancer Center, 1275 York Avenue,
   New York 10021, USA
Nečas, E., Katedra Patologické Fyziologie, Fakulty Všeobecného Lěkařstvi,
   Univerzity Karlovy, Unemocnice č. 5, 12853 Praha-2, CSSR
Netzel, B., Institut für Hämatologie, GSF, Abteilung Immunologie, Landwehr-
   straße 61, 8000 München 2, Federal Republic of Germany
Neuwirt, J., Katedra Patologické Fyziologie, Fakulty Všeobecného Lěkařstvi,
   Univerzity Karlovy, Unemocnice č. 5, 12853 Praha-2, CSSR
Niethammer, D., Hämatologische Abteilung der Universitäts-Kinderklinik, 7400
   Tübingen, Federal Republic of Germany
Nissen, C., Medizinische Universitätsklinik, Kantonsspital, 4002 Basel, Switzer-
   land
Porzsolt, F., Department für Innere Medizin, Abteilung Innere Medizin und
   Hämatologie, Steinhövelstraße 9, 7900 Ulm (Donau), Federal Republic of
   Germany
Sabbe, L. J. M., Department of Immunohematology, University Medical Centre,
   2300 RC Leiden, The Netherlands
Schmücker, H., Department für Innere Medizin, Abteilung Innere Medizin und
   Hämatologie, Steinhövelstraße 9, 7900 Ulm (Donau), Federal Republic of
   Germany
Schofield, R., Paterson Laboratories, Christie Hospital and Holt Radium Institute,
   Manchester, M 20, 9 BX, Great Britain
Shadduck, R. K., University of Pittsburgh, School of Medicine, Montefiore
   Hospital, 3459 Fifth Avenue, Pittsburgh, Penns. 15213, USA
Singer, J., The Fred Hutchinson Cancer, Research Center, 1124 Columbia Street,
   Seattle, Wash. 98104, USA
TeVelde, J., Pathologisch Laboratorium, Universitair Medisch Centrum, Postbus
   9603, 2300 RC Leiden, The Netherlands
Thierfelder, S., Institut für Hämatologie, GSF, Abteilung Immunologie, Land-
   wehrstraße 61, 8000 München 2, Federal Republic of Germany
Thomas, D., Division of Oncology, The Fred Hutchinson Cancer Research
   Center, 1124 Columbia Street, Seattle, Wash. 98104, USA
Thorsby, E., Tissue Typing Laboratory, The National Hospital, Oslo 1, Norway
Tigges, F.-J., Medizinische Universitätsklink, Schnarrenberg, Otfried-MüllerStra-
   ße, 7400 Tübingen, Federal Republic of Germany
Weber, W., Medizinische Universitätsklinik, Kantonsspital, 4002 Basel, Switzer-
   land
Wickramasinghe, S. N., Department of Haematology and MRC Experimental
   Haematology Unit, St. Mary's Hospital Medical School, University of London,
   London W 12, Great Britain

Foreword

Research on aplastic anaemia has until recently been limited to clinical description, morphology and epidemiology. New methods to culture haemopoietic cells, and advances in our knowledge of proliferation and differentiation in the haemopoietic cell system opened a new area of scientific interest for this "prototype" of haemopoietic failure. In addition, bone marrow transplantation became not only a clinical method of treatment, but also a source of data useful for the discussion of pathophysiological models of aplastic anaemia.

This situation prompted us to arrange an international conference on aplastic anaemia, with particular emphasis on its pathophysiology and the rationals of the current therapeutic approaches. This conference was held at Schloss Reisensburg from July 20–22, 1978 with the participation of both experimental and clinical scientists active in this field or in related areas of research. The proceedings of the symposion reflect the present knowledge as well as the many new questions which arose from the discussions.

The editors are gratefully indebted to the participants of this meeting, to Gerlinde Trögele and all the co-workers of the University of Ulm engaged in preparation of this symposium and of this volume, and last not least to all sponsors who provided the financial basis for this scientific event.

Ulm, September 1979

H. Heimpel
E. C. Gordon-Smith
W. Heit
B. Kubanek

Acknowledgement

The International Symposium on Aplastic Anemia – Pathophysiology and Approaches to Therapy – was organized by the Departments of Internal Medicine and Haematology, of Clinical Physiology, University of Ulm, the Department of Haematology, Royal Postgraduate Medical School, London and the Sonderforschungsbereich 112 (Cell System Physiology), University of Ulm.

The symposium was realized by grants from the University of Ulm, Royal Postgraduate Medical School, London, the Deutsche Forschungsgemeinschaft and the Kultusministerium Baden-Württemberg.

The Organizing Committee gratefully acknowledges the support of the following pharmaceutical companies:
Bayer AG, Leverkusen, Biotest-Serum-Institut GmbH, Frankfurt, Byk-Essex GmbH, München, Ciba-Geigy GmbH, Wehr (Baden), Deutsche Wellcome GmbH, Burgwedel, Eli Lilly GmbH, Bad Homburg, Dr. E. Fresenius KG, Bad Homburg, Hoechst AG, Frankfurt, Johann A. Wülfing, Neuss, Medizinisch-pharmazeutische Studiengesellschaft e. V., Frankfurt, Pfizer GmbH, Karlsruhe, Sandoz AG, Nürnberg.

1 Introduction

Introduction

H. Heimpel

We are happy to welcome all of you on behalf of the Organizing Committee and the University of Ulm at this symposion on aplastic anemia. When we asked some of you one year ago whether there would be interest in such a meeting, we met a positive response and felt that the progress of the last five years made it useful to discuss the problems of this disease between clinicians and experimental investigators. The fact that, at a time overloaded with meetings and congresses, so many prominent and busy scientists followed our invitation demonstrates the current interest in this model of hemopoietic failure. I hope that, together with the special atmosphere of Schloss Reisensburg, this common interest will produce many stimulating and fruitful discussions. We appreciate that you made the effort to come here from many countries, and at this time we would like to thank the public and private sponsors who enabled us to arrange this meeting.

To introduce the subject, I want to give you a short summary on the evolution of research on aplastic anemia. Aplastic anemia was first described in 1888 by one of the real pioneers of biomedicine, Paul Ehrlich in Berlin, who outlined in his paper the clinical essentials and basic pathophysiology of the disease still valid today. 10 years later, Santesson from USA, in a paper written in German, detected that the aplasia may be due to an external factor, benzene. After the detection of Vitamin B12 and iron as therapeutics in anemia, it was shown in the thirties that these drugs were not effective in aplastic anemia, suggesting a basic difference from other types of hemopoietic failure.

An important stimulus for further research was the observation, that the antibiotic chloramphenicol could induce aplastic anemias, but did so only in a minority of exposed individuals. This pointed to the importance of intrinsic, possibly genetic predisposition. Despite the efforts of many investigators to elucidate a possible genetic background, the question of individual hypersensitivity has remained open to the present day.

When autoallergy was detected as a mechanism of disease, autoantibodies to blood cells were described in different forms of pancytopenia and thought to be responsible for the hemopoietic aplasia; however, this theory did not stand later critisism, and today we are inclined to believe that the reactions observed were due to HLA-allo-antibodies not known at this time.

A third external factor, the hepatitis virus, was described in 1955 and this relationship was later supported by many observations. A large review of such cases was made by Dr. Camitta who is going to speak about the viral induction of aplasia at this meeting. As in the case of chloramphenicol or phenylbutazon, we have been able to identify the external noxious agent, but we still do not know the link between virus infection and stem cell injury, and we do not know why only very few people with virus infection develop aplasia. Is the primary target actually the hemopoietic stem cell? Or is the primary target stromal tissue? Does viral infection trigger immunological reactions? The problem of "soil or seed", that means the primary role of hemopoietic stem cells or their supporting microenvironment had been raised earlier, for example by Krospe and Crosby 1971. This question has not yet been answered and there may be aplastic anemias of both types.

For a long time, no successful treatment was available for aplastic anemia except blood transfusions and supportive care. In 1959, androgens were found by Shahidi and Diamond to be useful in childhood aplastic anemia, mainly of the congenital types. After many years of experience with androgens, their therapeutic value and their mechanism of action are still a matter of discussion. Further therapeutic alternatives were opened by the work of Dr. D. Thomas on allogeneic bone marrow transplantation, inducing a new area of research activities on aplastic syndromes.

In recent years, many new data were obtained by the application of clonal stem cell assays after Pike and Robinson in 1971 adapted the techniques to human material. The last and most fashionable data come from the immunologists and suggests suppressor T-cells to be of pathogenetic importance. These data were recently published from a group in the Sloan Kettering Institute and soon followed by similar ones from other centers. I personally believe that it has not been unequivocally shown that suppression of hemopoietic stem cells exists in aplastic anemia and the relevance of these findings should be discussed very intensively at this meeting.

In spite of the progress made by all these observations, one has to admit that the pathogenesis of aplastic anemia is not very well understood. The goal of our symposium is to discuss recent advances and to establish working hypotheses from results obtained in the different fields of research. We should particularly try to find out to what extent the many and sometimes confusing experimental data are in agreement or disagreement with clinical observations.

The lectures of the first session are thought to summarize the most important clinical data. One of the main problems to be discussed may be whether aplastic anemia is one disease, or rather the non-specific end results of different underlying pathological mechanisms. This is an important question for the clinician as well as for the experimental researcher who works with material from such patients.

The following session on stem cells leads to the central topics of aplastic anemia research. We hope that in addition to the conventional clonal assays, culture techniques evaluating proliferation capacity of stem cells will be included, and that application of such techniques to human clinical research will be stimulated. One important aspect is the use of animal models. We invited Dr. A. Morley to contribute at this point, and we are sorry that he is not able to participate. In the general discussion on stem cells we hope to identify what we have really learned and what we are going to learn from experimental models and stem cell techniques for the human situation.

The session on therapeutic approaches continues the analysis of the natural course of the disease. For this symposion, we would like to use therapy as a clinical experiment done on human aplastic anemia. The main emphasis will therefore be placed on how and why certain treatments are effective or ineffective, rather than on clinical details. At the end of this symposium we hope to come to a better understanding of the results of various treatment modalities. This is particularly true for bone marrow transplantation, which in spite of its actual merits may still be regarded as experimental therapy.

The question of immun-mechanisms seems to be at present the most controversial matter in the pathogenesis of aplastic anemia. We particularly appreciate the presence of Dr. Thorsby from Oslo, who agreed to introduce this session, even though he only received our invitation shortly before the symposion. We expect the chairmen to use their experience in various in vitro and in vivo systems to look rather critically on the many data, which have arisen from experiments with serum and lymphatic cells from aplastic anemia patients. The end point of such experiments is usually a clonal stem cell assay, and in this respect the results of the previous stem cell session may be useful when the validity of the immunological data is discussed.

In the final general discussion, at the end of the second day, we would like to bring together the data and conclusions of the previous sessions, in an attempt to recognize which questions were to be answered and which problems may be subjected to further investigation in the near future. I am afraid that after two days we will not have solved the riddles of aplastic anemia, however, I hope that after this symposion clinicians, stem cell researchers, transplanters and experimental immunologists will have better mutual understanding of their work, and that we will know better what kind of future research projects may help to understand the disease aplastic anemia in the forthcoming years. I believe that better knowledge of the pathophysiological mechanisms in the only way for relevant progress in the prevention and therapy of this disease.

2 Clinical Observations in Aplastic Anemia

2.1 Clinical Features of Aplastic Anaemia

E. C. Gordon-Smith

The purpose of this Symposium is to collect and discuss data from divers groups of workers, data which may throw some light on the pathogenetic mechanisms of aplastic anaemia. It is desirable to understand the pathogenesis of disease, perhaps even more than to know the cause, for from such knowledge may come more rational and perhaps more successful treatment for that disease. We should try in discussion and in formal papers, to maintain an open, critical attitude to the problem and not interpret data in the light of preconceived hypotheses. We should keep in mind that there may be more than one pathway to aplastic anaemia and that the bone marrow, like most other organs of the body has a limited number of ways in which it can react to many assaults.

It is my task in this opening talk to set the scene for later papers without, I hope, pre-empting the contributions to follow. I will try, as far as possible to concentrate on features of the natural history of the disease which I believe need to be explained by speculations on the pathogenesis.

Definition: The definition of aplastic anaemia is not easy. It is characterised by a peripheral pancytopenia associated with a depletion of all haempoietic cell lines in the marrow, with replacement of the cellular marrow by fat cells, an absence of abnormal cells in the marrow and no increase in marrow reticulin. It is tempting to include a time element in the definition of aplastic anaemia. Cytotoxic drugs and irradiation will produce a similar picture to the one described but in the main clinicians would distinguish between the marrow failure produced in that way and the disease we are discussing here. However, to exclude all marrow failure resulting from cytotoxic drugs would be to ignore the development of prolonged aplastic anaemia which sometimes follows busulphan therapy. Busulphan-induced aplastic anaemia in rats has been proposed as an animal model for human aplasia [2] so it is important to consider similarities and differences between "conventional" and Busulphan-induced aplasia in man.

Aetiology: It is accepted that aplastic anaemia may follow exposure to certain agents. Sulphonamides, heavy metal compounds, organic arsenicals and amidopyrine were all implicated in the 1930's. Quinacrine hydrochloride was identified as a possible cause by epidemiological studies on American soldiers in the 1939–45 war [1]. It was predicted in 1949, on the basis of its structure, that chloramphenicol would cause haematological disorders [4] and the first case report of aplasia was published the following year [3].

Nitro-groups attached to or substituted in benzene rings are present in sulphonamides, amidopyrine, phenylbutazone, chloramphenicol, quinacrine. In nitrofurantoin the nitro group is associated with a furan ring. On the other hand, gold salts, amongst the well recognised causes of aplasia, are not aromatic salts and it should be emphasised that the Benzene (benzol) solvent which causes marrow failure is not the chemical benzene.

I do not want to make too much of the chemistry of aplasia producing substances but neither should their structure be completely ignored in constructing hypotheses. It is perhaps of passing interest that all the compounds mentioned produce granulocytopenia, almost certainly on an immune basis, much more often than they produce aplastic anaemia. The circumstances under which they produce the more permanent aplasia is what concerns us here.

The incidence of aplastic anaemia amongst people exposed to these drugs or to the other well documented cause, infectious hepatitis, is, of course, very small. Taking the most gloomy estimate of chloramphenicol-induced aplasia the incidence cannot be more than 1 in 10,000 [5]. We will, I hope, hear much more about epidemiology later but I would throw out for thought now that such rare personal idiosyncrasy may, in other instances, be based on an unusual immunological reaction, as with immune haemolytic anaemias, or with genetic abnormalities of enzyme systems as, for example, with the deficiency of pseudocholinesterase activity in otherwise normal people homozygous for an abnormal variant which will produce scoline apnoea.

Latent period of aplastic anaemia: One of the difficulties in assessing whether a drug or chemical is involved in the aetiology of aplasia is the delay in onset of symptoms following exposure to that agent. Sometimes there is a clear cut cause, exposure to which may be timed with reasonable accuracy in patients who later present with aplastic anaemia. Figure 1 shows the latent period in 12 patients with aplastic anaemia, 6 of whom had aplasia following infectious hepatitis and 6 following exposure to a single drug or chemical agent. The median latent period is six and a half weeks for the group as a whole.

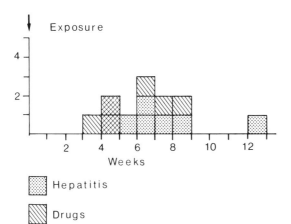

Fig. 1. Time interval between exposure to supposed aetiological agent and onset of pancytopenia in twelve patients who had serial blood counts between the two events

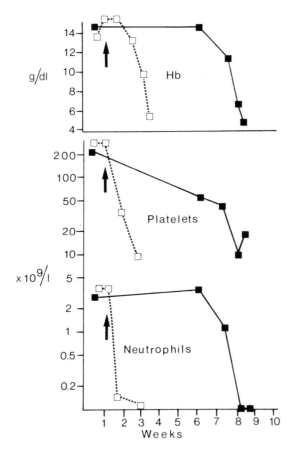

Fig. 2. Onset of pancytopenia in a patient who developed aplasia following hepatitis (■–■) compared with a haematologically normal marrow destroyed by cytotoxic drugs (□---□)

Occasionally one is fortunate enough to have serial blood counts during the production of aplasia and Figure 2 shows a comparison between a patient who developed aplastic anaemia following infectious hepatitis with a patient who was rendered aplastic following cytotoxic drug therapy for bone marrow transplantation – a patient with sex-linked agammaglobulinaemia who had a haematopoetically normal marrow before exposure to the drugs. This comparison emphasises two features of the development of aplasia, first the latent period and secondly the rate of development of pancytopenia once the count begins to fall, the latter being similar in each type of disorder.

There must be many speculations on the reason for the latent period. When cytotoxic drugs are given, all cells in division, that is the committed cell pool, are affected and there is a rapid depletion of the pool. In acquired aplastic anaemia the committed cells present may be spared but damage to the stem cells leads to production of defective cells, though proliferation continues until the pool is exhausted. Evidence to support this idea, that the main problem is one of stem cell inhibition or destruction rather than total loss of proliferative capacity, may be derived from further consideration of the haematological changes which occur in aplasia.

Proliferative Changes in Aplastic Anaemia

1. The bone marrow: We will hear later detailed descriptions of the bone marrow in aplastic anaemia but I would like to mention now three features of the aplastic marrow which I find of interest. First, in the early stage of aplasia, particularly, though not exclusively, after infectious hepatitis there is a striking increase in macrophage activity. Phagocytosis of all cell lines is marked (Fig. 3) and may contribute to the rapid development of pancytopenia. As the disease progresses this macrophage activity becomes less obvious and, as we all know, the typical appearance of the aplastic marrow is of a fatty marrow with a suggestion of increased lymphocyte infiltration. The second point about the aplastic marrow is the patchy development of aplasia. Even in severe aplastic anaemia foci of apparently normal marrow may continue to be present, both at microscopic and macroscopic level. One of the mysteries of aplastic anaemia is why these "normal" foci of marrow do not repopulate the fatty marrow, when it is known that syngeneic marrow, given intravenously, more often than not leads to recovery. Thirdly, in recovery, islands of a particular cell type, erythroblasts or granulocytic cells, develop, not mixed clones.

2. Red cell proliferation: Abnormalities of red cell proliferation are well recognised in aplastic anaemia. Morphological abnormalities have been described and may be associated with characteristics of damage or alteration to the red cell membrane as indicated by an increased sensitivity to cold antibody lysis, by macrocytosis and by increased expression of I and i antigens on the red cell

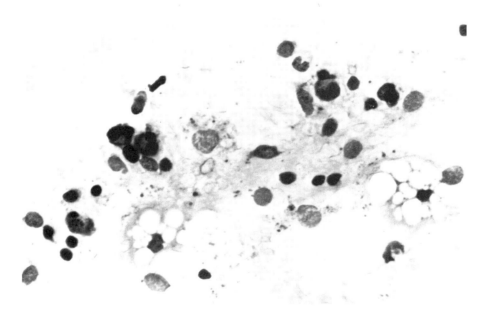

Fig. 3. Phagocytosis in the marrow during development of aplastic anaemia after infectious hepatitis

surface. These changes suggest that proliferative abnormalities are present in aplasia as well as quantitative changes.

3. *Development of abnormal clones:* The association of PNH with aplastic anaemia, particularly the development of PNH clones during the course of the disease, has led to the idea that there may be favoured clones of cells which are able to survive the pathogenic mechanisms of aplasia but, of course, as with the development of acute leukaemia in aplastic anaemia this may be an expression of increased mutagenic activity in the disordered marrow. The increased proportion of HbF containing cells, seen in some patients, may possibly be an expression of mutagenic activity but may also represent a failure of the normal control mechanisms, that is, gene derepression.

4. *Recovery in aplastic anaemia:* Spontaneous remission may occur in aplastic anaemia, even in those patients whose disease is considered severe by any of the accepted criteria. The tantalising problem with aplasia is whether all patients have the ability to recover in time or whether in some damage to the marrow is permanent. Recovery may occur within months of the onset or may be delayed for several years; there seems to be no way of predicting recovery. Of special interest has been the recent observation that autologous recovery of marrow function may occur in patients who have been prepared for marrow transplantation, some of whom have an apparently successful, though temporary graft. This observation, togehter with the reports of successful treatment of aplastic anaemia with antilymphocyte globulin, raises the possibilities of immune processes at work in the genesis of aplasia but I must not presume upon other territories at this stage.

I will close by summarising the main clinical features of aplastic anaemia as I see them and ask the rest of you here to explain them!

Table 1. Clinical features in the natural history of aplastic anaemia

Delayed onset
Proliferative as well as quantitative marrow changes
Development of abnormal clones of cells
Some spontaneous remissions
Evidence of persistent marrow abnormality after apparent recovery

References

1. Custer, R. P.: Aplastic anemia in soldiers treated with atabrine (Quinacrine). Am. J. Med. Sci. *210*, 211 (1946)
2. Morley, A. and Blake, J.: An animal model of chronic aplastic marrow failure. I. Late marrow failure after busulphan. Blood *44*, 49 (1974)
3. Rich, M. L., Ritterhof, R. J. and Hoffman, R. J.: A fatal case of aplastic anaemia following chloramphenicol therapy. Ann. Intern. Med. *33*, 1459 (1950)
4. Smadel, J. E.: Chloramphenicol (Chloromycetin) in the treatment of infectious diseases. Am. J. Med. *7*, 671 (1949)
5. Yunis, A. A.: Chloramphenicol-induced bone marrow suppression. Seminars in Hematology, *10*, 225 (1973)

2.2 Histology of Bone Marrow Failure, a Follow-up Study in Aplastic Anaemia

J. te Velde, H. L. Haak

Introduction

Aplastic anaemia (AA) constitutes only a small minority of the pancytopenic syndromes. It is defined as a peripheral pancytopenia due to bone marrow failure, which is not to be caused by a series of more or less well circumscribed conditions, such as malignancies, deficiencies, metabolic disorders, chemotherapy or radiotherapy. Such a "definition by exclusion" obviously leads to a heterogenous remainder of pancytopenic syndromes, called AA. Its heterogeneity will greatly depend on the accuracy, with which the various disorders to be excluded are sought for.

In the Leiden University Hospital, retrospective and prospective studies on AA have been initiated. The retrospective study concerned 40 adult patients admitted between 1964 and 1976. Results have been published before [1, 2]. During this retrospective study, the urgent need for an extensive protocol for patient investigation was realized. Thus, since 1972 all newly admitted patients and those still alive have been studied with a series of tests for morphological, biochemical and cytogenetical abnormalities with additional studies on the amount and properties of the stem cells and the inflammatory infiltrate in the bone marrow. Results on these studies have been reported on, as for instance by Haak et al. [3].

In the definition of AA, the peripheral cytopenia must be due to an absence of haematopoiesis in the bone marrow. Thus, we have considered it of primary importance to document this marrow failure by means of sufficient marrow histology. Since 1972, histological bone marrow investigations were performed on large bone marrow biopsy specimens, embedded in methyl-methacrylate [7]. This technique had been chosen for the excellent preservation of the cytological detail of all remnant bone marrow elements. Our findings on the prognostic value of the amount and distribution of the inflammatory infiltrate have been published elswhere [8, 10, 3].

Studies on repeated follow-up biopsy specimens in corroboration with the other repeated tests and clinical studies confirmed the central role of marrow histology for the diagnosis of AA. This report will therefor mainly concern the histological diagnosis of AA and the differential diagnosis with other pancytopenic syndromes. It is now based on long term follow-up studies in about 60 patients from the Leiden University Hospital, diagnosed as suffering from aquired aplastic anaemia, and on additional experience with slides sent in for consultancy or studied in several haematological centres elswhere in Europe.

Materials and Methods

Severe aplastic anaemia in adults is studied according to an extensive protocol. Peripheral blood cell values should be below 6.0 mM/L Hb with reticulocytopenia, and below 1.0 and 50/L for granulocytes and thrombocytes respectively. At physical examination or isotope studies hepato-splenomegaly or adenopathy is absent. Biochemical tests reveal no gross abnormalities, except for disorders in the liver function or during episodes of sepsis. Deficiencies, metabolic or storage diseases should be absent. Any suspicion of an occult malignancy warrants further investigations. Bone marrow histology should be performed on sufficiently large and adequately prepared biopsy specimens.

Bone marrow specimens are obtained with the trephine according to Burkhardt or the Jamshidi needle [10]. The specimens are fixed, dehydrated and embedded in methyl-methacrylate. Two micron thick sections are routinely stained with Giemsa after Gallamine etching, PAS, Gomori's reticulin stain, a combined stain for iron and DNA/RNA (Turnbull-Trevan), and a stain for osteoid and mineralized bone (Goldner). The histotechnical procedure has been described extensively elswhere [7].

Evaluation of the biopsy specimens was only performed, when the sections contained at least 30 mm^2 of bone marrow and spongious bone.

The patients studied have been collected from the retrospective and he prospective studies in the Leiden University Hospital, with special attention to the 161 biopsy specimens from 35 patients with aquired AA, diagnosed since 1974 when the protocol for patient investigation was routinely applied. Additional studies are being performed on 50 biopsy specimens from 31 children under the age of 15 years.

From our files, we also collected about 30 patients, who had been referred to our hospital as suffering from AA, but in whom the diagnosis had been changed on several grounds, partly on the basis of incompatible bone marrow histology. On the other hand, we collected sections of patients without AA, but with completely fatty marrow or with histological pictures, which might resemble AA.

In AA, patients were only evaluated, when they had been followed for at least 6 months, when still alive.

Results

Histology

Haematopoiesis

In our definition of AA, bone marrow aplasia or severe hypoplasia is an absolute requirement. Complete or almost complete absence of haematopoiesis is found in all our patients with severe AA at the time of diagnosis or at the time of referral to our center. During remissions, the marrow may become normal or even hypercellular, but unless repeated marrow examinations have demonstrated a previous absence of haematopoiesis, we do not accept an haematopoiesis of more than one third of normal to be compatible with the diagnosis AA. Extra consideration should be given to the age of the patient and the site of the biopsy. In the elderly, it is completely normal to find fatty marrow in the zone of $1/2-1$ cm width just beneath the cortical bone, but it is very unusual to find inflammatory cells in these normal atrophic marrow areas. The structure of the bone trabecula should be described, as very coarse trabecula may indicate, that the biopsy specimen has been taken paralel to the bone surface, just beneath the cortical surface.

At diagnosis, the megakaryocytic and the granulocytic lineages are most affected. A few, morphologically normal megakaryocytes may still be found, but they are completely absent in most of our patients. An increase or normal amount of megakaryocytes generally indicates peripheral platelet consumption or the presence of a myeloproliferative disorder. In the granulocytic lineage, when it is still present, one finds some scattered areas with polymorphonucleated cells and a few younger elements at the endosteal surface. If young myelo-monocytic cells are recognized in the areas away from the endosteal surface of the trabecula, one should be very cautious not to be dealing with an early or smouldering myelo- or monocytic leukaemia. Generally, erythropoiesis makes up the bulk of the remnant haematopoiesis. In some of the adults but more often in children, erythropoiesis is seen as loosely scattered young and more differentiated elements intermingled with inflammatory infiltrate. But in a few patients, large and compact fields of erythropoietic cells show a remarkable uniformity in the stage of differentiation. Clumps of proerythroblasts and fields of more differentiated elements are usually separated; ripe normoblasts are often seen as fields of scattered elements, as if part of the cells have already left the area for the periferal blood stream. Especially in these cases, the normoblasts may show severe nuclear abnormalities with muclear bridging, lobulation, and chromatin clumping, as depicted in our previous paper [8]. At diagnosis, the erythropoiesis is PAS-negative. PAS-positive granules in normoblasts were present in a patient, diagnosed as AA, but already at the time of diagnosis the buffy coat smear revealed sporadic erythroblasts, heralding the onset of erythroleukaemia and abolishing the clinical diagnosis of AA. It should be noted, that the bone marrow histology was completely compatible with AA, except for the PAS-positivity of the erythropoiesis.

Bone Marrow Stroma

As AA has been suggested to be a defect of the marrow stroma, we have paid special attention to the vascularization and stromal cells. In our first communication [8] we had been impressed by the presence of sinusoidal wall disruption in some of the 15 patients described. Since that time, we have observed specimens in equally severe AA with very little vascular defects. Reviewing a larger series of patients now, conspicuous oedema and sinusoidal wall leakage seems to be present in only a small minority of the patients. Condensation of reticulin fibres in these oedematous areas is at most very focal; more than sporadic increase of reticulin fibres or any collagen fibrosis is absent in our AA patients. Sinusoidal wall necrosis with oedema, fibrosis and a strong inflammatory infiltrate is generally seen in pancytopenic patients with collagen diseases or auto-immune disorders, like in rheumatoid arthritis or SLE. It may be found in patients with high titres of circulating immune complexes. It is questionable whether these patients should be regarded as suffering from AA, as they usually have additional peripheral blood cell destruction leading to pancytopenia with rather high amounts of residual haematopoiesis in the bone marrow. In the two patients with PNH, we have studied, the marrow showed a similar amount of vascular destruction with even proliferation of capillaries, a feature, which is completely absent in AA.

Reticulin or collagen fibrosis with and without conspicuous oedema is regarded by us with great suspicion, as we have observed it especially in patients with slowly developing myeloproliferative disorders. One patient serves as a good example: she was seen in extreme pancytopenia with complete disappearance of marrow haematopoiesis, diffuse reticulin fibrosis and moderate oedema. The spleen was barely palpable. She responded dramatically to prednison therapy, and after 1½ years her peripheral blood picture was completely normal. Only the spleen remained just palpable. A follow-up biopsy at that time revealed the classical picture of beginning osteomyelosclerosis. In the series, described by Heimpel et al. [4], three patients showed marrow fibrosis; two died and one survived 6 years, developing evident splenomegaly. In our series, we consider an increase of fibres as incompatible with the diagnosis AA.

Inflammatory Infiltrate

A large biopsy specimen completely filled with fatty marrow also contains at least foci with an abnormal agglomeration of inflammatory cells. Only in rare instances in very long-standing moderate or even severe AA, the specimen fails to show areas with a concentration of lymphoids, plasma cells, mast cells and macrophages within the fatty marrow or in the remnant haematopoietic areas. At the time of diagnosis in severe AA, all adult patients show an absolute increase of these inflammatory cells, compared to normal. In children, we have not been able to quantify these inflammatory cells, as young children normally have high numbers of lymphoids in aspiration smears, and histological controls of normal children are not available in our files.

The composition of the inflammatory infiltrate varies from patient to patient, but in a single patient it is constant, unless influenced by immune suppressive treatment. In some of the patients, one single cell type dominates, usually consisting of lymphocytes. Generally, the infiltrate consist of a mixture of lymphocytes with mast cells and plasma cells with macrophages.

In our previous communication [8], the amount and the distribution of the inflammatory infiltrate seemed to be correlated with survival during conventional, i.e. non-immune suppressive treatment. We had divided our patients in three grades: in grade I only a slight increase of inflammatory cells was found; in the grades II and III a conspicuous increase of inflammatory cells was seen, at least focally. Grade I biopsy specimens are uniformly associated with survival for at least 6 months, mostly for several years, but it is generally found in patients in clinical remissions of long duration. We have observed it only once in a patient with severe AA and once in moderate AA, both patients surviving at least one year. Grade II and III are both found in severe AA. In both grades, a marked increase in inflammatory cells are found, but the difference between grade II and IIII lies in the distribution of the infiltrate over de bone marrow. In grade III it is found in every microscopical field, whereas in grade II one must find at least one marrow area without an icreased amount of inflammatory cells. This difference implies, that the marrow biopsy specimen must be of sufficiently large dimensions before it is diagnosed as being grade III. Grade II may be diagnosed on a smaller specimen by finding already one or more areas without increased infiltrate.

Prognostic Value of the Grading According to the Inflammatory Infiltrate

In our previous communication [8], we reported on our efforts to find a correlation between survival and any of the histological features, scored semiquantitatively. We found no correlation, except for a significant correlation of survival during conventional treatment and the histological grades I and III. Grade II, although mentioned, was only represented by one patient in that report. Since then, more patients grade II could be followed: the survival appeared to be similar to the patients grade I, the greater majority of the patients survived at least 6 months (Table 1) [10].

Table 1. Aplastic anaemia, survival during conventional treatment (in brackets total number of biopsies)

	Grade I	Grade II	Grade III	Total
Alive more than 6 months after biopsy	7 (12)	5 (13)	–	12
Died due to aplasia within 6 months after biopsy	–	–	7 (8)	7

The number of patients, treated with supportive care alone, is still small, as bone marrow transplantation and various immune suppressive treatments are trought to improve survival. ATG infusions have been given to the majority of our patients, diagnosed in the last years. In this group of patients, a similar striking difference between patients grade II and grade III has been observed. The only responders to ATG infusions were found within grade III; no patient grade II has showed any clinical improvement, compared to the patients grade II under conventional treatment [10, 11]. In a review study of 36 patients with sufficiently long follow up and treated with supportive care and/or immune suppression by ATG or high dose prednison, the distinction between patients grade II and grade III attains an extra dimension. In Table 2, the results of a follow-up study covering 1 to 6 years for the patients still alive are given.

Table 2. Aplastic anaemia, follow-up study (in brackets after treatment with immunosuppression)

	Grading of the initial biopsy specimens			
	I/II	II	III	Total
Deminishing infiltrate + return of haematopoiesis	–	–	6 (6)	6
No change in follow-up biopsy specimens	4	4 (4)	1 (1)	9
Appearance of atypia in follow-up specimens	2	3 (2)	–	5
Development of atypia, followed by leukaemia	–	3 (2)	–	3
Death due to aplasia within 6 months	–	1 (1)	12 (3)	13
Total	6	11	19	36

Follow-up Study

As described above, no patient diagnosed as AA with compatible histological bone marrow features has shown any signs of beginning or smouldering leukaemia at the time of diagnosis. Yet, in 3 patients bone marrow histology has changed as seen in repeated biopsy specimens over a period, varying from ½ to 4 years. The bone marrow cellularity slowly increased, mainly by an increase of atypical megakaryocytes and erythropoiesis. Megakaryocytes tended to lie in clusters with large irregular forms, intermingled with small megakaryocytes with one round nucleus. In these areas an increase in reticulin fibres appeared, especially around the smaller megakaryocytes. In two instances, the ripe erythroblasts began to show PAS-positive intracytoplasmatic granules. Over a period of ½ to 1 year, blasts appeared in the periferal blood, and frank leukaemia ensued. This observation led us to review all specimens for similar atypia of the haematopoiesis in AA patients without evidence for leukaemia, and found another 5 patients. The clinical data on these patients will be delt with in another report (Haak et al., in preparation). For this report, we want to stress, that this atypia is only found in the patients, who have been graded as grade II. Two patients with longstanding AA and partial remissions were seen alternating with biopsy specimens graded I and II over periods ranging from 4 to 6 years. In both, atypical erythropoiesis and megakaryocytes were observed in follow-up biopsy specimens. Analyzing the data given in Table 2, it seems probable, that AA is made up of at least two different groups of patients. Grade III patients die within 6 months due to the aplasia (12/19) or survive and show clinical as well as histological improvement, but only under immune suppressive treatment (6/19). Patients with severe AA at diagnosis or with longstanding severe AA with and without partial remissions grade II and I generally survive for long periods with and without immune suppression (16/17). The immune suppression in these patients does not alter the amount of infiltrate in their bone marrow specimens. But in this group, 8/17 developes atypia of the haematopoiesis and 3 of these 8 have developed frank leukaemia.

Discussion

Diagnosis

The diagnosis of AA is reached by excluding several pancytopenic syndromes on clinical grounds, such as following previous chemotherapy or when an association is found with storage diseases or malignancies. Within the definition, the word aplasia means, that the cytopenia must be due to absence of haematopoiesis, which therefore must be documented sufficiently. If the diagnosis AA is made without an adequate bone marrow biopsy, then one must be aware, that a variable number of patients with pancytopenia of different origin may be included. Hellriegel et al. [5] give some indication about the selective value of bone marrow histology. In their report, an unknown number of patients with peripheral pancytopenia were screened for the clinical criteria of AA, and patients with a variety of disorders, known to lead to secondary pancytopenia were excluded. Ninety-two patients with clinical AA were found. But from these 92, about one third had to be excluded further on the basis of the bone marrow histology, as only two-thirds of the patients actually showed marrow aplasia or hypoplasia.

If a bone marrow biopsy is performed, then criteria for AA are still painfully lacking. Some authors even accept hypercellular bone marrow as compatible with the diagnosis AA [as e.g. 6]. No control is performed on the quality of the biopsy specimen, ignoring the fact, that in all normal elderly a small superficial needle biopsy specimen will show fatty, physiologically atrophic marrow. In this paper, we have tried to describe the histological spectrum of the bone marrow in pancytopenic patients with deminished haematopoiesis. Our experiences were made in the fortunate setting of close cooperation between clinicians and pathologists, and we also had the possibility to follow the patients for a long time with repeated marrow biopsies. Our findings in patients with marrow fibrosis or severe vascular desarrangement clearly suggest, that these pictures are associated with separate pancytopenic syndromes. Well prepared marrow sections may additionally help to diagnose early or smouldering leukaemia, at a time when the patient still seems to be suffering from AA.

Follow-up Study

In our series of adult AA patients, meeting all rigid criteria according to the protocol of investigation, severe AA seemed to be composed of at least two different subgroups. These could be divided on the basis of the amount and the distribution of the inflammatory infiltrate in the marrow. In the group of patients grade II, the patients generally survived whatever treatment given, but a great proportion of them developed a regenerating haematopoiesis, which showed defenite atypia or even developed into frank leukaemia. If AA is caused by a primary stem cell defect, then these patients supply a great deal of support for this theory. The patients grade III, however, seemed to be suffering from an overwhelming inflammatory process, which in most patients under conventional treatment leads to death within 6 months, or which could be reversed by immune suppressive treatment. Whether this immune reaction was triggered by a primary stem cell defect, altered membrane properties by e.g. viral infections, or by an abnormal immune regulation, is still unknown. In the infiltrate, however, immune competent cells have been demonstrated to inhibit the outgrowth of CFU-c in in vitro experiments [3]. Further studies are in progress to elucidate the role of the different lymphocyte subpopulation and the plasma cells, mast cells and the macrophages.

In severe aplastic anaemia in adults, no clear clinical difference existed between grade II and III patients at diagnosis, when seen very early in the disease. As grade II patients tend to live longer, while grade III patients die in a very short time, the number of grade II and grade III patients in the various treating centres will vary according to the time lapse between diagnosis and referral. In the Leiden University Hospital this time lapse is significantly shorter than in comparable European centers (data derived from the cooperative group for European Bone

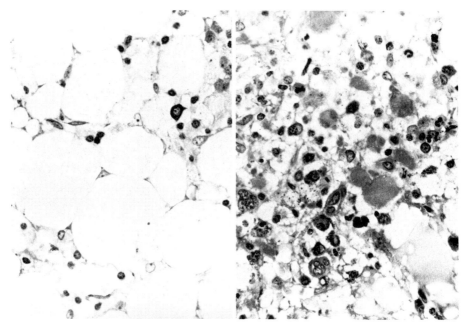

Fig. 1. Stroma: Moderate oedema and vessel wall leakage; no fibrosis *(left)*. Compare to pancytopenia in collagen diseases with severe stromal reactions and necrosis of haematopoietic cells *(right)*. Methyl-methacrylate, Gallamine-Giemsa

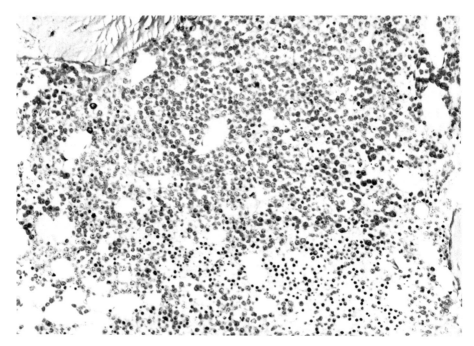

Fig. 2. Haematopoiesis: A rather large focus of haematopoiesis in AA, composed of large fields of erythroid cells in the same stage of maturation. Methyl-methacrylate, Gallamine-Giemsa

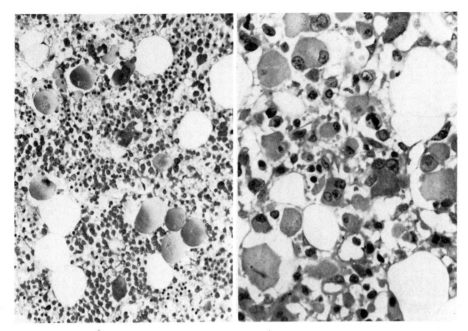

Fig. 3. Development of leukaemia: same patient as in Figure 2 after long standing AA grade II. *Left:* admixture of clustering and slightly atypical megakaryocytes to the irregular erythropoiesis. *Right:* follow-up biopsy specimen 6 months later: increase of very atypical megakaryocytes in the bone marrow, preceeding the appearance of blasts in the peripheral blood, followed by frank leukaemia. Methyl + methacrylate, Gallamine-Giemsa

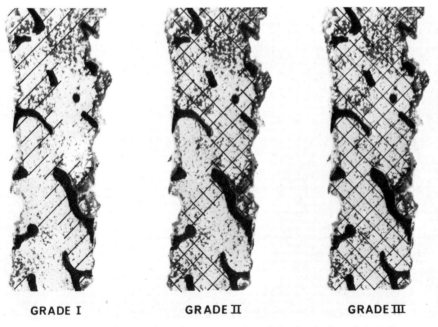

GRADE I   GRADE II   GRADE III

Fig. 4. Inflammatory infiltrate: schematic representation of the distribution of the inflammatory infiltrate in AA. Only slightly increased and focally absent infiltrate in grade I., strong infiltrate in grades II and III, but in grade II areas without increased inflammatory infiltrate

Marrow Transplantation). In reviewing slides from other University Hospitals, a striking preponderance of grade II biopsy specimens were observed (personal observations). This difference may explain the seeming controversy between various authors with regard to the question of the aetiology of AA: stem cell defect or "auto-immune disease". Our data suggest, that AA is a heterogenous group of syndromes, even when we try to apply a most rigid set of criteria. Obviously, the dilemma can only be solved by further experimental research together with long term clinical and pathological studies on well defined and adequately documented patient populations.

References

1. Haak, H. L., Hartgrink-Groeneveld, C. A., Eernisse, J. G., Speck, B. and van Rood, J. J.: Acquired aplastic anaemia in adults. I: A retrospective analysis of 40 cases. Acta Haemat. 58, 257–277 (1977)
2. Haak, H. L., Hartgrink-Groeneveld, C. A., Guiot, H. F. L., Speck, B., Eernisse, J. G. and van Rood, J. J.: Acquired aplastic anaemia in adults. II: Conventional treatment; a retrospective study in 40 patients. Acta Haemat. 58, 339–352 (1977)
3. Haak, H. L., Goselink, H. M., Veenhof, W., Pellinkhof-Stadelman, S., Kleiverda, J. K. and te Velde, J.: Acquired aplastic anaemia in adults, IV: Histological and CFU studies in transplanted and non-transplanted patients. Scand. J. Haemat. 19, 159–171 (1977)
4. Heimpel, H., Rehbock, C. and van Eimeren, W.: Verlauf und Prognose der Panmyelopathie und der isolierten aplastischen Anaemie. Eine retrospektive Studie an 70 Patienten. Blut 30, 235–254 (1975)
5. Hellriegel, K.-P., Züger, M. and Gross, R.: Prognosis in acquired aplastic anaemia. An approach in the selection of patients for allogeneic bone marrow transplantation. Blut 34, 11–18 (1977)
6. Suda, T., Omine, M., Tsuchiya, J. and Maekawa, T.: Prognostic aspects of aplastic anaemia in pregnancy. Experience on six cases and review of the literature. Blut 36, 285–298 (1978)
7. te Velde, J., Burkhardt, R., Kleiverda, K., Leenheers-Binnendijk, L. and Sommerfeld, W.: Methyl-methacrylate as an embedding medium in histopathology. Histopathology 1, 319–330 (1977)
8. te Velde, J. and Haak, H. L.: Aplastic anaemia in adults. Histological investigation of methacrylate embedded biopsy specimens; correlation with survival after conventional treatment in 15 adult patients. British J. Haemat. 35, 61–69 (1977)
9. te Velde, J. and Haak, H. L.: Aplastic anaemia in adults. Further evidence for the significance of the inflammatory infiltrate in methacrylate embedded bone marrow biopsy specimens. Bibl. Haemat. in press (1978)
10. te Velde, J., den Ottolander, G. J., Haak, H. L., van 't Veer, M., Hartgrink-Groeneveld, C. A., Spaander, P. J., de Koning, H., Bijvoet, O. L. M. and Schicht, I. M.: Iliac bone marrow trephine biopsy; some remarks on the technique. Neth. J. Med. 21, 221–227 (1978)
11. te Velde, J., Haak, H. L., Zwaan, F. E. and Spaander, P. J.: Severe aplastic anaemia in adults; histological observations in 13 patients under ATG-treatment. Proceedings of the Second European Symposium on Bone Marrow Transplantation, in press (1979)

Discussion

Camitta: Dr. te Velde, you called the cells you saw in aplastic marrows (lymphocytes, plasma cells, and mast cells) inflammatory infiltrates, suggesting an active process. Is it an active process or a passive one? Have these bone marrows been examined by electron microscopy to see if there is damage to endothelial cells? It might be germane to aplastic anemia because if you irradiate animals all the secondary changes are supposed to be due to damage to the microvasculature which then causes the eventual changes.

te Velde: One of the reasons you can call it an inflammatory infiltrate is that it is linked with destruction of sinusoids and endothelial cells though not to a large extent. The electron microscope shows destruction of the normal lining of endothelials. What the cells are doing there and whether they form an inflammatory process which is actively destroying things I don't know. As a pathologist I do not look at the cells in that way but whenever some immune reaction is triggered and the cells come in there, there will be a lot of innocent bystanders. It is also a general pathological description to say this is inflammation but we do not know if it is specifically directed against endothelial cells or whatever.

Lohrmann: When we interpret your data we always think that lymphocyte which accumulate in the marrow are in some way involved in the production of aplastic anaemia. I could imagine that it is the other way round, that due to disturbed regulatory factors lymphocytes passively accumulate in the marrow.

te Velde: I agree with you, at least in some patients. That is why we make the distinction between grade 2 and 3. In grade 3 there is a diffuse infiltrate which disappears if you kill off the lymphocytes with immunosuppression whilst in other patients if you give the same immune suppression, nothing happens. But I dont't think it is right to say it is just lymphocytes because it is a combination of lymphocytes, plasma cells and mast cells. I always have some trouble with the concept that all these cells ciculate and that in aplastic anaemia they have circulated in higher numbers into the marrow and are not doing anything there.

Singer: I would like to address a question to Dr. Velde. I believe there is a reasonable human model of immunological aplastic anemia in marrow graft rejection of allogeneic recipients. I was wondering if you have examined the marrow biopsies from some of those patients to see whether it correlated at all with your group III type patients.

te Velde: Yes, they do. We have not seen so much reaction but in patients who have rejected bone marrow, it is quite clear in the grade III marrow biopsies. I might say there is an enormous amount of edema and vascular damage especially in those cases. We see the inflammatory infiltrate all over the biopsies.

2.3 Epidemiology and Aetiology of Aplastic Anemia

L. E. Böttiger

Epidemiology of Aplastic Anemia

Epidemiological findings are of great importance as they may help us to understand aetiology better. My considerations of the epidemiology of aplastic anemia will be divided into two parts, the first being an analysis of the differences in prevalence of aplastic anemia between East and West, the second a short summary of a Swedish study.

Differences Between East and West

One of the most interesting and striking facts with regard to epidemiology of aplastic anemia is the difference in prevalence of the disease between the far East and the western world. In Japan, this has lead to the establishment of a Research Group on Aplastic Anemia, sponsored by the Japanese Government. Further, an international symposium on aplastic anemia was arranged in Kyoto, 1976. The Proceedings of the Symposium [3] have recently been published. The complete papers are more informative than the discussions during the symposium and point, not only to the already established differences in prevalence, but also to several possible reasons for it.

However, the epidemiological studies from the Kyoto Symposium are different in scope and structure. It is therefore difficult to analyze and compare their results, but the following conclusions seem warranted.

First of all *the prevalence* of aplastic anemia in the East is high. Figures from selected data, quoted by Whang [14], are given in Table 1. The mean annual number of patients with aplastic anemia in such hospital data in the West is 4, in the East no less than 22. True prevalence figures may by found for aplastic anemia in children. In Sweden, Böttiger [4] found 4 children with aplasia per million inhabitants below the age of 15, Nakayma and Nakagawa [12] as many as 22 in Japan. Both these comparisons demonstrate that aplastic anemia is *at least* 4–5 times as common in the East as in the West.

The further analysis brings to light a number of facts that might explain at least some of these differences:

1. *Age*. The age composition of the eastern data is entirely different from that of the western. In Sweden, e.g., 80 per cent of the patients with aplastic anemia are above 50 years of age [4]. Whang [14] from Korea has collected a large data of patients with aplastic anemia – only 12 per cent were above 50 years. This is a highly significant difference. The age composition is different, however, not only with regard to the oldest age groups – as shown in Figure 1 the Korean material has a very marked peak between 10 and 40 years of age.

West		East	
USA	2	China	40
USA	4	Japan	7
UK	5	Korea	18
Mean	4	Mean	22

Table 1. Annual numbers of patients with aplastic anemia in hospital materials

West	Male/Female	East	Male/Female
USA	1.3	China	4.2
USA	1.2	Japan	–
UK	1.3	Korea	2.4
Mean	1.3	Mean	3.3

Table 2. Sex ratio in hospital materials of patients with aplastic anemia

2. *Sex.* The sex ratio in the eastern data also differs from that of the western. Results taken from the same studies as in Table 1, show that whereas the western data contain approximately as many men as women – or as in other studies even more women –, the eastern patient groups are composed of a much larger proportion of men (Table 2).

3. *Occupation hazards.* The above findings demonstrate that while aplastic anemia in the West primarily is a disease of old age, in the East it affects young men in their active working age. This would imply that *toxic substances in the working environment* might be suspected as the cause of bone marrow damage and aplasia. And a further scrutiny of the eastern reports shows that such substances are in fact commonly implicated. Both insecticides and organic solvents seem to play an important role in causing aplasia in the East. It is true, that Aoki et al. [2] did not find any causal association between organic solvents

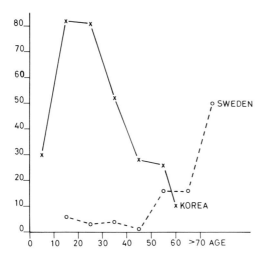

Fig. 1. Different age composition of Korean (×——×) and Swedish (○–––○) materials of aplastic anemia. The figures in both materials represent total number of patients during the time of investigation, not annual frequencies

and aplastic anemia in Japan. But, on the other hand, they list results from other Japanese studies in which organic solvents were suspected as causative factors in 6–12 per cent of patients with aplastic anemia. Whang [14] found a history of toxic exposure in 27 per cent of his Korean material, with chloramphenicol, insecticides and benzol as the most common offending substances (Table 3). Of those believed to have toxic exposure, no less than half had been exposed to insecticides and/or benzol. The insecticides, mainly of the organic phosphorous type (parathion etc.), were used in professional agriculture as well as in homes and private gardens.

Agents	No. of patients
Chloramphenicol	30
Insecticides	29
Benzol	7
Chloramphenicol and insecticides	4
Atabrine	2
Ephedrine	2
Terramycin	2
Salvarsan, quinine, chloroquine and hair dye – one patient each	4

Table 3. Drugs and chemicals associated with aplastic anemia in 81 patients [from Whang, 14]

4. *Differences in drug use.* Another explanation for the difference between East and West seems to be that drug use is different. Chloramphenicol, e.g., evidently is used extensively in the far East. The study by Yoshimatsu et al. [15] is especially elucidating. They performed a retrospective analysis of the drug record of a small group of patients with aplasia. They made a careful scrutiny of the past clinical records and, most important, they took direct and repeated contacts with every physician whom the patient previously had visited. They found that no less than 50 per cent of the patients had received chloramphenicol, a drug intake not suspected or listed in the primary hospital records. For comparison, they made the same analysis for a group of patients with acute leukemia – only 1 out of 16 had received a small amount of chloramphenicol.

5. *High incidence of viral infections (hepatitis).* The eastern studies also reveal that the number of patients with a preceding infection is higher than in the West. Komiya [10] and Takaku et al. [13], respectively, list 7–8 per cent of their aplasia patients as having had hepatitis before the onset of aplasia. Even if it is extremely difficult to find true figures for the occurrence of hepatitis before the onset of aplasia such figures are higher than those found e.g. in Sweden (2.5 per cent) [4]. Further, to my knowledge it is not known whether subclinical forms of hepatitis – without apparent jaundice or other signs of disease – might damage the stem cells of the bone marrow. An answer to such a question can only be found when large populations, and groups of patients with aplasia, have been examined for antibodies against hepatitis antigens.

6. *Ionizing radiation.* Ionizing radiation in a 1959 – survey in Japan was found responsible for 30 per cent of cases of *secondary* aplastic anemia. Later, however,

much lower figures were found and Aoki et al. [2] concluded "that radiation exposure at the low level common today does not cause excess incidence of aplastic anemia".

7. *Genetic influence.* There are indications that genetic factors also are of importance for the appearance of bone marrow disorders, including aplasia. Aoki et al. [2] have pointed out that the incidences of leukemia and aplastic anemia in many countries are inversely correlated. Countries with Caucasian populations such as e.g. the USA and Denmark have a high leukemia incidence and a comparatively low incidence of aplastic anemia, whereas, on the other hand, other countries such as Japan has a relatively high incidence of aplastic anemia and a correspondingly low incidence of leukemia. Another well known difference between East and West in the conspicuous lack of malignant disorders of the lymphatic system is the eastern populations. Further, Fujiki et al. [9] performed genetic studies and found increased tendency for consanguinity and other findings to indicate that heredity plays an important role in determining the suspectibility to hypoplastic anemia. Finally, even if many of the exogenic factors, such as e.g. insecticides used in homes and subclinical infections, could act also on children, the very high incidence of aplastic anemia in childhood would rather speak in favor of a genetic influence.

My conclusion from this analysis is that the difference in prevalence of aplastic anemia between East and West has a multifactorial explanation. Genetic differences in susceptibility form the background, against which acts a number of separate exogenic factors, that can be identified – and to a large extent probably also eliminated. In fact, the true importance of the various factors can only be ascertained when a number of the exogenic ones have been eliminated.

An Epidemiological Study of Aplastic Anemia in Sweden

Sweden, for many reasons, is ideal for epidemiological studies. In one of the seven health care regions, the Uppsala region with 1.2 million inhabitants, corresponding to 15 per cent of the entire population, all hospital discharge diagnoses are recorded on a computer. It is thus possible to retrieve and analyze hospital records for a specific disorder from all the hospitals within the region. The Swedish Adverse Drug Reaction Committee, which was established late in 1965, now receives more than 2000 reports of adverse drug reactions every year. Results from these two sources may be combined with such from continous drug prescription studies in one part of the Uppsala region and from statistics of nationwide drug sales data.

A detailed analysis of 80 patients with aplastic anemia, collected during a five-year period, has previously been published [4]. Only a few facts should be mentioned, as a basis for the discussion during the present symposium.

The overall prevalence rate was 13 per million inhabitants year. But the prevalence varies greatly with age (Fig. 2). It is only 4 per million for children below 15 years of age but goes up to 61 per million for people above 65. Aetiology also varies with age (Table 4). If the total material of 80 patients is divided into

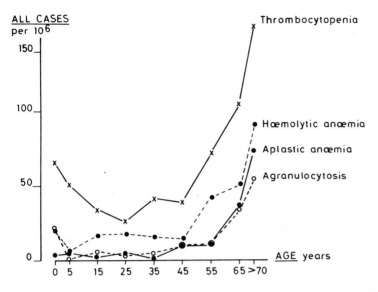

Fig. 2. Age-related incidence of all cases of aplastic anemia (as well as acquired hemolytic anemia, thrombocytopenia and agranulocytosis) occurring in the Uppsala region in Sweden during the 5-year period 1964–68. Curves represent annual incidence

a younger and an older group, with an arbitrarily chosen dividing line at 50 years of age, one finds that only 38 per cent of the young but as much as 81 per cent of the old patients have unknown etiology. This is a significant difference (chi square test: P<0.01). The prognosis is poor and only 13 of the 80 patients (16 per cent) were alive at follow-up, 3 to 7 years after the diagnosis of aplasia.

	Patients	
	below 50 years	above 50 years
Drugs	2	8
Congenital	3	–
Malignant lymphoma	1	3
Infectious hepatitis	1	1
Other viral infections	2	–
Immune mechanisms	1	–
	10	12
	P<0.01	
Unknown aetiology	6 (38%)	52 (81%)
Total	16	64
	80	

Table 4. Aetiology of aplastic anemia in Sweden

Aetiology of Aplastic Anemia

Many divide cases of aplastic anemia into primary and secondary types, according to whether a probable explanation can be found or not. However, clinical differences between the two types do not seem to exist, possibly with the exception for variations in prognosis (see below) [1]. Many studies, among others the analysis by Yoshimatsu et al. [15] already referred to, have demonstrated that the more carefully and detailed the medical history is analyzed, the more often a probable or possible reason for the bone marrow aplasia may be found. As also mentioned, it seems that the chances of finding an explanation for the aplasia is much smaller in the old patient. I have put forward the theory that aplastic anemia in the old might be the end result of combined processes, both circulatory and immunologic, connected with normal ageing [5].

Viral infections may be the cause of bone marrow aplasia. Infectious hepatitis is the best known example, but other viral infections have also been implicated. Böttiger [4] found two patients with viral infections of the upper respiratory tract, preceding the onset of aplasia. Recently a case of aplasia after infectious mononucleosis has been published [8]. Experiences especially from patients with pure red cell aplasia have demonstrated that *immunologic mechanisms* may also be responsible for bone marrow suppression and aplasia. This problem has been excellently reviewed by Cline and Golde [7].

Although effects of *toxic substances* such as organic solvents and insecticides seem to be comparatively common in the East (see above), *drugs* everywhere dominate among the known causes of aplastic anemia. And the single drug that for a long time has been in the center of interest with regard to bone marrow aplasia is chloramphenicol. Today, there can be not doubt that it causes a regular and dose-depending suppression of the bone marrow as well as a rare but unpredictably occurring aplasia. The pathogenic mechanisms for both these types have been studied by Yunis [16]. The incidence of chloramphenicol-induced aplastic anemia initially was regarded as low. Recent results, however, clearly

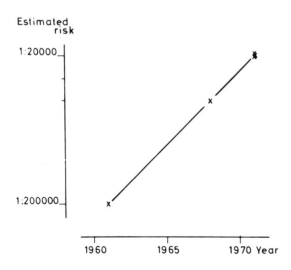

Fig. 3. Change in estimate of risk of developing chloramphenicol-induced aplastic anemia. The figure is composed from results by Leikin et al. (1961), Wallerstein et al. (1969), Hausmann (1974) and Böttiger (1972), respectively

	No. of patients	Whereof deaths
Oxyphenbutazone (Tanderil)	10	5
Chloramphenicol	5	4
Phenylbutazone (Butazolidin)	4	2
Sulfonamides	2	1
Metamizol (Dipyrone, Novalgin)	1	1
Ampicillin	1	1
Trimetadione	1	1
Indomethacin	1	–
Doxycycline	1	1
Griseofulvin	1	1
Amytryptiline	1	1
Diphenyldydantoine	1	1
	29	19 66%

Table 5. Drugs causing aplastic anemia in Sweden (1966–70)

demonstrate that the risk in the beginning was grossly underestimated. Four separate studies of the risk of getting chloramphenicol-induced aplastic anemia have been summarized in Figure 3. It shows that the estimated risk has gone up from 1:200,000 to 1:20,000 – a tenfold increase in ten years [5].

There are some indications to support the idea that aplastic anemia induced by chloramphenicol has a poorer prognosis than aplasia caused by other drugs. Such findings have been reported by myself [4] and later by Modan [11].

In recent years, however, chloramphenicol has been displaced from the top of the list of aplasia-inducing drugs in many countries. My study from Sweden [4] showed that already at the end of the 1960ies, a history of intake of phenylbutazone or oxyphenbutazone was found twice as often as of chloramphenicol in patients with aplastic anemia (Table 5). Now, the WHO Centre for Adverse Reactions to Drugs also shows the same findings, for the period 1968–73 and for approximately twenty countries that report to the Centre (Table 6). Interestingly enough, however, the latest analysis in Sweden [6] – for the years

Drugs	Case reports	Deaths
Phenylbutazone	53	31
Oxyphenbutazone	45	27
Chloramphenicol	18	11
Indomethacin	9	2
Sodium aurothiomalate	9	3
Trimetoprim-sulfametoxaxole	9	3
Phenytoin	7	3
Allopurinol	5	4
Chlorpropramide	5	2
Phenobarbital	4	2
Azathioprine	4	1
Ampicillin	3	1
Trimethadione	3	2

Table 6. Drugs most frequently reported as associated with aplastic anemia. Cases reported to the WHO Research Centre for International Monitoring of Adverse Reactions to Drugs, 1968–1973

Drug	No. of patients	Whereof deaths
Sulfonamides	9	1
Sulfapral (sulfa-metizol + sulfa-metoxipyridazine)	4	
Trimetoprim--sulfametoxazol	4	
Salazopyrin (sali-cylazosulfapyridine)	1	
Cytostatics	7	4
Acetazolamid	2	2
Chlorpropramide	2	1
Carbamazepine	2	
Nitrofurantoin	1	
Amitrypriline	1	
Diphenylhydantoine	1	
Ethosuximide	1	
Phenazone + kaffeine	1	
	27	8

Table 7. Drugs causing aplastic anemia in Sweden (1971–75)

1971–75 – shows that the anti-inflammatory drugs also have disappeared from the list and have been replaced by a number of sulfonamide preparations (Table 7) – Sulfapral (sulfametizol + sulfametoxipyridazine), a Swedish product for chronic urinary tract infections, Salazopyrin (salicylazosulfapyridine) and trimetoprim-sulfametoxaxol. The number of sulfonamide-induced cases of aplastic anemia has gone up from 2 cases in 1968–70 to 9 cases in 1971–75. Thus, there is a *real* increase and not only an unmasking by the disappearance of chloramphenicol and phenylbutazone/oxyphenbutazone.

One obvious reason for the shift among drugs causing side effects is difference in sale and use of the drugs. Figure 4 shows that the sale of chloramphenicol, after an early increase, has gone down remarkably in Sweden, as has, somewhat later,

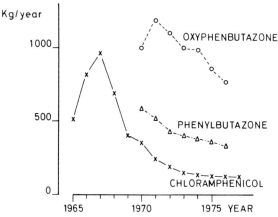

Fig. 4. Change in sales figures for chloramphenicol, oxyphenbutazone and phenylbutazone in Sweden

the sale of phenylbutazone and oxyphenbutazone. The decrease in sales figures for chloramphenicol is 87 per cent in ten years, for phenylbutazone and oxyphenbutazone 44 and 35 per cent, respectively, in a little more than five years. The sales situation for the sulfonamides is complex – among the three types mentioned, the sale for one drug has gone down, for another gone up and for the third remained remarkably constant.

An interesting fact is that although the drugs causing blood disorders have changed, the incidence of drug-induced aplastic anemia – and of other drug-induced cytopenias – has remained remarkably constant (Table 8). This might be taken as an indication that individual susceptibility plays an important role, perhaps more important than the drugs administered, at least at a low level of use of potentially toxic or dangerous drugs. Even if this would be true, it is of course always necessary to use only such drugs for which there are clear-cut indications and of which the prescribing physician has a thorough knowledge with regard not only to effects but also to adverse reactions. Many cases of aplastic anemia could have been avoided – and still may be avoided – by a more careful selection of therapeutic drugs.

	1966–70	1971–75
Aplastic anemia	0.7	0.6
Agranulocytosis	2.5	2.6
Thrombocytopenia	3.1	2.7
Hemolytic anemia – acquired	1.1	1.6

Table 8. Incidence of drug-induced cytopenias in Sweden. Number of cases per million inhabitants and year

Summary

Epidemiological data show that the incidence of aplastic anemia is a least 4–5 times as high in the far East as in the West. The reason seems to be complex – a number of exogenic factors act against a background of genetic differences in susceptibility.

A Swedish study of aplastic anemia shows the overall incidence to be 13 cases per million inhabitants and year, but also that the incidence varies greatly with age (range 4–61 cases per million).

Drugs are the single most important aetiologic factor. Chloramphenicol is the most common offender, but in many countries it now has been replaced by the anti-inflammatory agents phenylbutazone and oxyphenbutazone. In Sweden the latter also have disappeared and sulfonamide preparations dominate among drugs causing aplasia.

References

1. Abe, T. and Komiya, M.: Some Clinical Aspects of Aplastic Anemia. In: Aplastic Anemia. Proceedings of the First International Symposium on Aplastic Anemia, Kyoto 1976. Japan Medical Research Foundation (ed.). Tokyo: University of Tokyo Press, 1978, pp. 197–204
2. Aoki, K., Ohtani, M. and Shimizu, H.: Epidemiological Approach to the Aetiology of Aplastic Anemia. In: Aplastic Anemia. Proceedings of the First International Symposium on Aplastic Anemia, Kyoto 1976. Japan Medical Research Foundation (ed.) Tokyo: University of Tokyo Press, 1978, pp. 155–170
3. Aplastic Anemia. Proceedings of the First International Symposium on Aplastic Anemia, Kyoto 1976. Japan Medical Research Foundation (ed.). Tokyo: University of Tokyo Press, 1978
4. Böttiger, L. E. and Westerholm, B.: Aplastic Anemia. I–III. Acta Med. Scand. *102*, 315–326 (1972)
5. Böttiger, L. E.: Prevalence and Aetiology of Aplastic Anemia in Sweden. In: Aplastic Anemia. Proceedings of the First International Symposium on Aplastic Anemia, Kyoto 1976. Japan Medical Research Foundation (ed.) Tokyo: University of Tokyo Press, 1978, pp. 171–180
6. Böttiger, L. E. and Holmberg, L.: Drug-induced cytopenias in Sweden 1971–1975. Läkartidningen *76*, 611–613 (1979)
7. Cline, M. J. and Golde, D. W.: Immune suppression of hematopoiesis. Amer. J. Med. *64*, 301–310 (1978)
8. van Doornik, M. C., van t'Veer, E. T., Korthof and Wirenga, H.: Fatal aplastic anemia complicating infectious mononucleosis. Scand. J. Haematol. *20*, 52–56 (1978)
9. Fujiki, N., Nishigaki, I., Masuda, M., Hosokawa, K. and Kondo, M.: Clinicogenetic Studies on Hypoplastic Anemia. In: Aplastic Anemia. Proceedings of the First International Symposium on Aplastic Anemia, Kyoto 1976. Japan Medical Research Foundation (ed.). Tokyo: University of Tokyo Press, 1978, pp. 185–188
10. Komiya, M. and Okumura, H.: see 2. Aoki et al.
11. Modan, B., Segal, S., Shani, M. and Sheba, C.: Aplastic anemia in Israel: Evaluation of the aetiological role of chloramphenicol on a community-wide basis. Amer. J. Med. Sci. *270*, 441–445 (1975)
12. Nakayama, K. and Nakagawa, T.: Prevalence and Clinical States of Aplastic Anemia among Children in Japan. In: Aplastic Anemia. Proceedings of the First International Symposium on Aplastic Anemia, Kyoto 1976. Japan Medical Research Foundation (ed.). Tokyo: University of Tokyo Press, 1978, pp. 207–223
13. Takaku, F., Aoki, K. and Shimizu, H.: see 2. Aoki et al.
14. Whang, K. S.: Aplastic Anemia in Korea: A Clinical Study of 309 Cases. In: Aplastic Anemia. Proceedings of the First International Symposium on Aplastic Anemia, Kyoto 1976. Japan Medical Research Foundation (ed.). Tokyo: University of Tokyo Press, 1978, pp. 225–240
15. Yoshimatsu, H., Uetake, S., Miyao, S., Omine, M., Tsuchiya, J., Shirakura, T. and Maekawa, T.: A Retrospective Survey of Drugs and Chemicals Associated with the Development of Aplastic Anemia. In: Aplastic Anemia. Proceedings of the First International Symposium on Aplastic Anemia, Kyoto 1976. Japan Medical Research Foundation (ed.). Tokyo: University of Tokyo Press, 1978, pp. 189–193
16. Yunis, A. A.: Pathogenic Mechanisms in Bone Marrow Suppression from Chloramphenicol and Thiamphenicol. In: Aplastic Anemia. Proceedings of the First International Symposium on Aplastic Anemia, Kyoto 1976. Japan Medical Research Foundation (ed.). Tokyo: University of Tokyo Press, 1978, pp. 321–331

Discussion

Camitta: I would like to pose some questions to Dr. Böttiger. Firstly, we often say that a case of aplastic anaemia is due to a drug but how do we know that the aplastic anaemia is due to the drug per se, not to the underlying disease for which the drug was given? This would fit in with a viral etiology. Secondly, in looking at etiology, how can you distinguish between dose related effects of a drug such as chloramphenical where everyone develops pancytopenia versus the idiosyncratic effect? The former produces a transient pancytopenia, the latter a prolonged severe aplastic anaemia. Maybe you must separate the drug history of patients with mild disease from those with severe disease to get at the etiology of severe persistent aplasias. We must also think of individual metabolic differences in detoxication of drugs and also the possible importance of the route of administration. Perhaps giving a drug orally results in a metabolite produced by the intestinal flora which is causing aplastic anaemia. It may not be the parent compound and the route may be significant. That is always brought up in terms of chloramphenicol. Everyone has said that they never have seen a case of chloramphenicol induced severe persistent aplastic anaemia secondary to a non oral route of administration. Has there been a case of induced aplasia by chloramphenicol given i.v. or i.m.?

Böttiger: We have a way of classification for all adverse drug reactions that are being reported to the Swedish committee. They are actually classified by a group of physicians and pharmacologists. Where only one drug is given during a long period of time, it may be repeated dosage of that drug which causes aplasia but, of course, you can never be sure that the drug really causes the disease unless you can repeat the experiment. I think it is extremely important to evaluate all such reports before they are collected and worked up and commented upon, so that you can show a reasonable likelihood of a connection between the administration of a drug and the reaction. It is quite clear that some of them are what people call toxic and dose dependent, others are not, but I want to stress the fact that there are these differences.

Camitta: Perhaps there are differences between the patients with severe aplasia and patients who have transient aplasia in terms of etiology. Everyone lumps together all cases of aplastic anaemia with etiology, but there may be different etiologies for mild, for severe and for transient aplastic anaemia.

Böttiger: That is an important question. The prognosis may also be different in these cases. I have pointed out that chloramphenicol induced aplastic anaemia seems to carry poorer prognosis than, for instance, that caused by phenylbutazone. Of the question by Camitta about the route of administration, I can only say that all the Swedish cases have had oral administration. Cases of injection of drugs are rare in comparison with the amount that is orally administered. There are not many defined cases of aplasia. A comment on Dr. Heimpel's data: I think it is an artificial division to talk about primary and secondary cases. Idiopathic cases from the clinical point of view are the same as drug induced. The more carefully you examine the past history of drugs, the smaller the idiopathic group gets. I would like to comment on your statement about multiple exposures. In the small group of patients with chloramphenicol induced aplasia in Sweden 7 patients out of 4 had repeated administration of chloramphenicol which is of interest because one has always said that chloramphenicol is not sensitising but there may be sensitisation to the stepwise suppression of the bone marrow function.

Heimpel: If you look carefully at drug histories, you often find multiple exposures to one drug or exposure to multiple drugs. There may be individuals who are sensitive to different kinds of drugs and there may be one hit by one factor and secondary effects caused by other drugs.

2.4 The Role of Viral Infections in Aplastic Anemia

B. M. Camitta

Introduction

Ninety years have passed since Paul Ehrlich first described an illness now known as aplastic anemia [12]. During this time much has been learned about the natural history and treatment of marrow aplasia. However, underlying pathogenetic mechanisms remain obscure. A few patients develop aplastic anemia as part of inherited syndromes. Irradiation and chemotherapy for malignant diseases may produce severe, prolonged, focal or diffuse marrow aplasia. Certain drugs produce reversible dose-related marrow suppression or, less often, idiosyncratic dose-unrelated severe aplastic anemia. For other drugs or chemicals an association with aplastic anemia is supported only by statistical correlations or suggestively related clinical events. At least 50% of cases of marrow aplasia are of unknown etiology. Viruses are rarely mentioned as causes of aplastic anemia.

Aplastic anemia may result from direct damage to the pluripotent stem cell or from damage to the marrow microenvironment. Alternatively, aplasia could arise from damage to peripheral organs. There is evidence linking viral infections with damage to each of these tissues. Nevertheless, it has been difficult to correlate specific viral infections with development of marrow aplasia.

In this paper evidence implicating viruses in the etiology of aplastic anemia is reviewed. There is good, though incomplete, data linking infectious hepatitis viruses and marrow aplasia in mice and man. There is currently little evidence that other viral infections play a role in the pathogenesis of human aplastic anemia.

Marrow Aplasia and Nonhepatitis Viruses (Tables 1, 2)

Fikrig and co-workers [14] induced aplastic crises in CBA/T6 and NZB/B1 mice by injecting the animals with Coxsackie B4 virus. Female NZB animals aged 4–8 months and CBA mice of any age or sex developed reticulocytopenia. Bone marrow examination revealed decreased red cell precursors. Marrow suppression lasted 6–8 days. There was little change in the animals' hematocrits during that time. Events in this model are similar to aplastic crises seen during presumed viral infections in patients with hereditary hemolytic anemias [25].

Osborn et al. [27] injected four week old mice with $10^{5.0}$ plaque forming units of murine cytomegalovirus. Mild to moderate pancytopenia developed. The mean hematocrit decreased from 50% to 38%. Leukopenia due to decreased mononuclear cells was followed a few days later by mild leukocytosis. Platelet counts fell from $1-1.4 \times 10^6/mm^3$ to 0.5×10^6 within 4 days and returned to

Table 1. Viruses and marrow aplasia – murine and guinea pig studies

Virus	Observations	Reference
Coxsackie B4	Transient reticulocytopenia; decreased marrow red cell precursors	Fikrig
Cytomegalovirus	Transient pancytopenia; viral antigen in megakaryocytes	Osborn
Encephalomyocarditis	Progressive thrombocytopenia; variable morphologic changes in megakaryocytes	Modai
Friend	Thrombocytopenia; viral particles in megakaryocytes	Dalton
Hepatitis	Variable transient pancytopenias; focal areas of bone marrow necrosis	Hunstein
	Pancytopenia; bone marrow red cell hypoplasia plus disturbances in white cell/megakaryocyte maturation; abnormal morphology of lymphoid, reticuloendothelial tissues; virus cultured from bone marrow	Piazza
Newcastle	In vitro abnormalities of megakaryocytes; viral antigen in megakaryocytes	Jerushalmy

Table 2. Viruses and marrow aplasia – evidence in man

Virus	Observation	Reference
Cytomegalovirus	Intranuclear inclusions in megakaryocytes	Chesney
	Congenital infections with anemia, thrombocytopenia	Oski
Dengue	Variable mono and pancytopenias	Bierman
Hepatitis	0.1–0.2% of patients with hepatitis develop marrow aplasia	Böttiger
	2–10% of patients with aplasia have a history or laboratory evidence of hepatitis	Abe, Böttiger, Camitta
	Administration of virus to volunteers produces pan-leukopenia then relative lymphocytosis with atypical lymphocytes	Havens
	Worsening of aplasia during exacerbation of hepatitis	Hagler
	Systemic nature of infectious hepatitis	Conrad
	Variable cytopenias during infectious hepatitis	Kivel
Herpes Simplex	Neonatal infections with anemia, thrombocytopenia	Oski
Rubella	Congenital infections with anemia, thrombocytopenia; marrow with increased phagocytosing reticulum cells; virus isolated from marrow	Zinkham
Rubeola	Live (but not inactivated) vaccine causes transient thrombocytopenia with abnormalities of megakaryocytes; virus not isolated from marrow	Oski
Varicella	Intranuclear viral particles in megakaryocytes	Espinoza
Other	Hematolytic anemias: Aplastic crises post viral infections	
	Immune thrombocytopenia, anemia after a variety of viral infections	
	Early leukopenia in many viral illnesses	

normal within 7 to 14 days. During the thrombocytopenia megakaryocytes showed cytoplasmic and nuclear vacuolization. There was strong nuclear and weak cytoplasmic fluorescence of megakaryocytes for murine cytomegalovirus antigen. Chesney et al. [8] recently reported intranuclear inclusions in the megakaryocytes of a 4 year old child with congenital cytomegalovirus infection and thrombocytopenia (platelets $10^5/mm^3$). Other hematologic values were normal. Thrombocytopenia and Coombs negative anemia had been present neonatally in this child, as in other children with congenital cytomegaloviral infections [29].

Several other reports have linked viral infections with thrombocytopenia and/or megakaryocyte damage. Encephalomyocarditis virus infection in guinea pigs has resulted in progressive, fatal thrombocytopenia [26]. Megakaryocytes appeared normal in vivo but when infected in vitro, vacuolization and decreased granulation were seen. Jerushalmy et al. [21] noted failure of granulation and positive immunofluorescence for viral antigen in guinea pig megakaryocytes infected in vitro with Newcastle disease virus. Murine leukemia virus infection of mice may cause thrombocytopenia. Electron microscopy has demonstrated virus particles in megakaryocytes whether or not the mice were thrombocytopenic [10]. Intranuclear viral particles have also been identified in the megakaryocytes of a patient with varicella and thrombocytopenia [13]. Finally, live but not inactivated rubeola vaccine has been reported to cause transient thrombocytopenia [28]. Megakaryocytes were decreased and showed nuclear and cytoplasmic vacuolization. There was no evidence of increased platelet destruction. Virus was not isolated from marrow specimens. In all of these studies, overall bone marrow cellularity was normal or not mentioned.

Rubella virus can produce congenital infections with concomitant anemia and thrombocytopenia. In one study of these infants, bone marrow specimens showed increased phagocytosing reticulum cells [40]. Rubella virus was isolated from most marrow specimens but concurrent viremia was not excluded. Megakaryocytes were normal. Reports of decreased megakaryocytes in this and other congenital viral infections may represent difficulties in sampling neonatal marrow rather than a true decrease in megarkaryocyte numbers [29].

Finally, leukopenia is a common finding early in the course of many viral illnesses. More severe mono – or pancytopenias may develop later. These cytopenias are due to direct marrow infection [3] or to development of antibodies versus marrow constituents, e.g. anti-i cold agglutinin in mononucleosis [19], antiplatelet antibodies in ITP [16].

Marrow Aplasia and Hepatitis Viruses (Tables 1, 2)

Piazza et al. [30] injected NMRI mice with 100 LD_{50} of murine hepatitis virus (MHV-3). Progressive pancytopenia was detectable within 24 hours. Bone marrows developed red cell hypoplasia, a white cell maturation delay with appearance of abnormal forms and abnormal megakaryocyte maturation. Reticuloendothelial hyperplasia with obliteration of sinuses, karyorrhexis of lymphocytes, and endothelial hyperplasia were seen in lymphoid organs. High concentrations of virus were present in the bone marrow, lymph nodes and spleen but comparable titers were also present in blood. Hunstein et al. [20] reported focal marrow necroses in CF mice injected with smaller doses of MHV-3 than in the above work. After 4 to 8 days repair occurred and marrow histology returned to normal.

Major causes of human viral hepatitis include hepatitis A virus (infectious hepatitis), hepatitis B virus (serum hepatitis), cytomegalovirus and non A – non B hepatitis virus. There has been no definitive evidence suggesting that any of the latter three viruses cause human aplastic anemia. This may be due to the transient nature of hepatitis B antigenemia [18], the slow development of antihepatitis B surface antigen antibody [18], the absence until recently of serologic tests for non A – non B hepatitis viruses and the failure to perform antibody studies in most patients. Thus, although some cases of post hepatitic aplastic anemia may be due to other viruses, the remainder of this section will discuss presumed hepatitis A induced human marrow aplasia.

Infectious hepatitis is a systemic disease [9]. Virus is detectable in blood, urine and feces. Histologic changes occur in the intestines and kidneys as well as in the liver. Experimental administration of virus to volunteers causes panleukopenia followed by a relative lymphocytosis with the presence of atypical lymphocytes [17]. The leukocyte changes usually return to normal within 1 to 2 weeks of the onset of fever. Similar changes are seen in patients during epidemics of infectious hepatitis [23]. In addition up to 50% of patients develop mild anemia, usually accompanied by reticulocytopenia, macrocytosis, and slightly shortened Cr^{51} red cell survival [9, 23]. Mild thrombocytopenia has also been noted [15]. Approximately 0.1 to 0.2% of patients with hepatitis develop pancytopenia with marrow aplasia [5]. Conversely, 0 to 11% (median approximately 2.5 to 4.0%) of patients with aplastic anemia have a prior history or laboratory evidence of recent hepatitis [1, 5, 7]. Worsening of marrow aplasia has been reported following recrudescence of hepatitis [15].

More than 200 cases of hepatitis associated aplastic anemia have been reported [2, 6, 15]. More than half on these patients developed aplasia within 9 months of initial hepatitis infection and had no history of treatment with potentially myelosuppressive drugs or another underlying disorder. Characteristics of patients with posthepatitic marrow aplasia are summarized in Table 3. Hepatitis was not more severe than usual. The patient's age, sex and HLA types (increased HLA B8) are typical of large groups of patients with hepatitis per se. The severity of posthepatitic aplasia is reflected in the acute lethal course of most affected individuals (Table 4). Few patients recover completely. There is no evidence that modern medical care (i.e., transfusions, antibiotics, anabolic steroids) has significantly altered survival. In a recent study, none of five patients with posthepatitic aplastic anemia survived despite administration of androgen and other supportive meansures [7]. In contrast, 4 of 7 patients receiving early bone marrow transplants from histocompatible siblings are alive and hematologically normal. Despite small numbers, this study and previously published reports [6, 31, 33, 34] suggest that early bone marrow transplantation is the treatment of choice for severe aplastic anemia following hepatitis.

Table 3. Characteristics of posthepatitic aplastic anemia

		Reference 6[a]	Reference 15[b]
Age in years		18 (1–74)	20 ± 14
Sex		Male 49/Female 27/?4	Male 83/Female 59/?32
Maximum recorded:			
Bilirubin-total, mg/100 ml		17 (1–59)	–
SGOT: U/ml		915 (50–4300)	–
Weeks from:			
Hepatitis to aplasia		8 (0–36)	9.3 ± 9.0
% Patients with lowest recorded:			
PMN/mm³	<1000	85%	–
	<500	75%	–
Platelets/mm³	<50,000	90%	–
	<20,000	65%	–
Marrow cellularity:		92% moderate to marked decrease	Aplastic 75
			Hypoplastic 43
		8% normal or increased	? 54

[a] Criteria for inclusion: Hepatitis to aplasia <9 months, no drug ingestion, no other disease. Values are medians (range) unless otherwise noted

[b] Criteria for inclusion: Hepatitis to aplasia <12 months, aplasia to death <16 months, not chronic hepatitis. Values are means ± one standard deviation

Table 4. Outcome of posthepatitic aplastic anemia (See Table 3 for criteria for inclusion)

	Reference 6	Reference 15
Died	70 (87%)	148 (85%)
Alive	10 (13%)	26 (15%)
complete recovery	6	13
partial recovery	2	13
continued aplasia	2	
Weeks from aplasia to death	median 11 range 1–41	11.3 ± 12.5[a]

[a] Median ± one standard deviation

Viruses and Aplastic Anemia: Possible Mechanisms

Viral infections could produce aplastic anemia by many mechanisms (Table 5). Depletion of pluripotent or committed marrow stem cells by direct viral infection seems likely. However, if a decreased stem cell pool was the only problem, one would expect repair of aplasia as these cells first self replicated and then matured. Persistence of aplasia therefore implies an inability of the depleted stem cell pool to self renew or to mature. This concept has been expanded on elsewhere [4].

Damage to marrow supporting cells may also cause aplastic anemia [36, 37, 38]. However, the successful engraftment of histocompatible sibling marrow

Table 5. Possible mechanisms of hepatitis-induced bone marrow damage

A. Direct
Infection of pluripotent stem cells and/or their immediate progeny
Infection of marrow supporting cells

B. Indirect
Induction of antimarrow immune reactivity
Failure of damaged liver to remove toxic substance from the blood or to provide essential nutrient
Increased phagocytosis of marrow elements by hyperplastic reticuloendothelial cells
Depletion of helper lymphocytes necessary for normal hematopoiesis
Immune suppression with subsequent graft vs. host disease

C. Combined
Interaction of above with hereditary and acquired factors determining ability of host to resist and/or terminate viral infection plus repair damage

suggests that functional microenvironmental damage, if it does occur, is rapidly repaired [33, 34].

Marrow damage from viral infection might be mediated by injury to plasma or organelle membranes. Viral-induced damage to chromosomal DNA could interfere with cells' capacities to indefinitely replicate. Damage of this type is common after irradiation and chemotherapy with alkylating agents. Gross chromosomal changes are infrequent in post hepatitic aplasia [15]. The observation that chromosomal breaks can be induced in vitro by a factor in the plasma of infectious hepatitis patients has not been followed by similar observations in vivo [11]. Studies utilizing modern banding techniques are necessary to further evaluate whether chromosomal damage is present in hematopoietic cells of patients with hepatitis and posthepatitis aplastic anemia.

Viral infections may indirectly produce marrow aplasia by altering atigenicity of cell membranes. Immune thrombocytopenia and hemolytic anemia may follow viral infections. However, there is no evidence for red cell or platelet antibodies in the anemia and thrombocytopenia associated with congenital viral infections. Likewise, immune phenomena are exceptional in posthepatitic aplastic anemia, but have not been consistently sought [15]. Suppression of normal marrow colony forming capacity by aplastic patients' lymphocytes has been reported, but is rare if the target marrow is from an HLA identical sibling and the patient has never been transfused [32].

In vitro suppression of leukocyte mitosis by serum from patients with infectious hepatitis has been reported [24]. Karp et al. [22] recently reported a relationship between liver dysfunction (secondary to viral infection or chemicals) and the lack of serum stimulation of bone marrow growth in vitro. Whether this resulted from lack on an "essential nutrient" or presence of a "toxin" must be determined. It may be difficult to separate effects of hepatic dysfunction per se from a more direct action of underlying agents on bone marrow.

Increased phagocytosis of marrow elements by histiocytic cells has not been implicated in human aplastic anemia. Similarly, although recently implicated in the anemia of W/W^v mice [39], depletion of helper lymphocytes remains an intriguing but untested hypothesis in human aplastic anemia. There is little evidence supporting a role of graft versus host disease in viral induced aplastic anemia although concomitantly acquired immune definciency could lead to this complication [35].

In conclusion, viruses may induce aplastic anemia by a number of mechanisms. Hereditary (immune response genes, DNA repair capability) and acquired (previous infections and immunologic experiences) host factors may also be important to the extent that they enable a patient to resist, terminate, and repair a viral infection. Presumed infectious hepatitis is the only human viral infection that has been linked with sustained marrow aplasia. This association requires additional documentation. Pending further studies, other viral infections cannot be considered a major cause of aplastic anemia in man.

References

1. Abe, T., Komiya, M.: Some Clinical Aspects of Aplastic Anemia. In: Aplastic Anemia. Japan Medical Research Foundation (ed.). Tokyo: Univ. Tokyo, 1978, pp. 197–204
2. Ajouni, K., Doeblin, T. D.: The syndrome of hepatitis and aplastic anemia. Br. J. Haematol. 27, 345–355 (1974)
3. Bierman, H. R., Nelson, E. R.: Hematodepressive virus diseases of Thailand. Ann. Intern. Med. 62, 867–884 (1965)
4. Boggs, D. R., Boggs, S. S.: The pathogenesis of aplastic anemia: A defective pluripotent hematopoietic stem cell with inappropriate balance of differentiation and selfreplication. Blood 48, 71–76 (1976)
5. Böttiger, L. E.: Prevalence and Aetiology of Aplastic Anemia in Sweden. In: Aplastic Anemia. Japan Medical Research Foundation (ed.). Tokyo: Univ. Tokyo, 1978, pp. 171–180
6. Camitta, B., Nathan, D. G., Forman, E. N., Parkman, R., Rappeport, J. M., Orellana, T. D.: Posthepatitic severe aplastic anemia – An indication for early bone marrow transplantation. Blood 43, 473–483 (1974)
7. Camitta, B.: International Aplastic Anemia Study – unpublished data
8. Chesney, P. J., Taher, A., Gilbert, E. M. F., Shahidi, N. T.: Intranuclear inclusions in megakaryocytes in congenital cytomegalovirus infection. J. Pediatr. 92, 957–958 (1978)
9. Conrad, M. E., Schwartz, F. D., Young, A. A.: Infectious hepatitis – A generalized disease. Am J. Med. 37, 789–801 (1964)
10. Dalton, A. J., Law, L. W., Moloney, J. B., Manaker, R. A.: Electron microscopic studies of series of murine lymphoid neoplasms. J. Nat. Cancer Inst. 27, 747–791 (1961)
11. El-Alfi, O. S., Smith, P. M., Biesele, J. J.: Chromosome breaks in human leukocyte cultures induced by an agent in the plasma of infectious hepatitis patients. Hereditas 52, 285–294 (1965)
12. Erlich, P.: Über einen Fall von Anämie mit Bemerkungen über regenerative Veränderungen des Knochenmarks. Charite Ann. 13, 300 (1888)
13. Espinoza, C., Kohn, C.: Viral infection of megakaryocytes in varicella with purpura. Am. J. Clin. Path. 61, 203–208 (1974)
14. Fikrig, S. M., Berkovich, S.: Virus induced aplastic crisis in mice. Blood 33, 582–589 (1969)
15. Hagler, L., Pastore, R. A., Bergin, J. J.: Aplastic anemia following viral hepatitis: Report of two fatal cases and literature review. Medicine 54, 139–164 (1975)
16. Handin, R. I., Piessens, W. F., Moloney, W. C.: Stimulation of nonimmunized lymphocytes by platelet-antibody complexes in idiopathic thrombocytopenic purpura. N. Engl. J. Med. 289, 714–718 (1973)

17. Havens, W. P., Marck, R. E.: The leukocyte response of patients with experimentally induced infectious hepatitis. Am. J. Med. Sci. *212*, 129–138 (1946)
18. Hoofnagle, J. M., Seeff, L. B., Bales, Z. B., Zimmerman, H. J., Veterans Administration Cooperative Study Group: Type B hepatitis after transfusion with blood containing antibody to hepatitis B core antigen. N. Engl. J. Med. *298*, 1379–1383 (1978)
19. Hossaini, A. A.: Anti-i in infectious mononucleosis. Am. J. Clin. Path. *53*, 198–203 (1970)
20. Hunstein, W., Perings, E., Eggeling, B., Uhl, N.: Panmyelophthise und Virushepatitis. Acta. Haematol. *42*, 336–346 (1969)
21. Jerushalmy, Z., Kaminski, E., Kohn, A., de Vries, A.: Interaction of Newcastle disease virus with megakaryocytes in cell cultures of guinea pig bone marrow. Proc. Soc. Exp. Biol. Med. *114*, 687–690 (1963)
22. Karp, J. E., Schacter, L. P., Burke, P. J.: Humoral factors in aplastic anemia: Relationship of liver dysfunction to lack of serum stimulation of bone marrow growth in vitro. Blood *51*, 397–414 (1978)
23. Kivel, R. M.: Hematologic aspects of acute viral hepatitis. Am. J. Dig. Dis. *6*, 1017–1031 (1961)
24. Mella, B., Lang, D. J.: Leukocyte mitosis. Suppression *in vitro* associated with infectious hepatitis. Science *155*, 80–81 (1967)
25. Miesch, D. C., Baxter, R., Levin, W. C.: Acute erythroblastopenia: Pathogenesis, manifestations and management. A. M. A. Arch. Intern. Med. *99*, 461–473 (1957)
26. Modai, Y., Oren, R., de Vries, A., Kohn, A.: Thrombocytopenia in guinea pigs infected by encephalomyocarditis virus (EMC). Thromb. Diath. Hemorrh. *18*, 686–690 (1967)
27. Osborn, J. E., Shahidi, N. T.: Thrombocytopenia in murine cytomegalovirus infection. J. Lab. Clin. Med. *81*, 53–63 (1973)
28. Oski, F., Naiman, J. L.: Effect of live measles vaccine on the platelet count. N. Engl. J. Med. *275*, 352–356 (1966)
29. Oski, F., Naiman, J. L.: Hematologic Problems in the Newborn, 2nd ed. Philadelphia: W. B. Saunders, 1972, pp. 35–43, 294–296, 300–301
30. Piazza, M., Piccinino, F., Matano, F.: Hematologic changes in viral (MHV-3) murine hepatitis. Nature *205*, 1034–1035 (1965)
31. Royal Marsden Bone Marrow Transplantation Team: Bone marrow aplasia after infectious hepatitis treated by bone marrow transplantation. Br. Med. J. *1*, 363–364 (1974)
32. Singer, J. W., Brown, J. E., James, M. C., Doney, K., Warren, R. P., Storb, R., Thomas, E. D.: The effect of peripheral blood lymphocytes from patients with aplastic anemia on granulocytic colony growth from HLA matched and mismatched marrows: The effect of transfusion sensitization. Blood *52*, 37–46 (1978)
33. Storb, R., Thomas, E. D., Buckner, C. D., Clift, R. A., Johnson, F. L., Fefer, A., Glucksberg, H., Giblett, E. R., Lerner, K. G., Neiman, P.: Allogeneic marrow grafting for treatment of aplastic anemia. Blood *43*, 157–180 (1974)
34. Storb, R., Thomas, E. D., Weiden, P. L., Buckner, C. D., Clift, R. A., Fefer, A., Fernando, L. P., Giblett, E. R., Goodell, B. W., Johnson, F. L., Lerner, K. G., Neiman, P., Saunders, J. E.: Aplastic anemia treated by allogeneic bone marrow transplantation: A report on 49 new cases from Seattle. Blood *48*, 817–841 (1976)
35. Strauss, R. G., Bove, K. E., Lake, A., Kisker, T.: Acquired immunodeficiency in hepatitis associated aplastic anemia. J. Pediatr. *86*, 910–912 (1975)
36. Tavassoli, M.: Studies on hemopoietic microenvironments. Exp. Hemat. *3*, 213–226 (1975)
37. Umehara, S.: Bone Marrow Stroma in Aplastic Anemia. In: Aplastic Anemia. Japan Medical Research Foundation (ed.). Tokyo: Univ. Tokyo, 1978, pp. 109–121
38. te Velde, J., Haak, H. L.: Aplastic anemia: Histological investigation of methacrylate embedded bone marrow biopsy specimens; correlation with survival after conventional treatment in 15 adult patients. Br. J. Haematol. *35*, 61–69 (1977)
39. Wiktor-Jedrzejczak, W., Sharkis, S., Ahmed, A., Sells, K. W., Santos, G. W.: Theta-sensitive cell and erythropoiesis: Identification of a defect in W/W^v anemic mice. Science *196*, 313–315 (1977)
40. Zinkham, W. H., Medearis, D. N. jr., Osborn, J. E.: Blood and bone marrow findings in congenital rubella. J. Pediatr. *71*, 512–524 (1967)

2.5 Aplastic Anemia Terminating in Leukemia

K. P. Hellriegel, I. Fohlmeister, H. E. Schaefer

Aplastic anemia is characterized by well defined clinical and morphological features, although there are fundamental differences in pathogenesis. The cardinal symptom of aplastic anemia, that of bone marrow aplasia, may be constitutional, induced by radiation or cytostatics, by certain drugs, or even due to agranulocytosis; and in some cases its origin may be unknown.

The occurrence of acute leukemia as a complication of aplastic anemia has been reported on several occasions, especially following the use of anabolic steroids in aplastic anemia [6, 15]. In aplastic anemia, the incidence of leukemia is considerably lower than in patients with refractory anemia which are known to have a 30–50 per cent risk for developing acute leukemia. The relation between aplastic anemia and acute leukemia, however, is challenging in the following subgroups of patients:

Aplastic anemia due to chloramphenicol and phenylbutazone [3, 4, 7, 9, 10, 11, 18],

aplastic anemia following exposure to benzene [1, 5],

aplastic anemia – paroxysmal nocturnal hemoglobinuria syndrome [2, 8, 16, 18, 21] and

Fanconi's anemia.

The following patient initially presented with aplastic anemia and later terminated in blastic crisis of chronic myeloid leukemia (CML). The case report illustrates the complexity of problems and raises some questions. To our knowledge no similar case has been described.

Case Report

A 36-year-old Turkish workman without any drug history first presented with hemorrhages and severe pancytopenia (hemoglobin 7.2 g/dl, leukocytes 0.8×10^9/l, granulocytes 0.3×10^9/l, platelets 11×10^9/l) in January 1976. He had neither hepato- nor splenomegaly. The demonstration of positive HAM's and sucrose lysis tests was consistent with a red cell defect characteristic of paroxysmal nocturnal hemoglobinuria (PNH). Bone marrow biopsy demonstrated severe hypocellularity with large areas containing fat cells only and small nests with mainly granulopoietic cells (Fig. 1).

Following androgen treatment (metenolone 3 mg/kg/day) an increase of peripheral blood values was observed concomitant with leukocytosis and shift to the left (May 5, 1976): hemoglobin 13.1 g/dl, leukocytes 15.9×10^9/l, platelets 50×10^9/l. Bone marrow aspirate demonstrated hypercellularity with increased granulopoiesis and decreased incidence of megacaryocytes. The leukocyte alkaline phosphatase score was 4. Cytogenetic analysis revealed the caryotype 46,XY,Ph[1]; banding studies showed the translocation 9; 22. Anabolic steroids were withdrawn, with blood values remaining unchanged during the following 11 months; only the platelet count rose to 115×10^9/l.

In April 1977, splenomegaly developed; concomitantly a leukocytosis (63×10^9/l) was observed. Neither in the peripheral blood nor in the bone marrow (Fig. 2), there was an increase of blast cells at

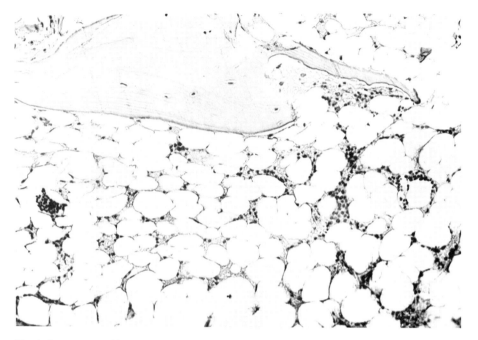

Fig. 1. Bone marrow biopsy of patient C. T.: Severe hypoplasia at diagnosis in January 1976

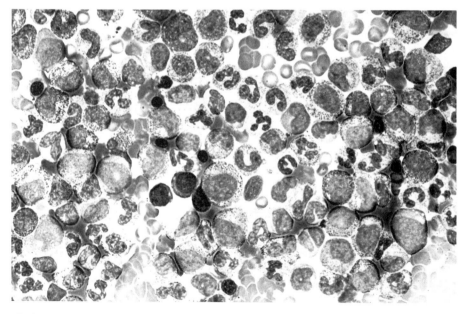

Fig. 2. Bone marrow aspirate of patient C. T.: Hypercellularity with increased granulopoiesis in April 1977

this time. Cytogenetic studies of bone marrow cells, however, exhibited mitoses with chromosome numbers ranging between 46 and 52. In the metaphases with 52 chromosomes, the supernumerary chromosomes were identified as number 8, 9, 12, 17 and 22 by Giemsa banding technique; additionally, a small, unidentifiable marker chromosome was present. By cytochemical studies, a peroxydase defect was demonstrated in 80% of neutrophils and monocytes.

Under treatment with thioguanine (120 mg/day), peripheral blood values improved, but thrombocytopenia persisted. The course was characterized by an increase of monocytes, progressively becoming more and more immature, in peripheral blood and bone marrow. By repeated cytogenetic studies, the hyperdiploid karyotypes were demonstrated to persist unchanged. After a two-year course and a 4-week-period of monoblastosis with cell counts up to 250×10^9/l, the patient died on February 6th, 1978.

At autopsy, a dense diffuse infiltration of the bone marrow, the liver (2800 g), the spleen (2400 g) and the lymph nodes was demonstrated.

Discussion

Aplastic anemia, irrespective of its origin, is associated with PNH or at least a proportion of red cells with the characteristic defect, in 10 to 20 per cent of patients [8]. In some cases of PNH, severe bone marrow insufficiency occurs within the course of the disease, whereas in others primarily aplastic anemia is present with subsequent development of the features of PNH. Thus, the term "syndrome of aplastic anemia-paroxysmal nocturnal hemoglobinuria" has been proposed [16]. Nowadays, PNH has been accounted to the preleukemic states, since several cases developing acute leukemia have been described [12, 13, 14, 18, 19, 21]. PNH, however, not only occurs in aplastic anemias and acute leukemias, but also in other myeloproliferative disorders and thus seems to be a syndrome rather than a specific disease entity [2]. An association of the red cell defect with CML has to our knowledge been described only twice [17, 20]. In both cases, the red cell defect has been observed when bone marrow cellularity was increased, not decreased as in this case.

Besides the atypical onset of the CML, there were additional extraordinary findings: Although being Ph[1] positive, the survival time of the patient was unusually short. Additional chromosome aberrations, obviously resulting from clonal evolution, appeared unusually early in the course of the disease, and the acquired, partial peroxydase defect of the mature neutrophils and monocytes as well as the final occurrence of monoblastosis is also atypical for CML. These observations would appear to make this case unique.

The main question this case has raised however is: Are aplastic anemia and the subsequent CML two distinct diseases, or is it one disease which exhibits different pictures at different times. By cytogenetic studies it is known that in the leukemic phase as well as in the preleukemic phase, e.g. of refractory anemia, identical clonal chromosome aberrations can mostly be found, leading to the conclusion that this may be one disease. In our case, the demonstration of the Ph[1] chromosome during the aplastic phase would have been the clue to the puzzle, but unfortunately chromosome studies were not done at this stage. It thus, can only be speculated whether the patient had CML from the onset, primarily presenting bone marrow hypoplasia which proceeded to overt leukemia, or whether the patient had true aplastic anemia with subsequent development of a leukemia characterized by a particularly bad course.

The observation that in the phase of "remission" the Ph[1] chromosome was already present, and the further course of the disease as well as the appearance of additional chromosome aberrations retrospectively characterize this phase as chronic stage of CML.

It remains questionable whether the improvement of values in the peripheral blood was due to the action of anabolic steroids or occurred spontaneously. It appears unlikely, however, that the anabolic steroids have induced the malignant transformation.

It thus seems more reasonable to assume that in this patient an unusual course of CML has been present beginning with an aplastic phase and subsequently resulting in a fatal course of blast crisis. And this case demonstrates that aplastic anemia may be defined as a syndrome which exhibits both a uniform clinical and morphological picture, but may have a divergent pathogenesis and final outcome.

With financial support of the Ministerium für Wissenschaft und Forschung des Landes Nordrhein-Westfalen.

References

1. Aksoy, M., Dincol, K., Erden, S.: Acute leukemia due to chronic exposure to benzene. Amer. J. Med. 52, 160–165 (1972)
2. Beal, R. W., Kronenberg, H., Firkin, B. G.: The syndrome of paroxysmal nocturnal hemoglobinuria. Amer. J. Med. 37, 899–914 (1964)
3. Brauer, M. J., Dameshek, W.: Hypoplastic anemia and myeloblastic leukemia following chloramphenicol therapy. Report of 3 cases. New Engl. J. Med. 277, 1003–1005 (1967)
4. Cohen, T., Creger, W. P.: Acute myeloid leukemia following seven years of aplastic anemia induced by chloramphenicol. Amer. J. Med. 43, 762–770 (1967)
5. DeGowin, R. L.: Benzene exposure and aplastic anemia followed by leukemia 15 years later. J. Amer. Med. Ass. 185, 748–751 (1963)
6. Delamore, I. W., Geary, C. G.: Aplastic anaemia, acute myeloblastic leukaemia, and oxymetholone. Brit. Med. J. 2, 743–745 (1971)
7. Dougan, L., Woodliff, H. J.: Acute leukaemia associated with phenylbutazone treatment: a review of the literature and report of a further case. Med. J. Austr. 1, 217–219 (1965)
8. Editorial: Paroxysmal nocturnal haemoglobinuria and leukaemia. Brit. Med. J. 3, 483–484 (1969)
9. Fraumeni, J. F.: Bone marrow depression induced by chloramphenicol or phenylbutazone. Leukemia and other sequelae. J. Amer. Med. Ass. 201, 828–834 (1967)
10. Hamer, J. W., Gunz, F. W.: Multiple aetiological factors in a case of acute leukaemia. N. Z. Med. J. 71, 141–142 (1970)
11. Hellriegel, K. P., Gross, R.: Follow-up studies in chloramphenicol-induced aplastic anaemia. Postgrad. Med. J. 50, Suppl. 5, 136–142 (1974)
12. Holden, D., Lichtman, H.: Paroxysmal nocturnal hemoglobinuria with acute leukemia. Blood 33, 283–286 (1969)
13. Jenkins, D. E., Hartmann, R. C.: Paroxysmal nocturnal hemoglobinuria terminating in acute myeloblastic leukemia. Blood 33, 274–282 (1969)
14. Kaufmann, R. W., Schechter, G. P., McFarland, W.: Paroxysmal nocturnal hemoglobinuria terminating in acute granulocytic leukemia. Blood 33, 287–291 (1969)
15. King, J. B., Burns, D. G.: Aplastic anaemia, oxymetholone and acute myeloid leukaemia. S. Afr. Med. J. 46, 1622–1623 (1972)
16. Lewis, S. M., Dacie. J. V.: The aplastic anaemia – paroxysmal nocturnal haemoglobinuria syndrome. Brit. J. Haematol. 13, 236–251 (1967)

17. Mehta, B. C., Bhatt, P. D., Iyer, P., Golwala, A. F., Pathel, J. C.: Paroxysmal nocturnal hemoglobinuria in a case of chronic myeloid leukemia. Ind. J. Med. Sci. *27*, 249–252 (1973)
18. Seaman, A. J.: Sequels to chloramphenicol aplastic anemia: Acute leukemia and paroxysmal nocturnal hemoglobinuria. Northwest Med. (Seattle) *68*, 831–834 (1969)
19. Tsevrenis, H., Pouggouras, P., Simos, A., Contopoulou, Ir., Papazacharis, A., Dariotis, A.: Evolution d'une hémoglobinurie nocturne paroxystique, maladie de Marchiafava-Micheli, en leucose aigue. Nouv. Rev. franc. Hématol. *10*, 274–277 (1970)
20. Tso, S. C., Chan, T. K.: Paroxysmal nocturnal haemoglobinuria and chronic myeloid leukaemia in the same patient. Scand. J. Haematol. *10*, 384–389 (1973)
21. Wasi, P., Kruetrachue, M., Na-Nakorn, S.: Aplastic anemia-paroxysmal nocturnal hemogobinuria syndrome – acute leukemia in the same patient. The first record of such occurrence. J. Med. Ass. Thailand *53*, 656–662 (1970)

2.6 General Discussion

Moderators: B. M. Camitta, E. C. Gordon-Smith

Gordon-Smith: I will start by putting a question to the floor. Occassionally one is lucky enough to see a patient actually developing aplastic anemia before the marrow becomes empty. This can happen in the post hepatitis patients where they may have serial blood counts before aplasia starts. In our experience with 5 such patients, phagocytosis of hemopoietic cells by macrophages is very striking but this phenomenon disappears as the marrow becomes more empty. I would like to ask Dr. te Velde if he has seen such a phenomenon and if he thinks that the reticuloendothelial system plays a part in the pathogenesis of aplastic anemia.

te Velde: You must have been very lucky to have the opportunity to look at patients' marrow as they develop aplasia. I have not had the chance to study patients at this time. In the acute phase when a patient is first diagnosed as having aplasia, erythrophagocytosis in the marrow may be conspicuous but I wonder if you have to give it a special role in pathogenesis. Erythrophagocytoses is seen in other hematological disorders and is probably not specific.

Wickramasinghe: We studied the ultrastructure of the marrow in patients with drug induced aplastic anemia, though not especially early in the disease. The striking feature was not phagocytosis of cells but lipid loading of macrophages. There was a spectrum of cells from typical macrophages to large cells which looked like typical fat cells.

Gordon-Smith: Do you know where they came from?

Wickramasinghe: No. The marrow is a closed compartment and if you lose haemopoietic cells the space has to be filled up with something. Animal studies have shown the marrow space to fill with macrophages and the same seems to be happening in humans.

Camitta: In the cooperative aplastic anaemia study I reviewed all the biopsies from 100 patients, each of whom had been diagnosed within a few weeks, at most, after the onset of symptoms. There was no evidence of increased phagocytosis to any marked degree, though it could occur to some extent.

Gordon-Smith: It has been said that maybe there are many different sorts of aplastic anaemia but if we think that every case of aplastic anaemia that we are going to talk about today may have a different pathogenesis, we are not going to get very far. I think we must accept that patients who have what we understand by acquired aplastic anaemia have a specific disease and there is a common pathogenesis. There may be a few exceptions but in the main it is the same. Professor Heimpel would like to say something about drug induced and idiopathic anaemias.

Heimpel: When the etiology of aplastic anaemia is discussed, it is common use to distinguish cases induced by drugs or virus from so-called "idiopathic" cases. The incidence of such idiopathic cases varies between 24 and 54% in larger series[1]. From the analysis of our own material of more than 100 cases, we were not able to distinguish definitely between the two groups. First, the classification of any case as drug induced or idiopathic depends largely on the scrutiny of the drug history taken, and the criteria applied to decide whether or not any exposure is regarded to be of possible etiological relevance. Table 1 gives an example of how the history of exposure may change when the "normal" drug history is completed by a special search, inducing data obtained not only from the patient himself, but also from his relatives, his physician and his pharmacist. Second, when we analyzed the main clinical date and course of disease in a smaller number of recent cases, no difference could be detected between patients exposed to potentially harmful drugs within 6 months prior to the first symptoms and the patients without such exposure (Table 2). It has to be stressed that these observations relate only to patients with prolonged, histologically proven aplasia, but not to transient pancytopenia without clear

---

1 Heimpel, H. and Kern, P.: Arzneimittelbedingte Panmyelopathien. Blut *33*, 1 (1976)

Table 1

"Normal" drug history
1971:   Cephalotin; Binotal
1974:   Sulfafuradantin; Iron, Folic-Acid and Vitamin-B$_{12}$; Combionta; Adumbran; Esberitox;
         Ruticalzon; Ultralan; Agiolax; Vitamin C

"Special" drug history
1970/71: Chloramphenicol
1972:   Mysteclin ovula; Senecion; Jellin-Ointment; Megacillin forte; Baycillin; Mallebrin;
         Sulfafuradantin
1973:   Tussipect; Gelondia; Dolo-Attritin; Senecion; Kendural; Gyno-Sterosan; Evazol; Ferro-
         sanol duodenal
1974:   Chloramphenicol; Venostasin; Dolantin; Macrodex; Ultracorten; Paraxin; Ubretid;
         Doryl; Nuran; Eryfer; Atosil; Valium; Sterofundin B; Kendural C; Allional; Itridal;
         Sulfafuradantin
         Albothyl; Ferrosanol duodenal; Methergin
         Duogynon simplex; Hirudoid-Ointment; Gelonida; Neurovegetalin; Dragee 19;
         Dihydergot

Table 2

	Aplastic anaemia	
	Drug induced	"Idiopathic"
Age (years): mean	46	44
range	15–70	15–74
Male/female	5/14	15/13
Peripheral blood:		
platelets/µl	24,800 ± 21,300	18,500 ± 23,750
leucocytes/µl	2,700 ± 1,800	2,460 ± 1,450
granulocytes/µl	1,120 ± 1,160	880 ± 820
reticulocytes/µl	23,100 ± 24,250	20,600 ± 24,350
hemoglobin [9%]	8.1 ± 2.0	7.5 ± 2.5
Mean survival	2.6 ± 2.0	1.5 ± 1.2
Patients alive two years after diagnosis	60%	50%

histology. From these observations we conclude that the term idiopathic anaemia may be merely a confession of our ignorance. Probably all cases of aplastic anaemia are at least triggered by external factors. The main unsolved question then remains the basis of the individual hypersusceptibility to a variety of chemical or microbiological factors.

Haak: Do you have any information on the hereditary or familial aspects of this disease? We have looked at it and we could not find any positive family history.

Heimpel: There is no increased frequency of aplastic anaemia in nonidentical siblings or relatives of affected people compared with the general population. A few instances of identical siblings both exposed to chloramphenicol who each developed aplastic anaemia have been reported.[1]

---

1 Best, W. R.: Cloramphenicol-associated blood dyscrasias. J. Amer. Med. Ass. *201*, 181 (1967)
  Nagao, T., Mauer, A. M.: Concordance for drug induced aplastic anemia in identical twins. New Engl. J. Med. *281*, 7 (1969)
  Nora, A. H., Fernbach, D. J.: Acquired aplastic anemia in children. Texas Med. *65*, 38 (1969)

Gordon-Smith: The association of two rare events, aplastic anaemia and identical twinning, is clearly important in relation to pathogenesis if it occurs with unexpected frequency but that does not seem to be the case.

Barrett: It is important to clarify whether we think aplastic anaemia is an ongoing process or a single event. Aplasia may be rare because few people suffer two or more shots at the marrow. The first hit may produce subclinical damage and only subsequent hits aplasia.

Heimpel: Yes, one must consider not only genetic disposition but also acquired disposition. The results we presented earlier showed that a long time after partial or complete marrow recovery there is a residual defect in the haemopoietic stem cells. It may be that subclinical damage exists also before the haemopoietic tissue is hit by the external factor.

Weber: Did you observe any increase of congenital malformations in patients with acquired aplastic anaemia?

Heimpel: No, we did not.

Lohrmann: The relationship between aplastic anaemia, PNH and leukaemia is interesting. It was suggested that the damage to the marrow which induces aplasia at the same time induces clonal proliferation. This hypothesis explains the observation that there is first remission in aplastic anaemia and then leukaemia develops by proliferation of that particular clone.

Heimpel: The question of leukaemia has been raised very often. If you have good biopsies and you study the morphology very carefully you can sort out cases which are, from the beginning, acute leukaemia. If these cases are excluded, the incidence of leukaemia is very low, less than 1%.

Camitta: The question aplastic anaemia and leukaemia raises the question as to whether aplastic anaemia has more than one etiology.

You can have empty bone marrow with leukaemia at a time when there are not many leukaemic cells crowding out the bone marrow. There may possibly be some humoral suppressor or other controlling factor of haemopoiesis that has gone awry. In this regard, aplastic anaemia and leukaemia are potentially very interesting.

te Velde: If you look at the bone marrow of patients with pancytopenia with a clinical diagnosis of aplastic anaemia, you can sort out patients with preleukaemia and we have several examples and I have shown you one. On the other hand, we had two patients we had diagnosed as aplastic anaemia who developed leukaemia after 4 to 6 years of aplastic anaemia and we had no parameter to say it was a preleukaemia at that time. In pathology, you have the development of atrophy and the development of hypoplasia before the onset of malignancy in all sorts of organs. If something is going awry in the epithelial system, for example, a dense inflammatory infiltrate occurs, possibly reacting to altered cells but you cannot classify these cells as malignant at that time. The rarity of leukaemia following aplastic anaemia may be because the patients die before they have a chance to develop leukaemia and perhaps, with better support keeping the patients alive, leukaemia will become a more frequent complication.

Camitta: In our study of 200 children with acute lymphoblastic leukaemia 2 to 3% presented with an aplastic marrow. It was not possible to distinguish these leukaemic patients from aplastic patients by marrow morphology alone but we did find that those who responded rapidly to corticosteroids were those who developed leukaemia. The increase in i-antigen on red cells and production of HbF in many of these patients probably reflects a return to fetal erythropoiesis and are not reflections of the pathogenesis.

Lohrmann: I think the observation that acute leukaemia may develop in aplastic anaemia supports the concept of a stem cell abnormality. It ist difficult to see how abnormalities of microenvironment could induce leukaemia.

Moore: It is a crucial point for us to understand whether or not aplastic anaemia is a clonal stem cell disorder, as are myelofibrosis, polycythaemia vera and CML. A clonal stem cell disorder which allows self-maintenance but fails to differentiate would give the morphological picture one sees in aplastic anaemia and also perhaps the rare progression to acute leukaemia. The work of Cotter & Essex[1] with feline leukaemia is relevant here. Following infection with the feline leukaemia virus only about 10% of cats go on to develop leukaemia – the remainder have a sprectrum of disease from myelofibrosis to aplastic anaemia. I do not think we should dismiss the leukaemic aspect of aplastic anaemia as a mis-diagnosis.

---

1 Cotter, S. M., Essex, M.: Animal model: Feline acute lymphoblastic leukemia and aplastic anemia, Amer. J. Path. *87,* 265 (1977)

Camitta: If aplastic anaemia were a clonal stem cell disorder in which there was a failure to replicate, how do you explain the persistence of aplasia? It seems that any remaining stem cells should repopulate the marrow.

Moore: I was saying that stem cells replicate but do not differentiate so they remain, ineffective, in the marrow.

Camitta: You have to postulate some agent that damages all stem cells or there would be normal stem cells left.

Moore: The normal stem cells would be differentiated out.

Gordon-Smith: The concept of stem cells suiciding by differentiation is very interesting. Perhaps this could explain some of the lag phase, as remaining stem cells differentiate out. One also has to explain spontaneous recovery, excluding the development of abnormal clones such as PNH and acute leukaemia. One thing that is interesting morphologically in the recovery phase is that, early on, one finds islands of haemopoieitic cells in which only one cell type is represented, it may be erythroblasts or an island of granulopoiesis. Does that fit with the idea of regenerating stem cells?

Moore: That is what is seen in the spleen of irradiated mice. That would seem to be imposed by this particular microenvironment, that is, that the stem cell is replicating and differentiating along a particular path rather than a non-commital one.

Heimpel: It is not only the fact that they are of the same cell line but, in such a cluster of cells, all cells have the same maturity. This could possibly be due to circulating stem cells settling in the marrow and giving rise to synchronised colonies.

Camitta: While we are on the subject of stem cells I would like to ask Dr. Moore if it is possible that aplastic anaemia is due to wearing out of stem cells. Perhaps some people's stem cells are not immortal and their stem cells lose their ability to maintain their numbers because they are unable to repair acquired damage.

Moore: Patients who receive intensive, long term chemotherapy for CML or other diseases do not show features of delayed type aplastic anaemia once they are in remission so I doubt whether failure of stem cell self-renewal, as a genetic variation, is important in aplastic anaemia. We looked at stem cell self-renewal as a function of age in experimental animals. We found one mouse strain (NCC), that ran out of stem cells and became totally aplastic at 12 months of age. It turned out this was caused by a high concentration of fluoride in the diet and there was also an excessive growth of teeth. High doses of fluoride given to other strains of mice produced a disease similar to aplastic anaemia.

Mangalik: We tested patients with pancytopenia for PNH and found only those patients with patchy or cellular marrows had PNH – those with hypocellular marrows did not have PNH clones. There is a difference between PNH positive and negative patients with pancytopenia.

Gordon-Smith: I think people might be able to show examples which would be exceptions to this rule but the PNH may be only a laboratory phenomenon and not produce clinical features.

3 Kinetics and Control of Hemopoietic Stem Cells

3.1 Aplastic Anemia – A Stem Cell Disorder?

B. Kubanek

The pathogenesis and often the etiology of aplastic anemia is still obscure [1, 4, 10]. However, it is generally recognised that there is a basic defect in the pluripotent stem cell compartment which results in a diminished feeding of later, more differentiated compartments, eventually leading to a failure of the production of mature cells.

The chain of events leading to haemopoietic failure has not been elucidated for several reasons, the most important being our lack of understanding of the organisation and control mechanism of stem cells, particularly in humans. In addition, there are technical difficulties, namely the lack of an assay for human pluripotent stem cells and the sampling of a well defined and representative cellular substrate.

Numerous experimental and clinical evidence implies that the hemopoietic stem cell is responding to a high demand for increased production, even over prolonged periods of time without failing. In addition there is evidence that the pluripotent stem cells are capable of restoring their compartment size after severe and even prolonged damage (e.g. chronic irradiation). Assuming that the human stem cells behave similarly to that of the mouse the question arises, why does the stem cell system fail to recover in aplastic anemia? Several possibilities have been discussed one being unbalanced differentiation and self-replication [1, 4]. Evidence exists that a critical compartment size of pluripotent stem cells must be accomplished after damage before differentiation can occur, otherwise a depletion of the stem cell compartment occurs.

A possible mechanism for this could be an undesirable ability of the pluripotent stem cells to differentiate at the expense of self-replication. It is very unlikely that this is an effect of an overriding strong stimulus, but rather an intrinsic abnormality of the stem cell compartment. A faster ageing of the stem cells induced by extensive proliferation has been discussed [8] as a possible cause. Again this event must be preceded by a loss of compartment size regulation.

In this context, the question of a latent stem cell defect should be discussed. This was suggested by Morley et al. [9] in myleran-treated mice and by our group [3, 7] for patients with aplastic anemia in remission. However, the pathomechanism which allows the latent stem cell defect to become an overt failure is unknown! There may be an acquired or congenital defect, possibly clonal, which renders the stem cell more vulnerable to a secondary extrinsic damage.

Recently humoral factors are again in discussion for regulation of differentiation of the pluripotent stem cell rather than a stochastically controlled process for the determination of early stem cells [5, 6]. Although this is at present a rather theoretical proposal, a failure or inhibition of such a regulating process should be discussed as a possible pathomechanism of aplastic anemia.

Aplastic anemia could be caused by an altered microenvironment which does not support an indefinite stem cell reproduction and a regulated differentiation. This possibility is often discussed as unlikely since the majority of patients with aplastic anemia can be cured by syngeneic or HLA- or MLC matched bone marrow transplants from siblings [11]. However, one cannot exclude the possibility that critical cell(s) for the re-establishment of the microenvironment are also transplantable. The long term culture system of Dexter [2] seems to be a useful system for further studies of the matrix supporting a regular structured haemopoiesis. Even assuming that the poorly defined control system of murine hemopoieses applies to human hemopoiesis there are a number of difficulties in interpreting the data collected from patients with aplastic anemia. Due to the lack of an assay for a human pluripotent stem cell we can only infer from the purely descriptive data of the committed erythroid and myeloid compartment that there must be a depleted earlier stem cell compartment. Interpretation of these data are hampered by the fact that we are only able to measure stem cell concentrations rather than absolute numbers. We do not know fluxes from one compartment to another and we cannot measure proliferation rates in these depleted compartments due to the very low incidences of the cells under study.

The riddle of aplastic anemia can only be solved with methods which enable us to measure early human stem cells and to characterize the selfreplication and differentiation of those in vitro. Dexter's long term culture system applied to human haemopoiesis may empower us to examine and elaborate the stem cell as well as the necessary microenvironmental factors.

Supported by SFB 112 of the Deutsche Forschungsgemeinschaft

References

1. Boggs, D. R. and Boggs, S. S.: The pathogenesis of aplastic anemia: a defective pluripotent hematopoietic stem cell with inappropriate balance of differentiation and self-replication. Blood 48, 71–76 (1976)
2. Dexter, T. M. and Moore, M. A. S.: In vitro duplication and "cure" of haemopoietic defects in genetically anaemic mice. Nature 269, 412–413 (1977)
3. Hansi, W., Rich, I., Heimpel, H., Heit, W. and Kubanek, B.: Erythroid colony forming cells in aplastic anemia. Brit. J. Haemat. 37, 483–488 (1977)
4. Heimpel, H. and Kubanek, B.: Pathophysiology of aplastic anemia. Brit. J. Haemat. 31, (Suppl.), 57–68 (1975)
5. Iscove, N. N.: Erythropoietin-Independent Stimulation of Early Erythropoiesis in Adult Marrow Cultures by Conditioned Media from Lectin-Stimulated Mouse Spleen Cells. ICN-UCLA Symposium on Hemopoietic Cell Differentiation, (1978) in press
6. Johnson, G. R. and Metcalf, D.: Pure and mixed erythroid colony formation in vitro stimulated by spleen conditioned medium with no detectable erythropoietin. Proc. Natl. Acad. Sci USA 74, 9, 3879–3882 (1977)
7. Kern, P., Heimpel, H., Heit, W. and Kubanek, B.: Granulocytic progenitor cells in aplastic anemia. Brit. J. Haemat. 35, 613–623 (1977)
8. Micklem, H. S., Ogden, D. A. and Payne, A. C.: Ageing, Haemopoietic Stem Cells and Immunity. In: Haemopoietic Stem Cells. Ciba Foundation Symposium 13. Amsterdam: Associated Scientific Publishers, 1973, p. 285–297

 9. Morley, A., Trainor, K. and Blake, J.: A primary stem cell lesion in experimental chronic hypoplastic marrow failure. Blood *45*, 5, 681–688 (1975)
10. Stohlman, F., jr.: Aplastic anemia. Blood *40*, 2, 282–286 (1972)
11. Storb, R. and Thomas, E. D.: Marrow transplantation for treatment of aplastic anemia. Clinics Haematol. *7*, 3, 597–609 (1978)

3.2 Mechanism of Damage to the Stem Cell Population

R. Schofield

Introduction

Bone marrow aplasia will be recognised clinically at any time there is a failure to produce mature cells or their recognisable precursors. This situation could conceivably arise in a number of ways: (a) failure to induce differentiation of committed precursor cells (eg. erythropoietin-responsive cells) or destruction of that population; (b) failure of the earliest committed precursor cells (e.g. erythroid burst-forming cells) to proliferate or loss of these cells; (c) failure of differentiation from the pluripotent stem cell; (d) total lack of haemopoietic stem cells.

In fundamental terms only (c) and (d) are truly aplastic anaemias (as opposed to mature cell aplasias) in that no cells whatsoever committed to a specified differentiation pathway are being produced. The situation arising in (d) will give rise to a total marrow aplasia and it is this aspect, and the less complete situation of reduction in the stem cell population, which this communication sets out to consider.

There is yet no way in which we can study the pluripotent cells of the marrow viz. The spleen colony-forming cells (CFC-S)[1] in any animal other than the mouse and even in that animal we need to clarify the relationship between the CFC-S and the haemopoietic stem cell, which have been equated and yet which offer evidence of non-identity.

An outstanding example of such evidence is found in the W/W^v mouse in which virtually no CFU-S can be detected but which nevertheless produce a near-normal supply of erythrocytes and a totally normal supply of other blood elements for a normal mouse life-span. Clearly haemopoietic stem cells must exist to feed the supply but they do not form spleen colonies.

The CFU-S in the mouse show no decline in either numbers or repopulating ability throughout a lifepsan of 2–3 years [9]. Harrison [4] showed that the decline of repopulating ability in the CFU-S of the W/W^v mouse transplanted with normal marrow shows little decline over a number of generations and has recently [5] offered evidence that the decline which does occur is due, not to aging of the CFU-S population, but to the transplantation of the cells. In other words, for all intents and purposes the stem cell shows no loss of potential through one or several generations and is, therefore, for all practical consideration, immortal.

---

1 The following abbreviations are used throughout: spleen colony-forming cell – CFC-S; spleen colony-forming unit (i.e. those spleen colony forming cells which actually form colonies in the spleen after injection into irradiated mice) – CFU-S

Yet evidence has been accumulating for many years to indicate that the CFU-S is a heterogeneous population insofar as its repopulating capacity is concerned. The loss of repopulating capacity as a result of serial transplantation was first observed by Siminovitch et al and published in 1964 [21]. Others have refined the procedure and tried to eliminate the decline by increasing the size of the grafts and extending the intervals between grafts but without success [9, 16]. The evidence produced suggests that the CFU-S actually run out i.e. exhibit mortality.

Worton et al. [23] as long ago as 1969 physically separated CFU-S having different reproductive capacities and I inferred the same conclusion from data obtained from comparison of the growth rates of CFU-S from different haemopoietic sources in the spleen of a standard, radiated mouse [18]. Our experiments with isopropylmethane sulphonate (IMS) [19] showed elimination of CFU-S with high reproductive capacity and suggested that those with low reproductive capacity had a greater probability of differentiation with consequent loss of CFU-S identity. This was probably the first clear suggestion of an age-structure in the CFU-S population. There has been a number of observations in which there has been an elimination or enrichment of CFU-S with higher than average reproductive capacity [12, 13, 8]; each consistent with the concept of an age-structure in the CFU-S population. Rosendaal et al. [17] equate aging and functional maturation.

Most cytotoxic treatments e.g. chronic or acute radiation, with or without marrow transplantation, and many drugs, of which myleran is a well-documented example, result in incomplete recovery of CFU-S in the marrow, both quantitatively and qualitatively [14]. On the other hand treatment with other compounds, of which IMS is the best example, will recover from a 99% CFU-S depletion back to normal, both quantitatively and qualitatively, within 10–14 days. The loss of repopulating ability of marrow CFU-S which have been depleted and therefore forced to repopulate has been attributed [6, 7] to the using up of reproduction potential as a result of forced "aging" in a population in which there is a limited capacity for self-reproduction. This cannot explain why aging is not apparent in recovery from IMS nor, indeed, as a result of the approximately 200 divisions (on average) undergone by a CFU-S during one normal life-span [20].

However, the existence for what appears to be an immortal stem cell, compelling evidence for an age structure in the CFU-S population and the dissociation (in W/W^v mice) of CFU-S and stem cells presents us with a fundamental problem in the understanding of the maintenance of haemopoiesis. Its solution will undoubtedly contribute to our understanding of residual marrow damage, drug-induced hypoplasia, some aplastic conditions etc.

Stem Cell – CFU-S Relationship

The paradox of an immortal stem cell and an age-structured CFC-S is best explained by considering them as two separate entities, both pluripotent and both capable of self-renewal, the stem cell indefinitely, the CFC-S to a limited extent. The corollary to the indefinite self-renewal of the stem cell is that the cell also

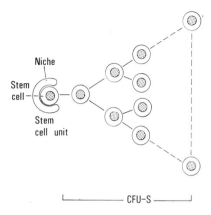

Fig. 1. A representation of the haemopoietic stem cell unit (stem cell and niche) and its progeny, the spleen colony-forming cells

never matures or differentiates to any other state. This is conceivable if the cell is held in a functional and, possibly, geographic situation in association with a controlling environment which would most probably be a cellular one (Figure 1). This environment I have called a stem cell "niche" and it is, at least partly, reproduced in vitro in the culture system described by Dexter and Lajtha [2] and Allen and Dexter [1] and discussed later in this book. Thus the only function of this cell is reproduction. Its progeny, dissociated from the parental niche, will become the producer of mature blood cells, being capable of reproduction and being separated from the environment which inhibits change. This cell, therefore, is the first generation colony-forming cell. The evidence of an age-structure in the CFC-S population is now understandable because, without any change occurring in the true stem cell, the colony-forming cells themselves can mature and during the maturation process undergo a number of divisions. With increasing maturity they approach the stage a which they will differentiate and cease to be CFC-S and therefore, also, with each division, they become less reproducible as CFC-S. In other words they age. Thus what is being suggested is that the CFC-S population consists of cells with a range of self-renewal potential and which is in a constant state of production, maturation and loss.

Quality of CFU-S

If either the primary stem cell or some essential part of the niche is absent or deficient then there will be a deficient production, or no production at all, of CFC-S and a hypoplasia or aplasia must result. Two genetic situations are presented to us in the W/W^v and Sl/Sl^d mice. In the former a normal environment (niche) enables normal stem cells transplanted from normal litter mates to function in a normal way, correcting the haemopoietic deficiency of the mouse. On the other hand the Sl/Sl^d defect cannot be corrected since its stem cell environment cannot support either its own (normal) stem cells nor those transplanted from normal littermates.

Many of the hypoplasias and aplasias that arise are the results of treatment with cytotoxic agents and these situations can be reproduced experimentally. The most complete way is by whole-body radiation at a dose which will kill all stem cells. Since the animal's marrow can be repopulated by transplanted bone marrow it has been assumed that the environment is relatively resistant to radiation or recovers very quickly from it. Thereafter, by measuring the CFU-S in, say, the femur of the mouse a measure can be made of the level to which recovery takes place. Nevertheless it has become clear that this is an incomplete index of the recovery, and therefore of residual damage. After recovery, a given number of CFU-S is less efficient in repopulating the irradiated mouse than the same number of CFU-S from normal marrow. Thus we must attempt to assess the qualitative defect in CFU-S.

The only objective way in which the quality of CFU-S can be measured is by examining how well they repopulate a depleted mouse or, in practical terms, what would be the minimum number of CFU-S required to permit survival in a heavily-irradiated mouse. This is obviously a procedure requiring such large numbers of animals and beset with so many variable influences that it cannot be considered feasible except, possibly, for an occasional measurement. The measure I have adopted involves assaying the CFU-S content of a number, usually 50, spleen colonies derived from any sample of marrow etc. 11 days after grafting it into an irradiated mouse. The number of CFU-S in a colony lies in the range from 0 to>100. If the percentage of colonies containing up to 1 (i.e. 0 and 1), 5, 10 etc. is plotted a cumulative distribution curve is obtained, examples of which are shown in Figure 2. From the data the probability of self-renewal, p, can be calculated and comparisons can readily be made between the colonies produced under standard conditions by CFU-S from different sources. This procedure, also, requires large numbers of mice (150–200 per assay) but we (Schofield et al., in press) have shown that, in general, a fairly good indication of probability of self-renewal can be obtained from a measure of mean CFU-S/colony alone and this can be obtained much more economically.

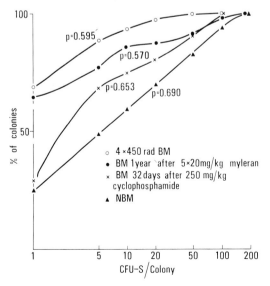

Fig. 2. Cumulative plots of the percentage of CFU-S per spleen colony in colonies produced from bone marrow of normal mice, ▲, mice treated with 250 mg/Kg cyclophosphamide 32 days earlier, x, mice treated 5 times with 20 mg/Kg myleran at 2-weekly intervals approximately 1 year earlier, ●, and mice given 4 whole-body exposures to 450 rad x-rays at 24 day intervals, the last 24 days before the marrow was assayed, o.

Figure 2 gives examples of the curves obtained by plotting the cumulative distribution of CFU-S content of colonies made from marrow of mice which have "recovered" after different treatments i.e. 5×20 mg/Kg myleran, 4×450 rad whole body x-rays, 250 mg/Kg cyclophosphamide. It is clear that in the treated and "recovered" mice a larger proportion of colonies produce small numbers of CFU-S. If the proportion containing up to 5 or up to 10 CFU-S/colony is read off this becomes obvious. The result is that the less reproductive the CFU-S the higher will the curve be above the curve obtained from normal marrow CFU-S.

The progressive deterioration of CFU-S as a result of serial transplantation can be diagnosed from such curves, which are worse after each transfer.

Repopulation of the Primary Stem Cell Population

If we accept the difference between the true stem cell and the CFC-S then we must also explain how the true stem cell population can recover from its own ablation. When bone marrow cells are injected the suspensions contain CFC-S of various "ages" and it appears to be these cells which effect the repopulation. However, these cells are only derivatives of the true stem cells. Two possibilities exist for the re-establishment of the stem cell population by CFC-S.
a) Only the first progeny of the stem cell can re-occupy the niche and assume the role of the immortal cell. This is feasible since the cell differs from the true stem cell only in its associations. If this is what occurs then the proportion of available niches filled must be quite small but each stem cell would be of the same quality as the original.
b) CFC-S of any age may home into niches and, because of the niche environment, be arrested from further maturation. It would seem highly unlikely, however, that these cells would retrace the maturation steps they had already taken and rejuvenate. The alternative is that they would retain the degree of "middle-age" they had reached before occupying the niche. In this way it is conceivable that all the niches could be filled but mostly with stem cells of lower proliferative capacity than the original stem cells. Figure 3 illustrates the effect of such a repopulation.

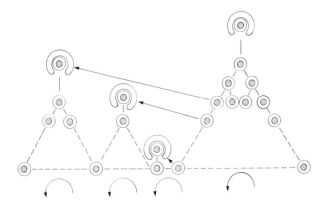

Fig. 3. A postulate of the repopulation of depleted stem cell niches by spleen colony-forming cells of different ages. This diagram illustrates the way in which loss of self-renewal capacity may occur as a result of such repopulation

In either of these situations a smaller number of CFC-S will be produced but in the second case a higher proportion of low-reproducing CFU-S will be found. The experimental data show this to be the case and therefore the evidence would seem to support the second possibility.

On the other hand there is evidence from other haemopoietic cell populations e.g. maturing erythroid-committed precursor cells that the number of divisions which can be achieved during the maturation process is variable. If this applies similarly to the CFC-S population then it would be expected that the CFC-S produced from a depleted stem cell population would undergo one or more extra divisions during their maturation in order to attempt to restore the numbers of CFC-S normally available for differentiation. Figure 4 illustrates the situation envisaged here. The result again would be a higher proportion of CFU-S of low proliferative capacity.

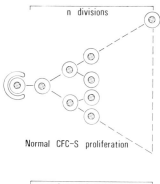

Normal CFC-S proliferation

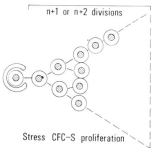

Stress CFC-S proliferation

Fig. 4. Increased production of spleen colony-forming cells may occur under conditions of high demand e.g. when the number of stem cells is reduced, by introducing extra divisions during the maturation (or aging) of the CFC-S

In any case a CFU-S/colony profile which is worse than the normal indicates either smaller numbers of primary stem cells or a normal number of lower quality stem cells – or some state between these extremes. The practical result is that the CFC-S population is qualitatively depressed. The new lower equilibrium state achieved may not be detrimental to the animal provided it is not subjected to further damaging treatment. One can see this in animals radiated and grafted, which can then produce a normal red cell, leucocyte, platelet etc. output for the rest of their lives. Nevertheless it cannot tolerate a second treatment as well as it did the first and if its marrow is grafted into a radiated host it is found to be less effective. In other words we can then see residual damage.

Preliminary Experimental Data

What has been presented here is an hypothesis on the basis of which a number of apparently anomalous observations can be explained. This does not mean the hypothesis is correct but it does act as a starting point for experiments to confirm, refute or modify it. Several such experiments are in progress but are only sufficiently progressed for preliminary data to be presented.

One of the questions which requires elucidation is the nature of the stem cell niche. The work of Dexter and Lajtha [2] and Allen and Dexter [1] can leave little doubt about the absolute requirements for a cellular environment for the maintenance of haemopoiesis in vitro. The existence of the $S1/S1^d$ mutant mouse can leave no doubt about the existence of an environment necessary for the complete functioning of stem cells in vivo. The nature and site of this environment (niches) are the subjects of current studies.

Histological evidence of early regeneration of radiation-depleted marrow would point to bone surfaces as the origin of regeneration and, by inference, of the primary stem cell sites. The experiments of Tavassoli and Crosby [22], Patt and Maloney [15] and Friedenstein et al, [3] on ectopic implantation of bone marrow indicate a requirement for bone itself before marrow can exist. Lord [11] has shown a disproportionate effect of injected ^{3}H-thymidine upon those CFU-S intimately associated with bone compared with the CFU-S in the lumen of the bone shaft. Furthermore the proportion of CFU-S in cells close to bone is higher than in axial bone marrow [10] and we (Schofield and Lord, manuscript in preparation) find that the self-renewal probability of bone-associated CFU-S and CFU-S form axial marrow are significantly different. Thus there is ample evidence that quantitative and qualitative differences exist between CFU-S of the axial marrow and those closely associated with bone. The functional significance of these differences is a problem which remains to be solved.

We have recently begun experiments, in collaboration with Drs. Humphreys and Thorne at the MRC Radiobiological Unit, Harwell, to study the effect upon CFU-S depletion and recovery of α-radiation of a narrow (~20 μ) band along bone surfaces with little radiation to the marrow as a whole. The radiation source being used is ^{239}Pu.

In studying bone marrow deficiences produced by cytotoxic agents the residual damage which often exists can be due to reduction either in numbers of stem cells or in their reproductive capacity. Reduction in numbers can occur either because the available number of CFU-S is low or because there has been a reduction in the number of niches available for occupation. Preliminary data on long-term bone marrow repopulation suggest the possibility of persistent environmental damage which may repair slowly. Injection of 10^6 or 10^7 normal bone marrow cells into lethally-irradiated mice results in faster repopulation in the group given 10^7 cells but, within 4 weeks of transplant there is no detectable difference between the number of CFU-S in the femur of either group. Both contain about 25% of the number found in control BDF_1 mice. During the next 4–6 weeks there is a gradual increase of the femoral CFU-S population to 50 or 60% of control irrespective of the size of the initial graft. Furthermore injecting a further 10^7 normal marrow cells 4 weeks after the first graft does not effect the level of CFU-S repopulation of the femoral marrow. The simplest explanation is that a limited number of sites (niches) is available for occupation and that no improvement in repopulation can be achieved by injecting more cells than the minimum number required to fill them.

If the transplantation of a standard number of bone marrow cells is delayed for 1 or 2 days after the radiation exposure the level of CFU-S recovery in the marrow is the same as when the cells are injected immediately after radiation exposure. One can deduce that there is no decay or atrophy of the unoccupied niches, nor is there any evidence of repair brought about by the presence or absence of injected cells.

Further data are being collected which, so far, are consistent with the age-structure concept of the CFU-S population. The experimental procedure is as follows: individual spleen colonies are dispersed into single cell suspensions and injected into heavily-irradiated recipients. The number of CFU-S in the colony is assayed in this way. Each secondary colony thus formed is assayed in the same way. To date no CFU-S has produced a colony containing as many CFU-S as were present in the colony from which it was taken. The inference is that secondary colonies are established from more aged cells than those which formed the primary colony.

In summary, then, a hypothesis is presented in which the haemopoietic stem cell is seen as an immortal cell, pluripotent and capable of indefinite reproduction. The immortal, non-maturing state is maintained by an environmental niche which the stem cell occupies. The progeny of the stem cell, outside the niche, is the first generation CFC-S which has some reproductive capacity as CFC-S, and pluripotentiality. As the CFC-S mature to a state in which they can respond to various differentiating stimuli they also divide. Hence as they approach the differentiable state the number of potential divisions, as a CFC-S, reduces i.e. the cell "ages". The CFC-S, if it can find a stem cell niche, becomes a stem cell, probably being arrested at the "age" it had reached. Its primary progeny will then have a lower reproductive capacity than the progeny of the original stem cell.

The aging CFC-S population is probably capable of expansion under situations of extra demand e.g. when the stem cell population is reduced. Under these circumstances the average reproductive capacity of the CFC-S will be reduced since the production of the last (extra) divison will have low self-renewal capacity as a CFC-S and these cells constitute approximately 50% of all CFC-S. This reduction in self-renewal probability can be found in many situations where recovery of haemopoiesis has followed e.g. radiation and cytotoxic therapy. Similarly low CFU-S self-renewal probability would be expected in cases of primary hypoplasia no matter how it had arisen.

Preliminary data are consistent with the concept that damage to "niches" by radiation limits bone marrow repopulation no matter how large the marrow graft. The minimal understanding of the selective effects of other cytotoxic agents so far prevents the assignation of sites of deficiency in hypoplasias and aplasias. Nevertheless the cure of primary aplastic anaemia may well be a more complex problem than simply supplying the animals with compatible haemopoietic spleen-colony forming cells or their equivalent.

Acknowledgements

This work was supported by grants from the Cancer Research Campaign and Medical Research Council.

References

1. Allen, T. D. and Dexter, T. M.: Surface morphology and ultrastructure in murine granulocytes and monocytes in long-term liquid culture. Blood Cells 2, 591–606 (1976)
2. Dexter, T. M. and Lajtha, L. G.: Proliferation of haemopoietic stem cells in vitro. Brit. J. Haematol. 28, 525–530 (1974)
3. Friedenstein, A. J., Chailakayan, R. K., Latsinik, N. V., Panasuk, A. F. and Keiliss-Borok, I. V.: Stromal cells are responsible for transferring the microenvironment of the hemopoietic tissue. Transplantation 17, 331–340 (1974)
4. Harrison, D. E.: Normal production of erythrocytes by mouse bone marrow continues for 73 months. Proc. Nat. Acad. Sci. 70, 3184–3190 (1973)
5. Harrison, D. E., Astle, C. E. and Delaitre, J. A.: Loss of proliferative capacity in immunohemopoietic stem cells by serial transplantation rather than by aging. J. Exp. Med. 147, 1526–1531 (1978)
6. Hellman, S. and Botnick, L. E.: Stem cell depletion: an explanation of the late effects of cytotoxins. Int. J. Radiat. Oncology, Biol., Phys. 2, 181–184 (1977)
7. Hellman, S., Botnick, L. E., Hannon, E. C. and Vigneulle, R. M.: Proliferative capacity of murine hematopoietic stem cells. Proc. Natl. Acad. Sci. 75, 490–494 (1978)
8. Kretchmar, A. I. and Conover, W. R.: A difference between spleen-derived and bone marrow-derived colony-forming units in ability to protect lethally-irradiated mice. Blood 36, 772–776 (1970)
9. Lajtha, L. G. and Schofield, R.: Regulation of Stem Cell Renewal and Differentiation: Possible Significance in Aging. In: Advances in Gerontological Research. B. L. Strehler (ed.) New York: Academic Press, 1971, p. 131–146
10. Lord, B. I., Testa, N. G. and Hendry, J. H.: The relative spatial distributions of CFU-S and CFU-C in the normal mouse femur. Blood 46, 65–72 (1975)
11. Lord, B. I.: Cellular and Architectural Factors Influencing the Proliferation of Haemopoietic Stem Cells. In: Differentiation of Normal and Neoplastic Hematopoietic Cells. 5th Cold Spring Harbor Conference on Cell Proliferation (in press)
12. Metcalf, D. and Moore, M. A. S.: In: Frontiers in Biology – Haematopoietic Cells. Neuberger, A. and Tatum, E. L. (eds.) Amsterdam: North Holland 1971
13. Micklem, H. S., Anderson, N. and Ross, E.: Limited potential of circulating haemopoietic stem cells. Nature 256, 41–43 (1975)
14. Morley, A., Trainor, K. and Blake, J.: A primary stem cell lesion in experimental chronic hypoplastic marrow failure. Blood 45, 681–687 (1975)
15. Patt, H. M. and Maloney, M. A.: Evolution of marrow regeneration as revealed by transplantation studies. Exp. Cell Res. 71, 307–312 (1972)
16. Pozzi, L. V., Andreozzi, U. and Silini, G.: Serial transplantation of bone marrow cells in irradiated isogenic mice. Current Topics in Radiation Research Quarterly 8, 259–302 (1973)
17. Rosendaal, M., Hodgson, G. S. and Bradley, T. R.: Haemopoietic Stem Cells are organised for use on the basis of their generation age. Nature 264, 68–69 (1976)
18. Schofield, R.: A comparative study of the repopulating potential of grafts from various haemopoietic sources. Cell Tissue Kinet. 3, 119–130 (1970)
19. Schofield, R. and Lajtha, L. G.: Effect of isopropyl methane sulphate (IMS) on haemopoietic colony-forming cells. Brit. J. Haematol. 25, 195–202 (1973)
20. Schofield, R.: Haemopoietic Cell Kinetics. In: Int. Cong. Series No. 349, Vol. 1. Cell Biol. and Tumour Immunol. Amsterdam, Excerpta Medica 1974, 18–23
21. Siminovitch, L., Till, J. E. and McCulloch, E. A.: Decline in colony-forming ability of marrow cells subjected to serial transplantation into irradiated mice. J. Cell. Comp. Physiol. 64, 23–31 (1964)

22. Tavassoli, M. and Crosby, W. H.: Transplantation of marrow to extramedullary sites. Science *16*, 54–56 (1968)
23. Worton, R. G., McCulloch, E. A. and Till, J. E.: Physical separation of hemopoietic stem cells differing in their capacity for self renewal. J. Exp. Med. *130*, 91–104 (1969)

Discussion

Neuwirth: Dr. Schofield, are there two different signals for stem cell differentiation? One for the stem cells to leave the niche and another one for the stem cells to differentiate?

Schofield: I would suggest that if the stem cell leaves this niche it loses its stem cell potential. I visualize the hemopoietic stem cell as a unit which is dependent on both the cell and its environment. If you separate these two then you have no longer a pluripotent stem cell.

Fliedner: If Dr. Schofield is right, then one could ask the question; what is the number of stem cell niches, because if the stem cell population is controlled by a factor that is determined among other things by what he calls this "half-moon niche", then the number of these niches which are not available for stem cells to seed in, is a very important consideration. The question is then, if there is any possibility to quantify these niches.

Schofield: I think this was something I was trying to say at the end of my talk: in irradiated mice it almost looks as if there is a limit to the size of the niche population after irradiation, because no matter how many bone marrow cells I put in, I can't fill it up beyond a certain point. Ten times the graft size doesn't give me ten times the recovery potential. It comes, in fact, during the first ten weeks, to the same size, no matter what the size of the graft.

Neuwirth: I would like to make a comment to this definition of the niche. It has been shown, that cyclohexamide is a protein synthesis inhibitor, when given in high doses to rats. As a result epithelial cells are lost from the gut, since a protein is needed for their cohesion. When we administered cyclohexamide to mice, cells were released from the bone marrow even at very low doses and the concentration of CFU-s rises in the peripheral blood. With very high doses of cyclohexamide (100 mg/kg) the rise of CFU-s in the peripheral blood is nine times that of normal. This would indicate that a protein, which has not yet defined, is needed for the cohesion of bone marrow cells. It could contribute to the function of the microenvironment.

Fliedner: Do you have any information of the effect of this mechanism on the cell surface and what is the time sequence?

Neuwirth: We have done these experiments very recently and have just information on the release of stem cells in the peripheral blood, which is happening very quickly peaking in a matter of hours.

3.3 Congenital and Induced Defects in Haemopoietic Environments, Stem Cell Proliferation and Differentiation

T. M. Dexter, R. Schofield, J. Hendry, N. G. Testa

The recent development of a system whereby the pluripotent haemopoietic stem cells (CFU-S) can be maintained *in vitro* for several months [4, 5] has greatly facilitated studies on cellular interactions in haemopoiesis. When mouse bone marrow cells are cultured in an appropriate media, an adherent layer of cells is formed which contains a variety of cell types, including "endothelial" cells, fat cells and macrophages [2, 5]. Such adherent layers seem to provide the necessary *in vitro* microenvironment for sustained haemopoiesis to occur, since when an eablished adherent layer is reinoculated with syngeneic or allogeneic bone marrow cells, proliferation and differentiation of stem cells occurs for many weeks [5, 6, 7]. In the absence of a suitable adherent layer, haemopoiesis declines rapidly and stem cells are lost within a few days. This system, therefore, lends itself for the study of induced or congenital defects in the stem cells or in the inductive haemopoietic environment, and the consequences of these defects in terms of self-renewal and differentiation of the stem cell and committed cell populations. Herein are reviewed some of our studies in this field.

Materials and Methods

The bone marrow culture techniques have been described in detail elsewhere [4]. To establish the long term bone marrow cultures, the contents of a single mouse femur are flushed into culture flasks containing 10 ml. of Fischer's medium + 25% horse serum (Flow Labs.) supplemented with antibiotics. No attempt is made to obtain a single cell suspension (marrow fragments being more effective than a single cell suspension for establishing the adherent layers). The cultures are maintained at 33° C [5] in air + 5% CO_2, and are fed weekly by removal of half the growth medium and addition of an equal volume of fresh medium. After culture for 2–3 weeks, an adherent layer is established containing a variety of cell types – and onto this established adherent layer is then added a further inoculation of 10^7 syngeneic or allogeneic cells. These cultures, maintained at 33° C are fed as before by demi-depopulation. The growth medium, removed at weekly intervals, contains cells in suspension, which after appropriate washing and dilution are assayed for the variety of haemopoietic colony forming cells [4].

Results and Discussion

1. Culture of Normal Bone Marrow Cells

Using the technique described above, we have reported that extensive cell proliferation occurs for several months. For the first few weeks in culture, these cells consist mainly of granulocytes, in all stages of maturation. Megakaryocytes are also being produced, but the cultures contain no morphologically recognisable erythroid or lymphoid cells [5, 1]. Extensive proliferation of normal stem cells and concomitant production of granulocytic (CFU-C), megakaryocytic (CFU-M) and erythroid (BFU-E) precursor cells is also occurring [5, 13, 14, 12].

The results are summarised in Table 1. Thus far, no evidence for lymphopoiesis in these cultures has been obtained.

Weeks cultured	CFU-S	CFU-C	CFU-M	BFU-E
2	880	8000	280	ND
5	680	7000	140	97
6	500	5500	185	180
9	230	2500	10	<10
13	60	800	0	ND

Table 1. Maintenance of stem cells and production of committed precursor cells in long term cultures of BDF_1 bone marrow

2. Culture of Congenitally Defective Bone Marrow

The role of the adherent layer in the long term maintenance of stem cells had been implied by studies where we had shown that that haemopoiesis was not sustained in siliconised culture flasks (where cell adherence is prevented) nor in culture flasks showing a qualitatively or quantitatively poor development of adherent cells. Similarly, we have emphasised the role of the adherent layer as a site of stem cell production and commitment to differentiation [8]. Fairly conclusive evidence for the role of the adherent layer in long term haemopoiesis, comes, however, from our studies using congenitally anaemic W/W^v and Steel ($S1/S1^d$) mice [6]. The lesion in W mice is assumed to be an intrinsic defect in the stem cell population, whereas in S1 mice the defect lies in the environment into which stem cells migrate [11, 3, 10]. Evidence for this view comes from the observation that the anaemia in W mice can be cured by a graft of normal stem cells (or stem cells from S1 mice), whereas the anaemia in S1 mice is only alleviated following a graft of neonatal spleen or bone marrow (from normal or from W mice), which can act as a suitable haemopoietic environment.

In a recent paper, we showed that these defects could be reproduced in vitro, in that $S1/S1^d$ adherent layers, inoculated with $S1/S1^d$ marrow cells and W/W^v adherent layers inoculated with W/W^v marrow cells, both showed a marked defect in the production of CFU-C and maintenance of granulopoiesis compared with cultures established from their respective normal littermates. Of particular interest was the observation that haemopoiesis rapidly *declined* in cultures using defective adherent layers from $S1/S1^d$ mice inoculated with defective W/W^v stem cell, but was *sustained* in cultures where a "competent" W/W^v adherent layer was incoulated with equally "competent" $S1/S1^d$ stem cells. Stem cell proliferation, therefore, depends upon interaction between the stem cells and an appropriate microenvironment. Since there is no gross qualitative or quantitative difference between the normal and the defective $S1/S1^d$ adherent layer and – in terms of the arrangement or types of adherent cells present (Allen, Dexter and Moore, unpublished observations) – we assume that the defect lies at the level of cell/cell (membrane) interactions, and studies along these lines are now in progress.

3. X-Ray Induced Haemopoietic Defects

The work outlined above raised the possibility that the long term marrow culture system could be a suitable model for studying environmental lesions involved in the genesis of the various haemopoietic aplasias. When mice are subjected to 4×450 rad X-rays, with 21 day intervals between each dose, it has been shown that CFU-S recovered to a plateau value only 10% that of normal mice, or of mice which had received a single dose of 800 rad X-rays followed by bone marrow reconstitution [9]. Is this low level recovery related to stem cell damage or damage to the haemopoietic environment? The finding that 3×450 rad mice, given a further dose of 800 rad X-rays and reconstitution with a graft of normal bone marrow cells, are able to sustain growth of the grafted marrow (indeed the CFU-S recover to values seen after reconstitution of mice which had received only 800 rad X-rays and marrow reconstitution) indicates that the radiation regime had not damaged the haemopoietic environment *any more than a single dose of 800 rad X-rays*. On the other hand, when stem cells from a 4×450 rad mouse are grafted into an irradiated host, they recover to a sub-optimal value compared with normal bone marrow cells [9] – indicating that the stem cell compartment is defective in reconstitution ability.

Using the long term marrow culture system, we have now investigated the potential of 4×450 rad bone marrow cells to produce a competent haemopoietic environment in vitro and concomitantly examined the selfrenewal capacity of 4×450 rad stem cells. The cultures were established from 4×450 rad BDF$_1$ mice, three weeks after the final irradiation, or from age matched control mice. To follow the competency of the environment or the stem cells, we utilised our previously published observation [7] showing that allogeneic stem cell culture chimeras could be successfully maintained in vitro, with no evidence of genetic resistance. In this way, we could selectively measure the efficacy of adherent cells derived from 4×450 rad mice to support proliferation of *normal stem cells,* and the ability of stem cells from 4×450 rad mice to replicate on an *adherent layer* established from *normal* mice. In the first case, we established adherent layers from BDF$_1$ mice previously given 4×450 rad X-rays or from normal age matched controls, and inoculated these established adherent layers with 10^7 normal C57Bl/6 bone marrow cells. The maintenance of proliferation of C57BL/6 CFU-S was then followed by injecting the cultured cells into irradiated C57BL/6 recipients. The results, shown in Table 2, demonstrate that adherent layers derived from 4×450 rad mice are very poor in their ability to sustain CFU-S proliferation. This environmental defect appears to be due to accumulated damage, since a single dose of 450 rad X-rays results in no detectable defect in the ability of marrow adherent cells to sustain stem cell proliferation (unpublished observation). In the second case, adherent layers established from normal C57BL/6 bone marrow cells were inoculated with 10^7, 4×450 rad BDF$_1$ marrow cells or 10^6–10^7 age matched control cells (to adjust for total number of *stem cells*

Table 2. Maintenance of normal CFU-S on adherent layers established from irradiated mice

Adherent layer	Cells inoculated	Total C57BL/6 CFU-S (weeks)				
		1	2	3	4	5
Normal BDF$_1$	10^7 normal C57BL/6	200	215	431	182	153
4×450 rad BDF$_1$	10^7 normal C57BL/6	15	22	5	0	0

inoculated or total number of *cells* added), and the production of BDF$_1$, CFU-S was followed by injecting the cultured cells into irradiated BDF$_1$ recipient mice. The results are shown in Table 3. While production of normal CFU-S is maintained for many weeks, the irradiated CFU-S undergo a rapid decline and are no longer detectable after the first two weeks in culture. We have also examined the self-renewal probability of stem cells in these cultures of normal and irradiated marrow. After one week in culture, the cells were injected into irradiated mice to produce 5–10 spleen colonies. Seven days later, the colonies in each group were analysed individually for their CFU-S content and the data used to calculate the mean CFU-S/colony and the self-renewal probability, p, of the CFU-S in the cultures. The mean CFU-S/colony in one week cultures of normal CFU-S was 10.9, with a p of 0.624. In contrast, the CFU-S in cultures of irradiated marrow had a low CFU-S/colony of 3.6 and a p of 0.540. Thus, while a normal environment has been provided in both cases, the quality of the CFU-S produced in the cultures seems to depend on the previous history of the marrow with which it is charged – clearly indicating instrinsic differences in the stem cell populations.

Table 3. Maintenance of irradiated CFU-S on adherent layers established from normal mice

Adherent layer	Cells inoculated	Total BDF$_1$ CFU-S (weeks)				
		1	2	3	4	5
Normal C57BL/6	10^7 normal BDF$_1$	296	1123	692	ND	191
Normal C57BL/6	10^6 normal BDF$_1$	133	97	207	132	204
Normal C57BL/6	10^7 4×450 rad BDF$_1$	70	33	0	0	12

A reasonable conclusion from these studies is that the defective haemopoiesis seen in 4×450 rad mice is associated with both environmental and stem cell damage. Since the defect in S1/S1^d mice lies only at the level of the environment, and in W/W^v mice only at the stem cell level, it is obvious that the variety of hypoplasias observed in experimental animals and man may be associated with single or multiple lesions. The determinations of these lesions is of practical importance in treatment of the human aplasias. While defective haemopoiesis due to stem cell "damage" or deficiency can be cured by transplantation of suitable genetically compatible marrow cells, a defective environment poses a greater problem. The transplantation of suitable haemopoietic stromal tissue is a possible course of treatment. Alternatively, isolation of the factors controlling stem cell proliferation and differentiation; their mechanism of action, synthesis and possible usefulness in treatment need to be explored.

Acknowledgements

The technical assistance of Ms. Gaynor Johnson and Stella Crompton was much appreciated. The work was supported by the Cancer Research Campaign and the Medical Research Council.

References

1. Allen, T. D.: Ultrastructural Aspects of in vitro Haemopoiesis. In: Stem Cells and Tissue Homeostatis. Lord, B. I., Potten, C. S. and Cole, R. (ed.), Cambridge: Cambridge University Press, 1978, p. 217
2. Allen, T. D., Dexter, T. M.: Cellular interrelationships during in vitro granulopoieseis. Differentiation 6, 191–194 (1976)
3. Bernstein, S. E., Russell, E. S., Keighley, G.: Two heridatory mouse anemias S1/S1^d and W/W^v deficient in response to erythropoietin. Ann. NY Acad Sci, 149, 475–485 (1968)
4. Dexter, T. M., Testa, N. G.: Differentiation and Proliferation of Haemopoietic Cells in Culture. In: Methods in Cell Biology. Prescott, D. M. (ed.). New York: Academic Press, Vol. 14, pp. 387–405
5. Dexter, T. M., Allen, T. D., Lajtha, L. G.: Conditions controlling the proliferation of haemopoietic stem cells in vitro. J. Cell Physiol. 91, 335–344 (1977a)
6. Dexter, T. M., Moore, M. A. S.: In vitro duplication and "cure" of haemopoietic defects in genetically anaemic W/W^v and S1/S1^d mice. Nature 269, 412–414 (1977)
7. Dexter, T. M., Moore, M. A. S., Sheridan, A. P. C.: Maintenance of hemopoietic stem cells and production of differentiated progeny in allogeneic and semiallogeneic bone marrow chimeras in vitro. J. exp. Med. 145, 1612–1616 (1977b)
8. Dexter, T. M., Wright, E. G., Krizsa, F., Lajtha, L. G.: Regulation of haemopoietic stem cell proliferation in long-term bone marrow cultures. Biomedicine 27, 344–349 (1977c)
9. Hendry, J. H., Testa, N. G.: Effect of repeated doses of X-rays or 14 MeV neutrons on mouse bone marrow. Radiation Research 59, 645–652 (1974)
10. Mayer, T. C., Green, M. C.: An experimental analysis of the pigment defect caused by mutations at the W and S1 loci in mice. Devel. Biol. 18, 62–75 (1968)
11. Russell, E. S., Bernstein, S. E.: Blood and Blood Formation. In: Biology of the Laboratory Mouse. Green, E. L. (ed.). New York: McGraw-Hill, 1966, pp. 351–372
12. Testa, N. G., Dexter, T. M.: Long-term production of erythroid precursor cells (BFU) in bone marrow cultures. Differentiation 9, 193–195 (1977)
13. Williams, N., Jackson, H., Rabellino, E. M.: Proliferation and differentiation of normal granulopoietic cells in continuous bone marrow cultures. J. Cell Physiol. 93, 435–440 (1977)
14. Williams, N., Jackson, H., Sheridan, A. P. C., Murphy, M. J., Elste, A., Moore, M. A. S.: Regulation of megakaryopoiesis in long-term murine bone marrow cultures. Blood 51, 245–255 (1978)

Discussion

Thierfelder: Dr. Dexter, do you know if you recover less stem cells if the feeder layer is genetically different or even incompatible?

Dexter: We have, if anything, more stem cells, if you establish a purely allogenic combination with different adherent layers and stem cells.

Kubanek: Did BFU-E develop in vitro from these genetically anaemic mice, as you have observed them from normal mice?

Dexter: We have not got to that yet.

Barrett: You have shown that there are three kinds of supportive cells. Do you still get maintenance of stem cells if you remove fat cells?

Dexter: In an adherent layer without fat cells there is no stem cell maintenance. We can't think of a way of selectively depleting for fat cells because they are being formed over a period of several weeks and it would be difficult to remove them without also displacing other cells.

Heit: As far as the monolayer or supportive layer function is concerned, using human fetal liver versus different preparations of human fetal bone marrow, the macroscopic pictures of these monolayers are different; although the fetal liver monolayer loses the capacity to produce erythroid cell differentiation, we tend to say that there is a difference between both, but we are not able to handle it correctly yet. Macroscopically and morphologically there is obviously a difference in man.

Wickramasinghe: It has been suggested that there is a short range or cell-to-cell interaction.

Dexter: I can't really answer this question, but the evidence for short-range interactions comes from in vivo experiments from which a local control of stem cell proliferation and differentiation was postulated. A point I did not raise is that the culture system is also deficient in colony stimulating factor. If you take conditioned medium from the cultures and attempt to stimulate CFU-c, you get no colonies and the dogma is that CSA is essential for granulocyte maturation. Then we thought that CSA could be membrane-bound and detected CSA in the cultures, by a radioimmunoassay as used by R. Stanley in USA.

Fliedner: I want to ask Dr. Dexter a question with respect to this phenomenon of using irradiated bone marrow to produce a feeder layer in the continuous culture which is running out of steam. You use 4×450 rad, is there anything magic about it? Does it happen, if one gives a single or a lower dose of irradiation?

Dexter: No, there isn't anything magic, it was just that we had those mice available at the time. We have also done low-dose irradiation in vitro. These results are extremely remarkable and I didn't want to bring that up, because I haven't any interpretation on those data, but if you first of all allow an adherent layer to be produced, and then irradiate that adherent layer with doses between 25 and 2000 rad, you have a consistent defect in the adherent layer to substain hemopoiesis for more than 5 weeks. Actually I didn't want to bring that up, because the point was raised to me, that we could be producing some sort of a radical by irradiating the adherent layer in the plastic culture flask. Although not giving detailed information, it seems that low-dose irradiation also causes defects.

Fliedner: Could we try to define the elements we define as microenvironment or what Dr. Schofield calls niche. What are cellular components in that, what are metabolic components, what are factorial components, what is meant by T-cell helper function? I have never found any clearcut definition. I think we should try to pick our brains in the general discussion to find out whether we can agree in the composition of that niche.

3.4 Independent Requirements for Early and Late Stages of Erythropoiesis

N. N. Iscove

Relatively primitive (BFU-E) and relatively mature (CFU-E) red cell precursors can be distinguished by their differing clonogenic properties in culture [1, 2]. Both are dependent on erythropoietin and also on serum for colony formation. For CFU-E, the erythropoietin requirement is in keeping with a similar dependence in the whole animal [1, 3]. Their serum requirement reflects only non-specific needs, namely albumin, transferrin, unsaturated fatty acid and cholesterol [4, 5]. The requirements of BFU-E are more puzzling. These cells show no dependence on erythropoietin in the whole animal [1, 3], while their proliferative activity is subject to other control mechanisms not involving erythropoietin [3]. In addition, they require serum for growth in culture even in the presence of albumin, transferrin and lipids. On the basis of these observations, it was suggested that the action of erythropoietin might be confined to later stages of erythroid maturation subsequent to BFU-E [3, 6, 7].

Recent observations by others have provided the basis for testing whether humoral factors distinct from erythropoietin might indeed influence early cells in culture. Johnson and Metcalf reported that medium conditioned by pokeweed mitogen-stimulated mouse spleen cells could induce formation of large erythroid colonies in cultures of mouse fetal liver cells. This effect was observed without addition of erythropoietin beyond that present fortuitously in the serum component of the cultures.

Conditioned media were therefore prepared in serum-free cultures of mouse spleen cells stimulated with either pokeweed mitogen or concanavalin A. The supernatants were tested in methyl cellulose cultures of adult mouse bone marrow containing erythropoietin, albumin, transferrin and lipids, and only 4% fetal calf serum. Under these conditions and without addition of conditioned medium, large erythroid colonies ("bursts") from BFU-E grew either at low efficiency or not at all. Conditioned medium increased their numbers in a dose-dependent fashion.

The active molecules in the conditioned media were bound by Concanavalin-A-Sepharose, and eluted with α-methylglucoside, indicating their glycoprotein nature. They displayed an apparent molecular weight of 35,000 on Sephadex G-150.

In methyl cellulose cultures of adult mouse marrow, this "burst-promoting activity" (BPA) reduced the amount of both serum and erythropoietin required for burst formation. In addition, it promoted formation of mixed colonies containing granulocytes and monocyte/macrophages in addition to erythroid cells.

   In order to determine the relative stages affected by BPA and erythropoietin, experiments were performed in which addition of one or the other principle to the methyl cellulose cultures was delayed. When cultures set up with BPA but without erythropoietin were incubated for varying periods of time before erythropoietin was added, CFU-E declined to very low levels within 24 hours. In contrast, BFU-E formed 11-day erythroid colonies normally even when addition of erythropoietin was delayed 5 to 7 days. No bursts formed if erythropoietin was never added. Marking of individual colonies on the 7th day, followed by addition of erythropoietin, established that recognizably erythroid 11-day bursts originated from colonies already present at day 7 but not yet recognizable as erythroid. Erythropoietin was therefore required only for the terminal maturation events which allow the erythroid character of the burst to become apparent. The converse experiment was also performed. Methyl cellulose cultures were set up with erythropoietin, and addition of BPA was delayed. Here, BFU-E declined rapidly, while CFU-E generated colonies normally, confirming results reported by others [8]. Thus, BPA is required for growth and survival of relatively primitive erythroid precursors (BFU-E), while erythropoietin is required for survival and growth only at later stages (CFU-E). These observations are all detailed elsewhere [9].

   The observations have led us to suggest a 2-stage model of humoral regulation hemopoiesis [7, 9]. It is proposed that BPA regulates pluripotential cells and their earliest committed progeny regardless of their prospective pathways. Later, as part of their maturation program, these cells would lose their receptors for BPA while simultaneously acquiring receptors for pathway-specific regulators such as erythropoietin. While the model can explain the present observations, it remains to be seen whether BPA will prove to have a regulatory role in the intact animal, and to what extent it will prove chemically separable from pathway-specific factors.

References

1. Axelrad, A. A., McLeod, D. L., Shreeve, M. M., Heath, D. S.: Properties of cells that produce erythrocytic colonies in vitro. In: Hemopoiesis in Culture (ed. W. A. Robinson), p. 26. U. S. Government Printing Office, Washington 1974
2. Iscove, N. N., Sieber, F., Winterhalter, K. H.: Erythroid colony formation in cultures of mouse and human bone marrow: analysis of the requirement for erythropoietin by gel filtration and affinity chromatography on agarose-concanavalin-A. J. Cell. Physiol. 83, 309–320 (1974)
3. Iscove, N. N.: The role of erythropoietin in regulation of population size and cell cycling of early and late erythroid precursors in mouse bone marrow. Cell Tissue Kinet. 10, 323–334 (1977)
4. Guibert, L. J., Iscove, N. N.: Partial replacement of serum by selenite, transferrin, albumin and lecithin in haemopoietic cell cultures. Nature 263, 594–595 (1976)
5. Iscove, N. N., Guilbert, L. J., Weyman, C.: Complete replacement of serum in primary cultures of erythropoietin-independent red cell precursors (CFU-E) by albumin, transferrin, iron, unsaturated fatty acid, lecithin and cholesterol. Cell and Tissue Kinetics, in press
6. Iscove, N. N.: Regulation of proliferation and maturation at early and late stages of erythroid differentiation. In: Cell Differentiation and Neoplasia, p. 195–209. (ed. Saunders, G. F.). New York: Raven Press 1978
7. Iscove, N. N., Guilbert, L. J.: Erythropoietin-independence of early erythropoiesis and a two-regulator model of proliferative control in the hemopoietic system. In: In vitro aspects of erythropoiesis. (ed. Murphy, M. J.), p. 3–20. New York: Springer 1978

8. Axelrad, A. A., McLeod, D. L., Suzuki, S., Shreeve, M. M.: Regulation of the population size of erythropoietic progenitor cells. In: Differentiation of normal and neoplastic hematopoietic cells. (eds. Clarkson, B., Marks, A., Till, J. E.), p. 155–163. Cold Spring Harbor Lab. 1978

9. Iscove, N. N.: Erythropoietin-independent stimulation of early erythropoiesis in adult marrow cultures by conditioned media from lectin-stimulated mouse spleen cells. In: Hematopoietic cell differentiation. (eds. Golde, D. W., Cline, M. J., Metcalf, D., Fox, C. F.), p. 37–52. New York: Acad. Press 1978

3.5 Estimation of the Pluripotential Stem Cells in Bone Marrow After Hydroxyurea

E. Nečas, P. Poňka, J. Neuwirt

Introduction

The method available at present for the measurement of the pluripotential stem cells (CFU_s) is based on the formation of the macroscopically visible colonies on the spleen surface of the lethally irradiated mice recipients of the bone marrow cell suspension [9]. This method demonstrates a fraction of the pluripotential stem cells injected [7].

Using this method a steady decrease of the CFU_s numbers in the bone marrow after a single injection of hydroxyurea was reported by several authors [4, 1, 6]. This decrease was observed despite the fact that hydroxyurea rapidly kills only CFU_s being in the S phase at the time of the drug administration, which represent only 10% of the whole CFU_s population [2, 5]. The decrease of CFU_s developing later has been tentatively attributed to the rapid and intensive differentiation of the pluripotential stem cells due to the severe depopulation of the bone marrow [1]. Measuring the efficiency of the Spleen Colony Technique for the bone marrow from animals treated with hydroxyurea we have arrived at data which shows a very different time course of the pluripotential stem cells in the bone marrow depopulated by hydroxyurea from that published previously.

Material and Methods

Spleen Colony Technique

Bone marrow cell suspension was obtained from mice femurs by flushing the bone marrow out of the femoral cavity and dispersing the cells in the tissue culture medium, the composition of which was very similar to that of M 199. The bone marrow from 7 to 10 femurs was pooled and the total cell number in the cell suspension was determined. This gave us the average bone marrow cellularity of the femur, and the suspension was used in the preparation of the diluted cell suspension, for CFU_s determination. 8×10^4 of the bone marrow cells in the volume of 0.2 ml were injected i.v. to the groups of 9 mice irradiated 2 to 3 hours before (using the ^{60}Co sourse, total dose 800 rad, dose rate 25 rad/min.). Eight days later the bone marrow recipients were killed, the spleens were removed, placed into Buin's solution and the colonies visible to the naked eye were counted on the spleen surface. Using the average bone marrow cellularity the results were expressed as CFU_s numbers per femur.

Seeding Efficiency of the Spleen Colony Technique

10^7 of the bone marrow cells from control or hydroxyurea treated mice were injected into the mice irradiated 2 hours previously with 800 rads. This cell suspension, appropriately diluted, was used at the same time for the determination of the CFU_s numbers. The recipients of the 10^7 bone marrow cells were killed 2 hours later and the cell suspension was prepared from 9 spleens obtained from mice which had received the same bone marrow sample. The average cellularity of the spleen was determined and 6×10^6 spleen cells were injected to a group of 9 irradiated mice for CFU_s determination. After CFU_s determination, performed 8 days later, 1. the total numbers of CFU_s injected to the recipient of the 10^7 of the bone marrow cells and 2. the total numbers of the CFU_s recovered in the spleen of these mice 2 hours later were calculated. The results were expressed as the percentage of CFU_s recovered in spleen from the total CFU_s injected. In each experiment the normal bone marrow was tested with 3 samples of the bone marrow from mice, receiving a single injection of hydroxyurea at various time intervals before. The percentage of the CFU_s recovered in the spleens of the recipients of the normal bone marrow ranged from 6 to 19% with the mean value of 11.2%.

Results

The depopulation of the bone marrow caused by a single injection of hydroxyurea (1000 mg/kg b.w. in 0.5 ml of saline intraperitoneally; Hydroxyurea, Calbiochem) is shown in Figure 1.

The time course of the CFU_s in femur is depicted in Figure 2 (CFU_s/femur curve). The data are mean values obtained in different experiments (variability of the assay method is not presented). At the time intervals where the data from more than two experiments were available, the arithmetic means from the results are presented by crosses.

The seeding efficiency of the Spleen Colony Technique for the CFU_s from hydroxyurea treated bone marrow is shown in Figure 3. The crosses indicate the mean values from more than two determinations.

The data from Figure 3 were used to calculate the pluripotential stem cell (CFC_s – colony forming cell-spleen) numbers in the bone marrow from hydroxyurea treated mice. The mean CFU_s levels detected by the Spleen Colony Technique

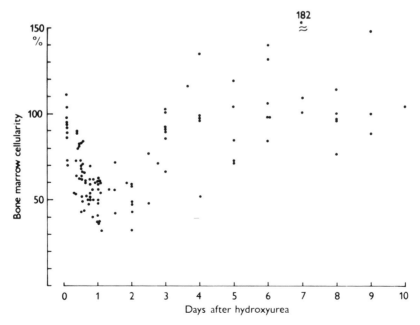

Fig. 1. Bone marrow cellularity measured in femur after a single injection of hydroxyurea

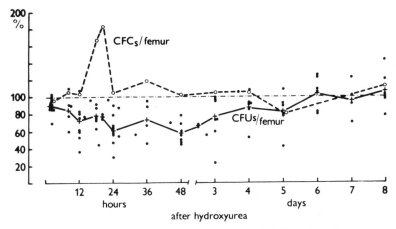

Fig. 2. Pluripotential stem cells in bone marrow after a single injection of hydroxyurea. CFUs/femur shows the data determined by means of the Spleen Colony Technique of Till and McCulloch (1961). CFCs/femur curve shows the CFUs data corrected for changes in seeding efficieny of the Spleen Colony Technique

(crosses on Figure 3) were multiplied by the factor 100/seeding efficiency of the hydroxyurea treated bone marrow at the particular time interval (crosses on Figure 3).

The results obtained in this way, reflecting the numbers of the pluripotential stem cells in the bone marrow depopulated by hydroxyurea, are depicted in Figure 2 (O – – – – O CFC$_s$/femur curve).

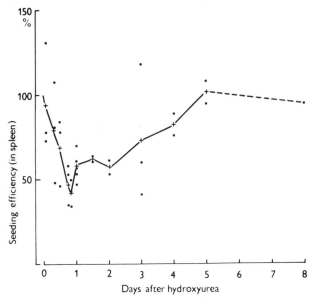

Fig. 3. Changes in seeding efficiency of the Spleen Colony Technique after a single injection of hydroxyurea

Discussion

From Figure 2 it is evident that after correcting the Spleen Colony Technique for the change in the seeding efficiency of this method, occuring in hydroxyurea treated bone marrow (Figure 3), a quite different time course of the pluripotential stem cells in the hydroxyurea depopulated bone marrow can be demonstrated. Almost no depletion of these cells in the bone marrow was observed. Eighteen and 20 hours after hydroxyurea there appeared to be an abundance of the pluripotential stem cells in the bone marrow, reaching values as high as 180% of the normal. This is in accordance with the well known effect of hydroxyurea on the CFU_s, i.e. triggering G_0 cells into cell cycle [10, 5]. The end result of this should be the production of new cells.

The reason for the change in the seeding efficiency of the Spleen Colony Technique elicited by hydroxyurea administration is not clear. It is known that hydroxyurea is metabolized in the course of 2 to 3 hours [3], so the lowered seeding efficiency cannot be caused by the direct effect of hydroxyurea on CFU_s. After the dose of hydroxyurea used in our experiments (1000 mg/kg b.w.) the majority of CFU_s were demonstrated to enter almost synchronously the cell cycle [5]. It is possible that the increase in the cell volume occuring during the progress of cells through the cell cycle [8] may have some effect on the seeding of CFU_s in spleens. On the other hand the effect of hydroxyurea on the seeding efficiency was present even after 3 to 4 days (Figure 3) which indicates that some other factors must participate in lowering the seeding efficiency too. One of those factors might be the immaturity of the cell membrane of the newly formed CFU_s.

The presented results indicate that in evaluating the pluripotential stem cell numbers after some pertubations in the hemopoietic tissue, caution should be taken regarding the possible effects on the seeding efficiency of the Spleen Colony Technique.

References

1. Hodgson, G. S., Bradley, T. R., Martin, R. F., Sumner, M., Fry, P.: Recovery of proliferating haemopoietic progenitor cells after killing by hydroxyurea. Cell Tissue Kinet. 8, 51–60 (1975)
2. Lajtha, L. G., Pozzi, L. V., Schofield, R., Fox, M.: Kinetic properties of haemopoietic stem cells. Cell Tissue Kinet. 2, 39–49 (1969)
3. Morse, B. S., Rencricca, N. J., Stohlman, F. jr.: The effect of hydroxyurea on differentiated marrow erythroid precursors. Proc. Soc. Exptl. Biol. Med. 130, 986–989 (1969)
4. Morse, B. S., Rencricca, N. J., Stohlman, F. jr.: Relationship of erythropoietic effectiveness to the generative cycle of erythroid precursor cell. Blood 35, 761–774 (1970)
5. Nečas, E., Neuwirt, J.: Control of haemopoietic stem cell proliferation by cells in DNA synthesis. Brit. J. Haemat. 33, 233–238 (1976)
6. Nečas, E., Poňka, P., Neuwirt, J.: Changes in stem cell compartments in mice after hydroxyurea. Cell Tissue Kinet. 11, 119–127 (1978)
7. Siminovitch, L., McCulloch, E. A., Till, J. E.: The distribution of colony – forming cells among the spleens colonies. J. cell. Physiol. 62, 327–336 (1963)
8. Terasima, T., Tolmach, L. J.: Growth and nucleic acid synthesis in synchronously dividing populations of Hela cells. Exp. Cell Res. 30, 344–352 (1963)
9. Till, J. E., McCulloch, E. A.: A direct measurement of the radiation sensitivity of normal mouse bone marrow cells. Radiat. Res. 14, 213–222 (1961)
10. Vassort, F., Winterholer, M., Frindel, E., Tubiana, M.: Kinetic parameters of bone marrow stem cells using in vivo suicide by tritiated thymidine or by hydroxyurea. Blood 41, 789–796 (1973)

3.6 CFU-C and Colony Stimulating Activity in Human Aplastic Anemia

A. Mangalik, W. A. Robinson, R. Bolin, Asha Mangalik, Maureen Entringer

Introduction

Aplastic anemia is a disease characterized by decreased production of cells of all the hematopoietic cell lines. Various mechanisms have been suggested as being important in the development of this disorder. Among these are loss of hematopoietic stem cells, inhibition of such cells by the humoral or cellular elements, decreased production of hematopoietic stimuli and alteration of the microenvironmental stroma [2, 17].

The studies described here were undertaken to determine whether specific mechanisms could be demonstrated in the alteration of granulocyte production in aplastic anemia. Patients with different forms of aplastic anemia have been studied using the in vitro semi-solid granulocyte colony technique in agar-gel. These studies have demonstrated that in the majority of patients with aplastic anemia, the granulocyte colony forming stem cells – colony forming units in culture (CFU-C), are reduced in the bone marrow. In these patients production of granulocyte colony stimulating factor (CSF), as evidenced by CSF activity (CSA) in serum is elevated. No evidence for circulating or humoral inhibitors, or inhibition of granulocyte colony formation by peripheral blood mononuclear cells has been demonstrated. These data suggest that the major abnormality in the usual form of aplastic anemia is a reduction in granulocyte committed stem cells through mechanisms as yet undefined.

In contrast, in patients with Fanconi's anemia CSF production by peripheral blood mononuclear cells appears to be defective. Thus, two separate mechanisms for the development of aplastic anemia as it affects the granulocyte stem cell line have been demonstrated.

Materials and Methods

All of the patients studied were seen at the University of Colorado Medical Center between 1970 and 1978. The studies were performed after written informed consent as approved by the Human Subject Committee, University of Colorado Medical Center. Bone marrow cells from patients and normal human volunteers were collected by aspiration through the posterior iliac crest into a heparinized syringe. After collection the cells were allowed to sediment by gravity for 1–2 hours. Plasma, containing bone marrow cells, was aspirated and the mature granulocytes were separated from the precursor cells by Ficoll-Hypaque gradient centrifugation [3] and monocytes removed by glass adherence [16]. The cells were washed twice in McCoy's 5A medium and then plated as described below.

Peripheral blood cells were, likewise, collected in heparin, sedimented by gravity and the white blood cell-rich plasma collected. Mononuclear cell fractions containing lymphocytes and monocytes

were prepared by centrifugation on Ficoll-Hypaque gradient centrifugation [3]. For preparation of
semi-purified populations of monocytes the Ficoll-Hypaque gradient fractions were incubated in glass
petri dishes for 1 hour at $37°$ [16]. The adherent cells were used as monocyte fractions and the
nonadherent cells as lymphocyte fractions. The culture techniques have been described in detail
elsewhere [21].

Outline of Experiments

1. Bone marrow CFU-C: Bone marrow cells were plated as overlays in 1 ml aliquots in 0.3% agar in
McCoy's medium in 35 mm petri dishes in concentrations of 5×10^4 cells after centrifugation on
Ficoll-Hypaque gradient and glass adherence. Stimulus for colony growth was peripheral white blood
cell feeder layers as described previously [21] or human placental conditioned medium (PCM) as
described by Burgess et al. [4]. The plates were incubated in a fully humidified incubator at $37°$ for 12
days at which time colony counts were done. Only aggregates containing 50 or more cells were
counted as colonies.
2. Serum CSA: Serum CSA values were determined as described previously [21]. 0.1 ml of serum was
plated in a 1 ml underlay of 0.5% agar and McCoy's 5A medium. Over this was added a 1 ml target
layer of 5×10^4 normal human bone marrow cells from which all CSF producing cells had been
removed by Ficoll-Hypaque gradient centrifugation and glass adherence as described above. Colony
counts were done after 12 days of incubation as described above.
3. Production of CSA by peripheral blood cells: Peripheral blood MNC, prepared by Ficoll-Hypaque
gradient separation, in a concentration of 2×10^5 /ml were incorporated in 0.5% agar and McCoy's
5A medium as a stimulus layer. The stimulus layer was overlaid with 5×10^4 separated normal human
bone marrow cells.
4. MNC conditioned medium: 1×10^6 peripheral blood MNC, in 1 ml of McCoy's 5A medium with
either no serum or with 50% human serum were incubated at $37°$ under full humidification with 7.5%
Co_2 in air for 72 hours. The supernatant conditioned medium was assayed for CSA using 5×10^4
normal human bone marrow cells devoid of CSF producing cells.
5. Cellular inhibition studies: Peripheral blood MNC's from 3 patients with aplastic anemia were
prepared by Ficoll-Hypaque gradient separation as described above. MNC's were added in increasing
concentrations ($0-4 \times 10^5$ per plate) to 5×10^4 separated normal bone marrow cells (devoid of CSF
producing cells) and plated over PCM in agar and McCoy's medium (Table 1).
6. Serum inhibition studies: 0.1 ml of serum from 11 patients and 14 normal subjects was incorporated
in feeder layers containing PCM in 0.5% agar in McCoy's 5A medium. These were overlaid with
normal bone marrow cells devoid of CSF producing cells.

Results

Figure 1 shows the number of colonies grown from the bone marrow of patients
with idiopathic or drug induced aplastic anemia, compared with normals from our
laboratory. It can be seen that the number of colonies grown from the bone
marrow of such patients was considerably lower than that of the normal standard
in our laboratory. Normal bone marrow cultured in our laboratory yields a mean
of 87 (range 19–219) colonies/5×10^4 bone marrow cells when stimulated with
PCM. Most patients with aplastic anemia grew less than 10 colonies under the
same experimental conditions.

Figure 2 shows the serum CSA values in these patients compared with that
obtained from normal human serum in our laboratory. In contrast to the CFU-C
levels noted above, serum CSA levels were elevated in patients suffering from
aplastic anemia. The sera from 11 patients and 14 normals were stored at $-20°$
C and were all tested against a single normal bone marrow. The normal sera
yielded between 1–15 colonies/5×10^4 bone marrow cells (median 5, mean 5.4)

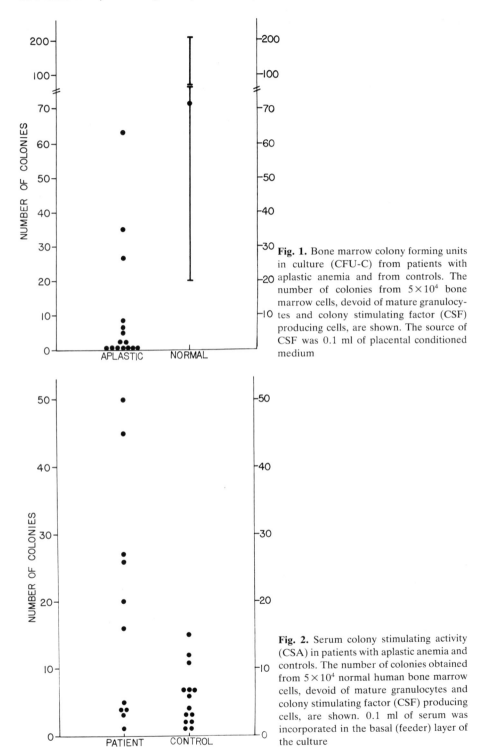

Fig. 1. Bone marrow colony forming units in culture (CFU-C) from patients with aplastic anemia and from controls. The number of colonies from 5×10^4 bone marrow cells, devoid of mature granulocytes and colony stimulating factor (CSF) producing cells, are shown. The source of CSF was 0.1 ml of placental conditioned medium

Fig. 2. Serum colony stimulating activity (CSA) in patients with aplastic anemia and controls. The number of colonies obtained from 5×10^4 normal human bone marrow cells, devoid of mature granulocytes and colony stimulating factor (CSF) producing cells, are shown. 0.1 ml of serum was incorporated in the basal (feeder) layer of the culture

while the aplastic anemia patients' sera yielded 1–50 colonies (median 15, mean 18.3). This difference is statistically significant (p=.02) using the Aspen Welch test for unpaired data with unequal variance. Also it appears that serum CSA from patients with aplastic anemia showed 2 subgroups – one with low values (5 patients below 5 colonies) and another with higher values (6 patients above 15 colonies). Analysis of the records on these patients did not reveal any major clinical or hematological differences between these subgroups.

Figure 3 shows the production of CSF by peripheral blood mononuclear cells from patients with idiopathic aplastic anemia compared with the production of CSA by normal human white blood cells. The results have been expressed as a percentage of the number of colonies produced by normal peripheral blood MNC in the same experiment. The levels of mononuclear cell production of CSA in such patients was similar to that in controls. These studies indicate that CSA production by peripheral blood cells of these patients is normal.

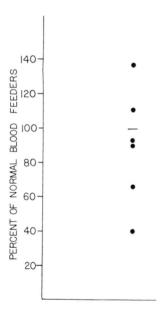

Fig. 3. Production of colony stimulating factor (CSF) by peripheral blood cells in patients with aplastic anemia as compared to control peripheral blood cells. The number of colonies from 5×10^4 normal human bone marrow cells, devoid of nature granulocytes and CSF producing cells, are shown as a percentage of colonies abtained when normal peripheral blood cells were used as the source of CSF. Each normal marrow was tested against patient and control peripheral blood cells incorporated in the basal (feeder) layer

Figure 4 shows the effect of combining normal and patient sera with PCM stimulus layers. It can be seen that with PCM alone 71 colonies were produced from 5×10^4 normal human bone marrow cells devoid of CSF producing cells. With addition of either normal or patient serum there was no significiant change in the number of colonies in either group.

Table 1 shows the effect of addition of peripheral blood lymphocytes from patients with aplastic anemia on granulocyte colony formation by normal human bone marrow cells, in the presence of a constant stimulus. No evidence of inhibition is noted, and, in fact, addition of mononuclear cells from these patients appears to enhance colony formation by normal human bone marrow when compared with studies in which normal human lymphocytes were added to the

	0	Granulocyte colonies/5×10^4 bone marrow cells lymphocytes added at concentrations noted			
		5×10^4	1×10^5	2×10^5	4×10^5
Case 1	79	128	104	117	99
Normal		114	131	129	86
Case 2	16	14	22	20	18
Normal		29	21	37	22
Case 3	78	92	98	107	139
Normal		92	104	114	107

Table 1. Effect of peripheral blood lymphocytes from normal humans and patients with aplastic anemia on granulocyte colony formation by normal human bone marrow stimulated by human placenta conditioned medium

same bone marrow preparation. These studies indicate that no inhibitory cells appear to be present in the peripheral blood of patients with aplastic anemia.

Figure 5 shows the ability of peripheral blood mononuclear cells from 2 patients with Fanconi's anemia to produce CSA. It can be seen that these cells produced little CSA either alone or in the presence of the patients' own plasma or normal human plasma. Normal MNC, on the other hand, produced no CSA in the absence of human serum but produced a large number of colonies when human serum (autologous or heterologous) was included in the incubation mixture.

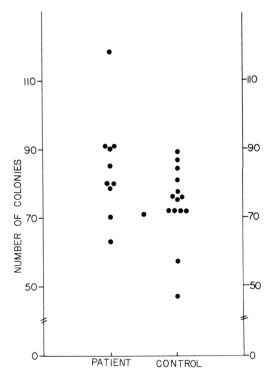

Fig. 4. The effect of serum on the number of colonies obtained from normal human bone marrow stimulated with placental conditioned medium (PCM). 5×10^4 normal human bone marrow cells, devoid of nature granulocytes and colony stimulating factor (CSF) producing cells, were stimulated with 0.1 ml of PCM. Sera from patients with aplastic anemia and controls was incorporated into the basal (feeder) layer. The central circle depicts the number of colonies obtained with PCM alone

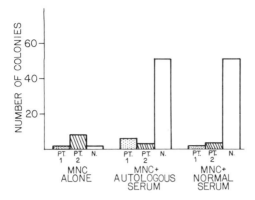

Fig. 5. Production of colony stimulating factor (CSF) by peripheral blood mononuclear cells (MNC). The number of colonies obtained from 5×10^4 normal human bone cells, devoid of nature granulocytes and CSF producing cells, are shown. Conditioned medium from peripheral blood MNC's from 2 patients with Fanconi's anemia and one control was incorporated into the basal (feeder) layer. The effect of addition of autologous and normal serum to the MNC's is shown

Discussion

Aplastic anemia is a clinical and hematological entity which may be the result of a variety of etiologic and pathogenetic mechanisms [2, 17]. The clinical and hematological picture and the course of the illness are variable. This, in some cases, can be related to etiologic factors, while in others a variety of etiologic factors produce similar clinical effects [2].

In addition to the large group of "classical" aplastic anemia a number of special subgroups of disease characterized by bone marrow hypoplasia are well known. These include Fanconi's anemia, Blackfan-Diamond syndrome, pure red cell aplasia and aplastic anemia associated with viral hepatitis and infectious mononucleosis [7, 9, 14, 24, 25]. A variety of laboratory findings have been noted and described in these disorders. Also the nature of some of these disorders and the associated abnormalities seen in them suggest that disorders like infectious mononucleosis and viral hepatitis may provide information which may unravel certain pathogenetic mechanisms. The universal applicability of this type of information may not be valid, however. Regardless of the cause and mechanism of production of the lesion in aplastic anemia the defect needs to be considered in terms of the stem cell compartment [2]. The defect, in a sense, lies in the failure of the stem cell compartment to provide and maintain adequate numbers of precursor and mature hematopoietic cells. In different conditions abnormalities in various aspects of this maturation process have been seen.

In the majority of our patients with idiopathic aplastic anemia the major problem would appear to be a decrease in numbers of effective hematopoietic stem cells to produce mature blood cells. The nature of this defect has not been determined but it does not appear to be the result of the production of circulating humoral substances or inhibitory cells as has been described previously using the erythroid colony forming technique.

Our studies have shown a marked decrease in bone marrow CFU-C's. Most patients yielded no CFU-C's when bone marrow was cultured. In a previous study using unseparated bone marrow stimulated with whole blood buffy coat feeders [15] and in the current study using separated bone marrow cells stimulated with

PCM the results have been similar. The results reported by Kern et al. [13] and Ragab et al. [20] have also been similar. Further, in Blackfan-Diamond congenital hypoplastic anemia the bone marrow and blood erythroid colony forming units (CFU-E) and erythroid brust forming units (BFU-E) have been found to be low [19]. Thus, one of the major abnormalities in aplastic anemia appears to be an absolute decrease in the committed stem cell population. The assessment of the pluripotent (uncommitted) stem cell population will have to wait development of an adequate assay system.

The other major finding in our patients was with regard to levels of serum CSA and the production of CSA by peripheral blood MNC's and peripheral blood buffy coat cells. The levels of CSA were found to be high or normal in patients with aplastic anemia. Similar findings have been reported by Kern et al. [13] and Ragab et al. [20]. Karp et al. [12] reported the findings of 26 patients with aplastic anemia. The incorporation of ^{3}H-Thymidine (^{3}H-TdR), ^{3}H-TdR labelling index, cell counts and morphologic studies were carried out on patients and normal marrow in the presence of patient serum. In 13 patients the serum was stimulatory and in the other half it was normal or inhibitory. Ten of the 13 stimulatory sera were from patients with idiopathic aplastic anemia while 10 of 13 sera that were normal or inhibitory were from patients with aplastic anemia associated with liver disease.

It would appear that in patients with idiopathic and drug induced aplastic anemia the quantitative defect of the stem cell population is associated with an increase in hormonal regulatory substances and it is reasonable to theorize that such a relationship is compensatory with the CSA rising in an attempt to stimulate granulopoiesis from the deficient stem cell compartment.

The addition of peripheral blood MNC's from 3 patients with aplastic anemia to normal bone marrow cells devoid of MNC's stimulated with PCM did not cause inhibition of colony growth (Table 1). This would suggest that the MNC's did not have a cell population that suppresses in vitro granulopoiesis directly or by elaboration of a humoral substance. Hoffman et al. [11] reported a marked reduction in erythroid colony formation in 5 of 7 patients when lymphocytes from patients with aplastic anemia were added in cultures. These workers also demonstrated a cell mediated immunity defect in patients with Blackfan-Diamond syndrome [10]. Ascensao et al. [1] reported 1 patient with aplastic anemia whose marrow cells inhibited CFU-C formation by normal bone marrow. Further, preincubation of patients bone marrow cells with antihuman thymocyte globulin and complement led to restoration of colony formation in soft agar cultures (CFU-C). Clinical data of Speck et al. [23] also suggest the possibility of an immunologic mechanism operative in patients with aplastic anemia. They compared the survival of patients transplanted with HLA mismatched bone marrow who were pretreated with antilymphocyte globulin (ALG) to that of patients transplanted with matched bone marrow and pretreated with cyclophosphamide. The ALG group did better despite the mismatched bone marrow transplant. In contrast to these reports 2 recent studies have demonstrated normal function of lymphocytes and no evidence of suppression of erythroid colony formation [8, 18]. The reasons for these differences are unclear. Recent preliminary data from Singer et al suggest that some of the inhibitory effects of lymphocytes from patients with aplastic anemia may be related to blood transfusions [22].

The incorporation of serum into a stimulus layer of PCM produced no change with normal or patient sera. Similar findings were reported by Ragab et al. [20]. In patients with pure red cell aplasia, on the other hand, serum inhibitors of erythropoiesis have been reported [14]. Serum inhibitors directed against CFU-C have been described in subpopulations of patients with a variety of hematologic disorders including some with atypical aplastic anemia [5].

The production of CSF by MNC of patients with Fanconi's anemia is deficient. Data from one patient suggest that the defect may be in the monocytes.

Since the granulocyte cell line is usually as abnormal as the erythroid cell line in aplastic anemia one might expect similar findings if such a mechanism were operative. Central to this argument is the nature of the colony forming cell in the granulocyte and erythroid colony forming techniques systems. It is conceivable that if the colony forming cell in the red blood cell technique was a multipotential hematopoietic stem cell that one might see suppression using that system while not observing it in the granulocyte colony forming technique in which the colony forming cell has been demonstrated to be a committed granulocyte precursor. All available evidence, however, indicates that the erythroid colony forming cell is comparable to the granulocyte colony forming cell and is a committed precursor rather than a true multipotential stem cell. A resolution of these diverse finding must await future studies therefore.

The nature of the defect in production of hematopoietic stem cells in this group of patients remains to be determined. One of the most interesting suggestions has been that this may result from an alteration in the microenvironmental stroma in which stem cells are nurtured. This microenvironment is not reproduced in the culture system used here. Recently developed techniques by Dexter et al. [6] for study of the microenvironment have not yet been applied to the study of human aplastic anemia but may yield interesting and valuable information in this regard.

The second mechanism described in aplastic anemia here is defective production of granulopoietic factors by peripheral blood mononuclear cells in two patients with Fanconi's anemia. In this disorder the number of granulocyte colony forming cells is normal or increased but CSA producing cells lack the ability to manufacture or release appropriate granulopoietic stimulus in the face of peripheral blood neutropenia. The nature of the defect in CSA production has not been determined. Likewise, it has not been determined how the defect in CSA production in this disorder relates to the anemia and thrombocytopenia also seen in such patients. Whether there is a concomitant alteration in production of erythropoietin and thrombopoietic factors remains to be determined.

References

1. Ascensao, J., Pahwa, R., Kagan, W., Hansen, J., Moore, M., Good, R.: Aplastic anemia: Evidence for an immunological mechanism. Lancet 1, 669–671 (1976)
2. Boggs, D. R. and Boggs, S. S.: The pathogenesis of aplastic anemia: A defective pluripotent hematopoietic stem cell with inappropriate balance of differentiation and self-replication. Blood, 48, 71–76 (1976)

3. Boyum, A.: Isolation of mononuclear cells and granulocytes from human blood. Scand. J. Clin. Lab. Invest. *21*, (Suppl) 97, 77–89 (1968)
4. Burgess, A. W., Wilson, E. M. A., Metcalf, D.: Stimulation by human placental conditioned medium of hemopoietic colony formation by human marrow cells. *Blood 49*, 573–583 (1977)
5. Cline, M. J., Herman, S. P., Golde, D. W.: Inhibitors of myelopoiesis. Trans. Proc. *10*, 99–102 (1978)
6. Dexter, T. M., Wright, E. G., Krizsa, F. R., Lajtha, L. G.: Regulation of haemopoietic stem cell proliferation in long term bone marrow cultures. Biomedicine *27*, 344–349 (1977)
7. Diamond, L. K., Wang, W. C., Alter, B. P.: Congenital hypoplastic anemia. Adv. Pediatr. *22*, 349–378 (1976)
8. Freedman, M. H., Saunders, E. F.: Diamond-Blackfan syndrome: Evidence against cell-mediated erythropoietic suppression. Blood *51*, 1125–1128 (1978)
9. Hager, L., Pastore, R. A., Bergin, J. J.: Aplastic anemia following viral hepatitis. Medicine (Baltimore) *54*, 139–164 (1975)
10. Hoffman, R., Zanjani, E. D., Vila, J., Zalusky, R., Lutton, J. D., Wasserman, L.: Diamond-Blackfan syndrome: Lymphocyte-mediated suppression of erythropoiesis. Science *193*, 899–900 (1976)
11. Hoffman, R., Zanjani, E. D., Lutton, J. D., Zalusky, R., Wasserman, L.: Suppression of erythroid-colony formation by lymphocytes from patients with aplastic anemia. New Eng. J. Med. *296*, 10–13 (1977)
12. Karp, J. E., Schacter, L. P., Burke, P. J.: Humoral factors in aplastic anemia: Relationship of liver dysfunction to lack of serum stimulation of bone marrow growth in vitro. Blood *51*, 397–414 (1978)
13. Kern, P., Heimpel, H., Heit, W., Kubanek, B.: Granulcocytic progenitor cells in aplastic anemia. Brit. J. Haemat. *35*, 613–623 (1977)
14. Krantz, S. B.: Pure red-cell aplasia. New Eng. J. Med. *291*, 345–350 (1974)
15. Kurnick, J. E., Robinson, W. A., Dickey, C. A.: In vitro granulocytic colony-forming potential of bone marrow from patients with granulocytopenia and aplastic anemia. Proc. Soc. Exp. Biol. Med. *137*, 917–920 (1970)
16. Mesner, H., Till, J. E., McCulloch, E. A.: Interacting cell populations affecting granulopoietic colony formation by normal and leukemic human marrow cells. Blood *42*, 701–710 (1973)
17. Metcalf, D.: Approaches to some unanswered problems concerning aplastic anemia. Trans. Proc. *10*, 151–153 (1978)
18. Nathan, D. G., Hillman, D. G., Chess, L., Alter, B. P., Clarke, B. J., Breard, J., Housman, D. E.: Normal erythropoietic helper T cells in congenital hypoplastic (Diamond-Blackfan) anemia. New Eng. J. Med. *298*, 1049–1051 (1978)
19. Nathan, D. G., Clarke, B. J., Hillman, D. G., Alter, B. P., Housman, D. E.: Erythroid precursors in congenital hypoplastic (Diamond-Blackfan) anemia. J. Clin. Invest. *61*, 489–498 (1978)
20. Ragab, A. H., Gilkderson, E., Christ, W. M., Phelan, E.: Granulopoiesis in childhood aplastic anemia. J. Pediat. *88*, 790–794 (1976)
21. Robinson, W. A., Pike, B. L.: Colony Growth of Human Bone Marrow Cells in vitro. In: Symposium on Hemopoietic Cellular Differentiation. Stohlman, F. jr. (ed.). New York: Grune & Stratton, 1970, pp. 249
22. Singer, J., Brown, J., Storb, R., Thomas, E. D.: The effect of lymphocytes from patients with idiopathic aplastic anemia on CFU-C growth from HLA matched and mismatched marrows. Blood *50*, (Suppl. 1) 260a (1977)
23. Speck, B., Cornu, P., Sartorius, J., Nissen, C., Groff, P., Burri, H. P., Jeannet, M.: Immunologic aspects of aplasia. Trans. Proc. *10*, 131–134 (1978)
24. Steier, W., VanVoolen, G. A., Selmanowitz, V. J.: Dyskeratosis congenita: Relationship to Fanconi's anemia. Blood *39*, 510–521 (1972)
25. VanDoornik, M. C., van'T Veer-korthof, E. T., Wierenga, H.: Fatal aplastic anemia complicating infectious mononucleosis. Scand. J. Haemat. *20*, 52–56 (1978)

3.7 Granulocytic and Erythroid Progenitor Cells in Recovering Aplastic Anemia

W. Heit, H. Heimpel, B. Kubanek

Introduction

The pathomechanisms which cause the obliteration of the hemopoietic system in aplastic anemia are poorly understood, the fate of an individual patient unpredictable and the overall prognosis bad.

More recently, the in vitro assay systems for clonal growing of human granulopoietic and erythroid progenitor cells (granulocytic colony forming cells, CFC's, erythroid colony forming units, CFU-e) have brought some insight into the functional structure of the hemopoietic stem cell compartments. They have helped to better define the targets for the possible pathomechanisms involved. However, they have so far failed to provide experimental evidence that allowed conclusions on the pathways which the etiologically relevant factors (e.g. various chemicals of unrelated structure, viruses) may follow.

It is well accepted from the work of Kurnick et al. (1971) [8] Howell et al. [4], Kern et al. [6], and Hansi et al. [2] that the defect may be located in the stem cell compartment, most probably on the level of uncommitted stem cells. Preliminary results from our group further indicate that the few patients who survive aplastic anemia with a recovered bone marrow and peripheral blood show quantitative und qualitative aormalities in the cultures for CFU-e and CFC's. The following presentation aims to critically evaluate the summarized data collected in our group.

Material and Methods

21 patients with aplastic anemia were included in the study. The diagnosis was based on pancytopenia, a hypo – or aplastic bone marrow in more than one biopsy and no recent exposure to chemotherapy or radiation. At time of investigation anabolic steroids and corticosteroids were in the history of all patients. None, however, received immunosuppressive agents (e.g. anti-theta cell-globulin). "Partial remission" (P.R.) was diagnosed in case of an improved clinical status and a recovery of at least one cell line in the marrow and peripheral blood; patients in "complete remission" (C.R.) showed normal marrow cellularity, normal or slightly subnormal peripheral blood cell counts and no need for supportive therapy.

The *preparation of the cell suspension* and the morphological evaluation of the bone marrow and peripheral blood specimen were performed as described in detail recently [6].

For the *agar cultures of granulopoietic colony forming cells* (CFC's) two techniques were used: in the double layer culture system as described by Robinson and Pike (1969) [9] the specific colony stimulating activity (CSA) was provided by feeder layers of a defined white blood cell composition (Heit et al., 1974 [3]). The single layer cultures, were employed according to the method of Burgess et al. (1977) [1] using conditioned media of cultured human placenta particles (CMHPL) as source of CSA. In essence, the cultures (1 ml total volume) consisted of alpha medium, 20% fetal calf serum,

20% CMHPL, and appropriate number of marrow cells and 0.3% bacto agar. The cultures were set up in triplicates and cultured in a fully humidified atmosphere of 5% CO_2 in air and 37° C for 10 days.

   To assay erythroid colony forming units (CFU-e) the methylcellulose modification (Iscove et al., 1974 [5]) was employed using a standard procedure previously described by Rich and Kubanek [5] and Hansi et al. [2]. The cultures were set up in duplicates consisting of a 1 ml mixture of alpha medium, cell suspension, 30% fetal calf serum, alpha-thioglycerol (10^{-4} M, final concentration), 0.8% Methylcellulose (Serva, primium grade, 4000 cps). During the study two different batches of a commercially available erythropoietin step III (Connaught Labs., Canada) had to be employed. Dose response curves from both batches of erythropoietin showed a maximum stimulatory activity at 0.4 units/culture. The cultures were incubated identically to the CFC cultures for a period of 7–8 days.

Results and Discussion

Culturing CFC and CFU-e in vitro results in a wide range of colony counts which may be ascribed to methodological pitfalls involved in any bioassay and the physiological inconsistency of cell suspensions. In our laboratory, the lowest

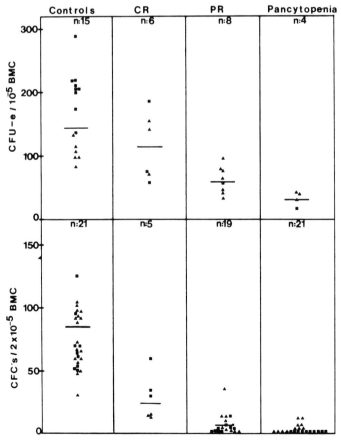

Fig. 1. Change in colony counts in the bone marrow of aplastic anemia patients. *Upper part:* CFU-e cultured in presence of the Lot-Nrs. of erythropoietin (0.4 units/ml). *Lower part:* CFC's grown in the double layer (▲) or single layer culture (■)

CFC counts in cultures of representatively collected marrow specimen of healthy normals range at about 50 per 2×10^5 nucleated cells. Approximately 80 colonies per 10^5 mononuclear cells is the corresponding value for the CFU-e. Patients with aplastic anemia constantly reveal a severe depletion of both precursor cell types (Fig. 1). No increase in CFC numbers can be observed in "partial remission". Furthermore, the majority of patients in "complete remission" show persistently decreased CFC-s in the bone marrow. In contrast, the CFU-e may recover, corresponding to the clinical course of the disease. There is little experimental evidence available to explain the discrepancy of the regenerative pattern of both cell types. Tentative conclusions favoring a preferential recovery of the erythroid cell line may be drawn from studies on stem cell differentiation in murine animal models. Besides the problem of species differences, the assumption that both stem cell types are functionally comparable entities is incorrect. Rich and Kubanek [8] clearly showed, that the CFU-e represents a "mature" stem cell which is closely related to the immature but morphologically recognizable red blood cell precursors.

No data are so far reported about the CFU-e precursor cells (Burst forming units, BFU-e) in CR of aplastic anemia. Preliminary results from two patients indicate, however, that the concentration of BFU-e in the bone marrow is decreased, thus confirming the observation of an impaired reconstitution of the CFC compartment in "complete remission" (CR) as shown in Figure 2: Compared to healthy controls all marrow specimen with a normal cellularity obtained from hematologically recovered aplastic anemia patients contained a disproportionally low number of CFC's. A comparison between the concentration of colony forming cells in agar (CFC's) and the relative numbers of more differentiated granulocytic precursor cells (myeloblasts, promyelocytes and myelocytes) shows (Fig. 3), that the more mature cell compartments are restored to normal values in regenerating aplastic anemia.

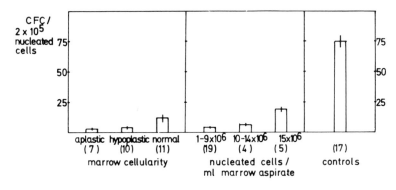

mean ± s.e.m.
number of cases in brackets

Fig. 2. Mean number of colonies ± SEM cultured from marrow suspensions of patients with aplastic anemia in relation to the marrow cellularity and the number of nucleated cells per millilitre aspirate. The number of cases falling within these categories are indicated in parentheses. The control marrows showed a normo-cellularity and a mean nucleated cell count of 30×10^6 per ml in the apsirate [6]

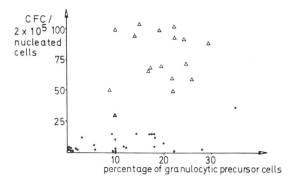

CFC /
2 × 10⁵ 100
nucleated
cells   75

50

25

10   20   30
percentage of granulocytic precursor cells

Fig. 3. Number of CFC plotted against the percentage of cells of the granulocytic mitotic pool (i.e. myeloblasts, promyelocytes and myelocytes) within the marrow suspension (●). Individual observations of the patients; (△) controls [6]

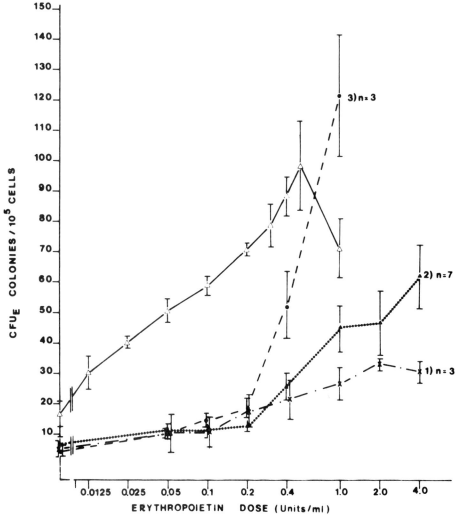

Fig. 4. Erythropoietin dose response on CFU-E colony formation. 1×10^5 nucleated bone marrow cells were plated per ml. The straight line shows the colony growth of seven control patients. Dotted lines show growth of the grouped patients with aplastic anemia. Bars indicate means + SEM [2]

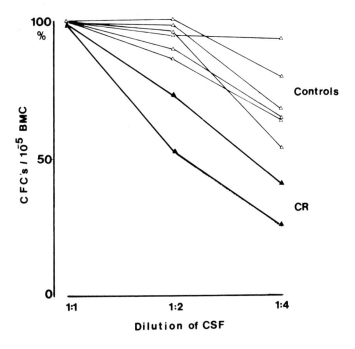

Fig. 5. CSA dose response on agar colony formation. 10^5 mononuclear bone marrow cells were cultured in presence of CMHPL, 20% v/v, ($\triangle$) normal controls, ($\blacktriangle$) patients in "complete remission"

Further information about a possible functional stem cell defect in aplastic anemia comes from experiments performed by Hansi et al. [2] who described an abnormally reduced responsiveness of CFU-e in vitro to low doses of erythropoietin in "partial" and "complete remission" marrows (Fig. 4). Similar results have been obtained for CFC's of two recovered patients which show a reduced response to low doses of colony stimulating factor (Fig. 5).

In summary, the data presented herein demonstrate a longterm impairment of two stem cell entities in hematologically recovered aplastic anemia. The discrepancy in the results obtained from the CFU-e and CFC cultures probably reflects the different grades of differentiation between both committed stem cell compartments. Two cell lines are affected, thus indicating that the pluripotent stem cells may be involved, too. The nature of the impaired stem cell regeneration in clinically recovered aplastic anemia, however, remains obscure. A persistent cellular defect of the pluripotent stem cell as well as an impaired local environment in the bone marrow or immunemechanisms may sustain the delayed reconstitution of the precursor cell compartments and cause an abnormal sensitivity to the humoral regulators. The long-term culture system of hemopoietic stem cells introduced by Mike Dexter may here provide a decisive contribution to a better understanding of the mechanisms which decide upon the prognosis of aplastic anemia.

Acknowledgements

The authors are indebted to Ms. G. Matzke for technical assistance throughout this work.

The work was supported by the Deutsche Forschungsgemeinschaft, SFB 112, Projects A2 and A5.

References

1. Burgess, A. W., Wilson, M. A., Metcalf, D.: Stimulation by human placental conditioned medium of hemopoietic colony formation by human bone marrow cells. Blood *49*, 573 (1977)
2. Hansi, W., Rich, I., Heimpel, H., Heit, W., Kubanek, B.: Erythroid colony forming cells in aplastic anaemia. Brit. J. Haematol. *37*, 483 (1977)
3. Heit, W., Kern, P., Kubanek, B., Heimpel, H.: Some factors influencing granulocytic colony formation in vitro by human white blood cells. Blood *44*, 511 (1974)
4. Howell, A., Andrews, T. M., Watts, R. W. E.: Bone marrow cells resistant to chloramphenicol in chloramphenicol-induced aplastic anaemia. Lancet *65, I*, (1975)
5. Iscove, N., Sieber, F., Winterhalter, H.: Erythroid colony formation in cultures of mouse and human bone marrow: analysis of the requirement for erythropoietin by gel filtration and affinity chromatography on agarose concanavalin. A. Journal of Cellular Physiology *83*, 309 (1974)
6. Kern, P., Heimpel, H., Heit, W., Kubanek, B.: Granulocytic progenitor cells in aplastic anaemia. Brit. J. Haematol. *35*, 613 (1977)
7. Kurnick, J. E., Robinson, W. A., Dickey, C. A.: In vitro granulocytic colony-forming potential of bone marrow from patients with granulocytopenia and aplastic anaemia. Proceedings of the Society for Experimental Biology and Medicine *137*, 917 (1971)
8. Rich, I., Kubanek, B.: Erythroid colony formation in foetal liver and adult bone marrow and spleen from the mouse. Blut *33*, 171 (1976)
9. Robinson, W., Pike, B.: Colony growth of human bone marrow cells in vitro. In: Stohlman, F., Jr. (ed.) Symposium on Hemopoietic Cellular Proliferation, pp. 249. New York: Grune & Stratton 1969

3.8 High BPA in Aplastic Anemia, a Possible Indicator of Immune Pathogenesis

Catherine Nissen, B. Speck

Some, but not all patients with idiopathic acquired aplastic anemia get long lasting remissions and are potentielly cured after treatment with antilymphocyte globulin (ALG) [10, 11]. This suggests that the diagnosis covers disorders of various origin. Since the pathogenesis is unknown, prediction of response to ALG is impossible.

Progress in treatment is to be expected from improved understanding of early events in hemopoiesis: by the use of tissue culture techniques it has recently been shown, that responsiveness to the pathway-specific regulators, CSA and erythropoietin (epo) is acquired at a relatively late stage of maturation; very primitive cells are regulated by another group of substances [4, 7, 8, 9, 12, 13]. This was first demonstrated in vitro for the red cell line: Colony formation by immature precursors (BFU-E) depends on the presence of "burst promoting activity" (BPA) in addition to epo. With maturation, the cells gradually loose their responsiveness to BPA, while sensitivity to epo increases.

BPA can be supplied by human leukocyte conditioned medium (HLCM) [1], and also – to a variable extent – by fetal calf serum, and by normal human serum. Plastic adherent cells are involved in making BPA available. It is not known, how BPA operates in vivo.

For detection of the activity, a culture assay poor in intrinsic BPA has to be provided. This is obtained by plating normal human non-plastic-adherent bone marrow cells in methyl-cellulose, essentially as described by Iscove [5, 6], at 1 unit of epo/ml in the absence of HLCM. Under these conditions, bursts will not grow unless BPA is added.

Using such an assay, we tested human sera and observed a striking increase of BPA in serum from many, but not all patients with acquired aplastic anemia. Compared to normal serum (Fig. 1) an excess of burst – and macrophage promoting activity was observed in patient's serum (Fig. 2). Colony formation by granulocyte precursors was not stimulated (Fig. 2). One high BPA serum was tested for inhibition of granulocyte precursors: reduction of colonies to 10% was observed in the presence of patient's serum. In addition, high BPA sera inhibited colony formation by the more mature erythroid precursors (CFU-E). One could assume that this inhibition reflects a pathophysiological mechanism operating in aplastic anemia. Since human leukocyte conditioned medium – supernatant of a culture of normal leukocytes – also strongly inhibits CFU-E (Fig. 3), it may be that there is an excess of a physiological regulator substance in aplastic anemia.

30 of our 31 patients had severe aplastic anemia as defined by Camitta et al. [3]. They were severly anemic and likely to have elevated epo. Unlike normal serum, patient sera in fact promoted the formation of hemoglobinized colonies

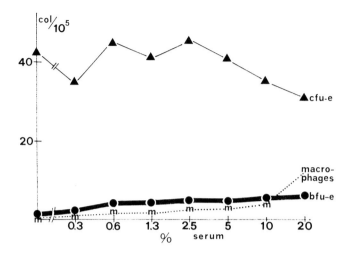

Fig. 1. Dose related effects of normal human serum on BFU-E, CFU-E and macrophage precursors: Mean values of 5 experiments (donor serum with ABO-matched normal bone marrow cells)

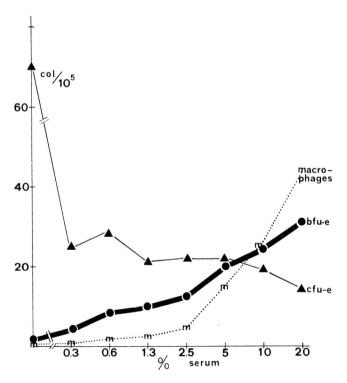

Fig. 2. Dose related effects of serum from patients with aplastic anemia on BFU-E, CFU-E and macrophage precursors. Mean values of 9 experiments (patient serum with ABO-matched normal bone marrow cells)

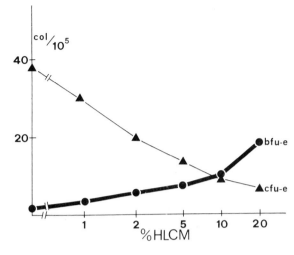

Fig. 3. Dose related effects of human leukocyte conditioned medium (HLCM) on BFU-E and CFU-E. Mean values of 4 experiments

without added epo. The stimulating effect of patient sera is, however, not sufficiently explained by high serum-epo, because
– there was no correlation between patient's hemoglobin values and serum-BPA.
– 5 patients with anemia of other causes (iron-deficiency, Blackfan-Diamond anemia, anemia of chronic disease, pure red cell aplastia and congenital hemolytic anemia) did not stimulate burst formation excessively.
– the stimulating effect could not be mimicked by addition of high dose epo (20 U/ml) to BPA-poor cultures.
– maximal precipitation of BPA, and epo respectively, occurred at different concentrations of ammonium sulphate.
– the two activities eluted separately on G-150 Sephadex-chromatography, BPA eluting with albumin.

Thus, high epo does not account for the stimulating activity described. We assume that BPA in these sera is elevated in addition.

Interpretation of this finding is merely speculative, since the physiological role of BPA is unknown. It seems unlikely that high BPA reflects a feed-back mechanism in response to peripheral demands. An alternate interpretation comes from clinical observation in the patients studied.

High BPA appears to characterize a certain patient category. As shown in Fig. 4, high BPA was confined to patients with either idiopathic acquired or chloramphenical induced aplasia. BPA was low in posthepatitic and in congenital aplasia. In one case of classical Fanconi's anemia [2], high BPA without excess macrophage stimulating activity was observed.

18 of these patient recovered autologous marrow function after either treatment with antilymphocyte globulin (ALG) or cyclophosphamide (Cy) given for attempted bone marrow transplantation (Table 1). All of them had high pretreatment BPA. One patient with high BPA did not respond to ALG. On the other hand, 4 patients treated with ALG and mismatched marrow had low

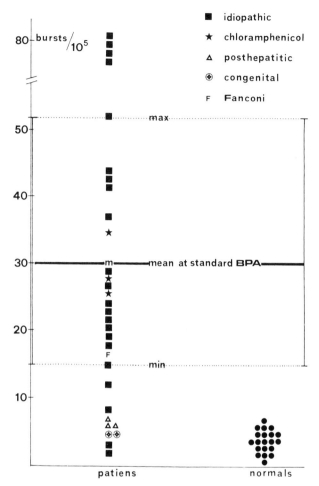

Fig. 4. Effect of 20% serum from 31 patients with aplastic anemia of various causes, and 20 normals on BFU-E in ABO-compatible normal marrow. Standard BPA = growth at 20% human leukocyte conditioned medium (mean, maximum and minimum values of 13 experiments)

pretreatment BPA, none of them responded to the treatment (Table 1). One low-BPA-patient recovered spontaneously from posthepatitic pancytopenia.

BPA values returned to the normal range in patients with complete autologous bone marrow reconstitution. Decline of BPA preceded hematological normalization (Fig. 5), suggesting that high BPA may play a primary role in the development of marrow damage.

Hence, high BPA appears to be of value in the prediction of autologous marrow recovery after ALG or cyclophosphamide. A positive response to such therapy suggests that an immune mechanism was involved in the pathogenesis of the disease. One may therefore argue that high BPA reflects an ongoing immune-reaction. There is evidence to support this hypothesis. In the superna-

Table 1. Clinical course of 30 patients in relation to pretreatment-BPA. Patients listed in descending order ob BPA as in Figure 4

Patient	Age		Treatment and course		Autologous marrow reconstitution	BPA before treatment
S.C.	♂	15	ALG + BMI		++	↑
L.M.	♂	15	ALG + BMI		++	↑
I.St.	♀	28	Cy + BMT	rejected	++	↑
R.L.	♂	18	Cy + BMT	rejected, died early	?	↑
B.S.	♂	29	ALG		++	↑
E.M.	♀	20	ALG + BMI		++	↑
I.B.	♀	16	Cy + BMT	rejected	+	↑
J.K.	♀	8	ALG		++	↑
S.E.	♀	20	ALG + BMI		++	↑
I.O.	♀	17	ALG	died of hemmorrhage after myeloid reconstitution	+	↑
R.T.	♂	35	ALG		–	↑
H.M.	♂	28	ALG + BMI		+	↑
E.K.	♂	22	ALG + BMI		++	↑
P.J.	♀	6	Cy + BMT	rejected, died after 2nd BMT	?	↑
A.B.	♂	14	ALG + BMI		++	↑
J.C.	♂	10	ALG + BMI		++	↑
A.C.	♂	18	ALG + BMI		++	↑
A.B.	♀	14	ALG + BMI		++	↑
M.R.	♂	10	ALG + BMI		++	↑
A.M.	♀	45	Cy + BMT	rejected	++	↑
A.E.	♀	10	ALG + BMI		?	↑
F.A.	♀	14	Cy + BMT	rejected, died after 2nd BMT	?	↑
M.W.	♀	6	Cy + BMT	permanent graft	?	↑
M.P.	♀	55	ALG + BMI		–	n
M.M.	♂	17	none	recovered from posthepatitic pancytopenia	++	n
A.R.	♂	14	ALG + BMI		–	n
G.S.	♂	19	Cy + BMT	died day 1	?	n
A.W.	♀	9	ALG		–	n
,, ,,		10	Cy + BMT	permanent graft	?	n
M.B.	♂	5	ALG + BMI		–	n
K.K.	♀	28	Cy + BMT	permanent graft	?	n

ALG = Antilymphocyte globulin
BMI = Bone marrow infusion (HL-A haplotype matched)
Cy = Cyclophosphamide
BMT = Bone marrow transplantation (HL-A matched)
+ = Partial autologous marrow reconstitution
++ = Complete reconstitution (patient does not need supportive care)
? = Possibility of autologous recovery not evaluable
– = No improvement of marrow function after 3 months
↑ = BPA elevated before treatment
n = BPA in the normal range treatment

tant of positive mixed leukocyte cultures (MLC) the BPA-activity was 2–4 times higher than in the supernatant of control cultures.

We conclude that the measurement of serum burst- and macrophage promoting activity allows us to caterogize patients with aplastic anemia as "high" and

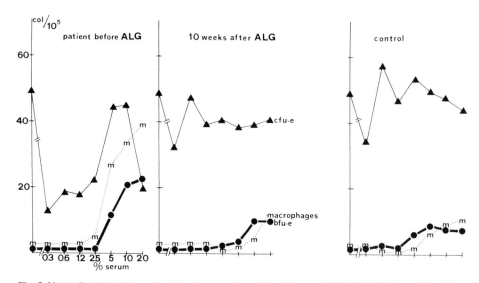

Fig. 5. Normalization of serum burst and macrophage stimulating and CFU-E inhibiting activity after treatment with ALG and bone marrow. (CFU-E-inhibition by serum before treatment is not evident at 5–10%, possibly because it is counteracted by epo at these concentrations) Patient's serum before and after treatment, and control serum tested on an ABO-matched bone marrow sample. Peripheral blood values:

	before ALG	10 weeks after	6 months after
Hb g/dl	6.7	9.6	15.5
Granulocytes/mm³	20	400	2500
Platelets/mm³	5000	26000	200000

"low" stimulators. Our data suggest that high stimulators have a greater chance of autologous recovery after ALG or cyclophosphamide. If this experience can be reproduced in a larger number of patients, improved choice of the appropriate treatment modality for individual patients could result.

Acknowledgements

Patient survival was improved by dedicated work of Pierre Cornu, Walter Weber and Alois Gratwohl (clinical management) and skilled technical assistance of Miss Beatrice Rubin and Miss Axenia Levak (supportive care). We wish to thank Mrs. Irène Lehnherr for organization and secretarial work. This contribution would not have been possible without advice on culture technique and interpretation of data by Norman N. Iscove.

Supported by Grants of the Swiss Cancer League FOR.080.AK.75 and FOR 101.AK.77 [2] and the Swiss Science Foundation 3.3320.74 and 3.890.0.77

References

1. Aye, M. T.: Erythroid colony formation in cultures of human marrow: effect of leukocyte conditioned medium. J. cell Physiol. *91*, 69, (1977)
2. Beard, M. E. J.: Fanconi anemia. In: Congenital Disorders of Erythropoiesis, Ciba Foundation Symposium 37 (new series). Amsterdam: Elsevier, 1976, p. 103
3. Camitta, B. M., Thomas, E. D., Nathan, D. G., Santos, G. W., Gordon-Smith, E. C., Gale, R. P., Rapaport, J. M. and Storb, R.: Severe aplastic anemia + a prospective study of the effect of early marrow transplantation on acute mortality. Blood *48*, 63 (1976)
4. Iscove, N. N.: Erythropoietin-Independent Stimulation of Early Erythropoiesis in Adult Marrow Cultures by Conditioned Media from Lecitin-Stimulated House Spleen Cells. In: Proceedings, ICN-UCLA Symposium on Hemopoietic Cell Differentiation, 1978, in press
5. Iscove, N. N., Senn, J. S., Till, J. E. and McCulloch, E. A.: Colony formation by normal and leukemic human marrow cells in culture: effect of conditioned medium. Blood *37*, 1 (1971)
6. Iscove, N. N. and Sieber, F.: Erythroid progenitors in mouse bone marrow detected by macroscopic colony formation in culture. Exp. Hemat. *3*, 32 (1975)
7. Johnson, G. R. and Metcalf, D.: Pure and mixed erythroid colony formation in vitro stimulated by spleen conditioned medium with no detectable erythropoietin. Proc. natl. Acad. Sci. (USA) *74*, 3897 (1977)
8. Metcalf, D. and Johnson, G. R.: Production by spleen and lymphnode cells of conditioned medium with erythroid and other hemopoietic colony-stimulating activity. J. cell. Physiol. *96*, 31 (1978)
9. Nathan, D. G., Chess, L., Hillman, D. G., Clarke, B., Breard, J., Merler, E. and Housman, D. E.: Human erythroid burst-forming unit: T-cell requirement for proliferation in vitro. J. exp. Med. *147*, 324 (1978)
10. Speck, B., Cornu, P., Nissen, C., Groff, P., Weber, W. and Jeannet, M.: On the Pathogenesis and Treatment of Aplastic Anemia. In: Experimentel Hematology Today. Ledney, G. D. and Bau, S. J. (eds.). New York: Springer, 1978, p. 143
11. Speck, B., Gluckman, E., Haak, H. L. and van Rood, J. J.: Treatment of aplastic anemia by anti-lymphocyte globulin with and without allogeneic bone marrow infusion. Lancet *II*, 1145 (1977)
12. Wagemaker, G.: Cellular and Soluble Factors Influencing the Differentiation of Primitive Erythroid Progenitor Cells (BFU-E) *in vitro*. In: In vitro Aspects of Erythropoiesies. Murphy, M. J. jr. (ed.). New York: Springer, 1978, in press
13. Wagemaker, G., Brouwer, A., Bol, S. J. L., Visser, T. P.: Analysis of factors in vitro inducing erythropoietin- and CSF-responsiveness in primitive hemopoietic progenitor cells of the mouse. Experimental hematology, *Vol 6*, Supplement No. *3*, 31 (1978)

3.9 Early and Late Effects of Adjuvant Chemotherapy (Adriamycin/Cyclophosphamide) on the Human Granulopoiesis

H.-P. Lohrmann, W. Schreml

Introduction

In clinical practice, the hematopoietic toxicity of chemotherapeutic agents is routinely assessed by following the changes in the concentration of peripheral blood neutrophilic granulocytes. However, for two reasons peripheral blood granulocytes do not directly reflect the extent of toxicity exerted by cytotoxic drugs on the human granulopoiesis:

a) since these drugs exert their cytocidal effects on the proliferating granulocytic compartments, this damage will appear in the peripheral blood with the appropriate delay determined by the kinetics of the granulocytic system

b) the presence of the bone marrow granulocyte storage pool will dampen any injury to the early granulocytic compartments.

In the literature, little information may be found on the effects of cytotoxic drugs on a normal (i.e., non-leukemic) human granulopoiesis. For that reason, we have initiated a series of studies designed to characterize the pattern of reaction of the human granulopoiesis to cytotoxic drugs, and to answer, among others, the following questions:

a) what are the changes of an unperturbed human granulopoiesis after exposure in vitro to cytotoxic drugs?

b) are there quantitative and/or qualitative changes persisting after discontinuation of cytotoxic drug therapy?

c) is the cytotoxic drug-induced marrow hypoplasia a useful model for other types of human marrow failure?

The data reported in this paper have been obtained in patients undergoing adjuvant chemotherapy for primary breast cancer. An adjuvant chemotherapy protocol appears particularly suited to answer the questions raised, for two reasons: here, patients without evidence of disease and with a presumably normal hematopoiesis are exposed to cytotoxic drugs; therefore, the changes observed are those of an unperturbed human granulopoiesis after exposure to the drugs. Furthermore, these patients are exposed to the chemotherapeutic agents for just a limited period of time, after which no further chemotherapy is administered; therefore, follow-up studies allow to detect possible late effects of such therapy.

Patients

44 female patients (age 32–64, median 46 years) were entered into the study. Selection of patients has been previously described [5]. Adjuvant chemotherapy consisted of six courses of cyclophosphamide

500 mg/sqm plus adriamycin 50 mg/sqm, given at 4 week intervals. At the time of this analysis (May 31, 1978), 33 patients had completed this chemotherapy protocol. Hematological studies, including repeated bone marrow aspirations, were performed during the first and during the sixth course, and again two months after completion of chemotherapy. In addition, peripheral blood was obtained before each course of chemotherapy.

Methods

Differential counts on peripheral blood and bone marrow smears were obtained by counting at least 300 cells. The functional bone marrow granulocyte reserve was determined from the maximal increment of neutrophils in the peripheral blood following rapid intravenous infusion of 200 mg hydrocortison i.v. [1]. Committed granulopoietic stem cells (colony-forming units in culture = CFU-C) were assayed utilizing a single layer agar culture system with human placenta conditioned medium as source of colony-stimulating activity [6]. Use of this conditioned medium provided the constant culture conditions necessary for the longitudinal studies performed in the present work. The proliferative activity of bone marrow CFU-C was assessed using a modification of the thymidine suicide technique [7].

Results

1. Peripheral Blood Neutrophils

Changes in peripheral blood segmented neutrophils (PMN) are shown in Figure 1. During the first 3 days, there were no changes in the peripheral blood PMN count; thereafter, neutrophils decreased progressively to their nadir around day 14, after which they recovered to pretreatment levels by day 28. With repeated administration of the cytotoxic drugs, a slight (not significant) decrease in the peripheral blood PMN counts was noted. After the sixth course, the reaction of peripheral PMN was similar to that seen during the first course. However, after completion of chemotherapy a persistent neutropenia was noted ($p < 0.05$ to < 0.001).

2. Peripheral Blood Mononuclear Cells

Changes of peripheral blood lymphocytes may be seen from Figure 2. Following discontinuation of chemotherapy, there was a transient lymphopenia; recovery to pretreatment levels occurred around day 200 after the last chemotherapeutic course. Peripheral blood monocytes had their nadirs around day 10 after chemotherapy, and thus showed earlier recovery to pretreatment levels than peripheral blood granulocytes (lower curve in Fig. 2).

After discontinuation of chemotherapy, there was a persistent reduction of peripheral blood monocytes, so far without evidence of return to pretreatment levels.

Peripheral blood reticulocytes and peripheral blood platelets were in the pretreatment range after discontinuation of chemotherapy.

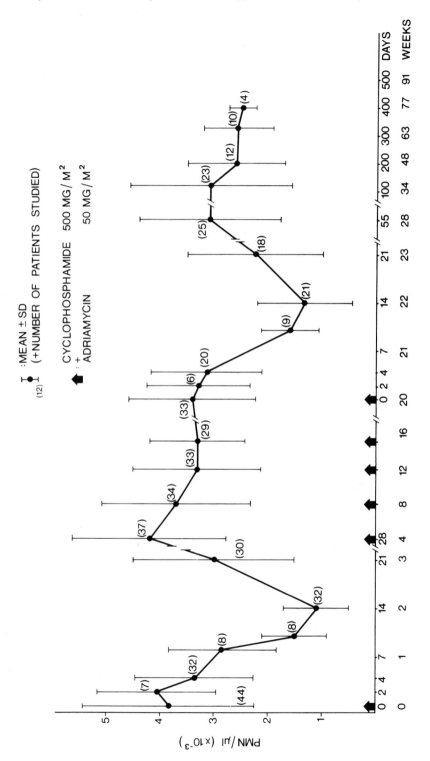

Fig. 1. Changes of the peripheral blood neutrophilic granulocytes during and after intermittent adjuvant chemotherapy with cyclophosphamide/adriamycin. Arrows indicate administration of chemotherapy

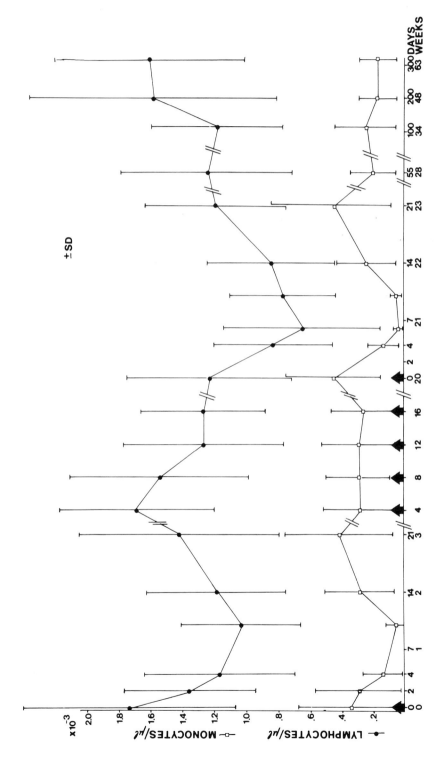

Fig. 2. Peripheral blood lymphocytes (*upper curve*) and monocytes (*lower curve*) during and after intermittent adjuvant chemotherapy with cyclophosphamide/adriamycin

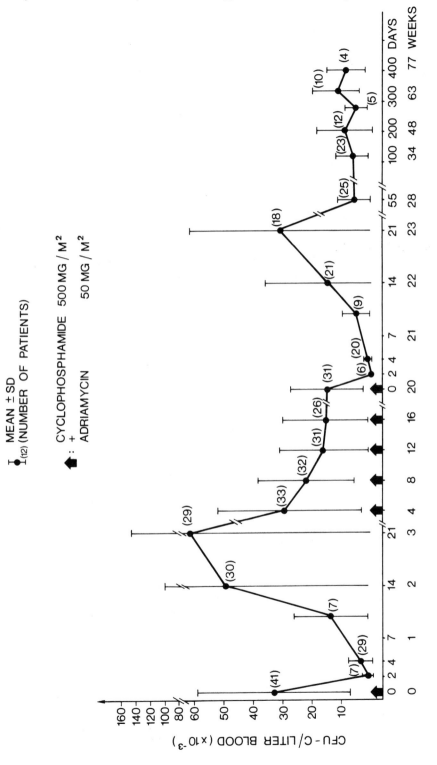

Fig. 3. Peripheral blood CFU-C during and after intermittent adjuvant chemotherapy with cyclophosphamide/adriamycin

3. Peripheral Blood CFU-C

Changes of the peripheral blood CFU-C pool (expressed as CFU-C per liter of whole blood) are demonstrated in Figure 3. Immediately after administration of the cytotoxic drugs, the peripheral blood CFU-C pool size was greatly reduced, with nadirs around day 2.

Thereafter, the peripheral blood CFU-C pool gradually returned to its pretreatment size. With repeated intermittent chemotherapy, a progressive reduction of the peripheral blood CFU-C pool size was noted. After the sixth (last) administration of the cytotoxic drugs, depletion of the peripheral blood CFU-C pool size was similar to that seen during the first course, whereas regeneration appeared to be delayed. Thereafter, a second and long-lasting reduction of the peripheral blood CFU-C pool size (to about 20% of pretreatment levels; $p < 0.001$) could be noted. So far, there has been no tendency of the peripheral blood CFU-C pool size towards return to pretreatment values.

4. Functional Bone Marrow Granulocyte Reserve

Changes of the functional bone marrow granulocyte reserve were not different during the first and during the sixth course (Fig. 4). Immediately after chemotherapy, there was a slightly increased granulocyte mobilisation following hydrocortisone infusion; thereafter, the functional marrow granulocyte reserve dropped to nadirs around day 14, and recovered to pretreatment values between days 21 and 28. After completion of chemotherapy, the functional marrow granulocyte reserve was not different from pretreatment values.

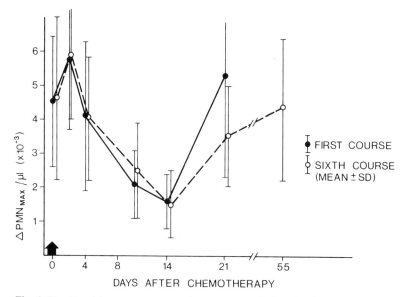

Fig. 4. Functional bone marrow granulocyte reserve during the first and sixth course of adjuvant chemotherapy. *Abscissa:* days after chemotherapy; *ordinate:* maximum peripheral blood granulocyte increments after 200 mg hydrocortison i.v.

5. Bone Marrow Differentials

Details of the changes in the cytological bone marrow composition are listed in Table 1. There was an initial rise in the M/E ratio, caused by an early, almost complete disappearance of cells of the erythroid series. Erythroid regeneration was marked around day 8, leading at this time to very low M/E ratios. With normalisation of marrow composition, M/E ratios returned to pretreatment values.

Shortly after chemotherapy, granulopoiesis was right-shifted. Granulocytic regeneration was morphologically identifiable around day 10, leading to a left-shifted granulopoiesis. Around day 14, the maturing granulocytic compartment was almost depleted, and segmented neutrophils were only rarely seen in marrow smears.

During the sixth course, the pattern of reaction was not different from that observed during the first course. However, after completion of chemotherapy bone marrow did not return to pretreatment marrow composition: the M/E ratio was reduced ($p < 0.001$), percentages of the cells of the proliferating granulocytic pool were reduced ($p < 0.01$), and percentages of segmented neutrophils were also reduced ($p < 0.001$).

6. Bone Marrow CFU-C

Values are expressed as CFU-C/10^5 marrow cells, and thus indicate relative numbers of bone marrow CFU-C. Immediately after chemotherapy, there was a sharp reduction of CFU-C, lowest values were recorded on day 2 (Fig. 5).

Table 1. Bone marrow differential counts during and after adjuvant chemotherapy

Course/day	% of bone marrow cells[*a]						
	M/E ratio	p[b]	Proliferating pool	p[b]	Maturing pool	PMN	p[b]
1/0	1.41±0.43		10.6±4.3		36.8± 5.8	11.7± 7.5	
1/2	6.36±3.8		6.9±2.3		52.9± 6.1	17.4± 5.5	
1/4	6.28±3.9		7.0±4.4		50.4±10.6	23.1± 9.7	
1/8	0.73±0.42		11.2±6.3		21.5±10.2	9.4± 5.1	
1/14	0.50±0.27		16.0±5.5		12.8± 5.4	1.7± 1.7	
1/21	1.95±0.82		13.0±4.6		43.5±10.4	5.2± 3.2	
6/0	1.67±1.02	NS	8.2±3.6	<0.05	40.1± 8.5	10.4± 4.1	NS
6/2	8.6 ±3.6		6.6±4.1		62.0±12.2	22.6±11.4	
6/4	8.2 ±6.3		6.4±2.2		56.3± 8.0	21.1± 6.7	
6/10	0.32±0.16		5.5±1.9		15.4± 8.0	3.4± 2.4	
6/14	0.33±0.13		10.0±4.3		11.1± 4.1	1.7± 1.7	
6/21	1.06±0.32		11.0±4.7		37.8± 4.7	4.2± 2.2	
6/55	1.06±0.41	<0.001	7.7±2.8	<0.01	31.2± 6.9	7.1± 2.8	<0.001

[a] Means±S.D.
[b] When compared to pretreatment values (day 1/0). n.s.: not significant

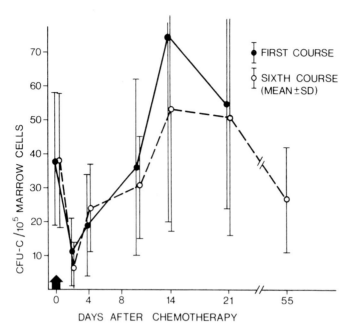

Fig. 5. Changes of bone marrow CFU-C during the first course, and during and after the sixth course of adjuvant chemotherapy

Thereafter, relative numbers of CFU-C increased rapidly, to somewhat overshooting values around day 14 before returning to pretreatment values. Changes observed during the sixth course were not significantly different from those observed during the first course (Fig. 5). However, after discontinuation of chemotherapy, a reduction in the relative numbers of bone marrow CFU-C was noted (means $\pm$ SD: before treatment 38 ± 21; after chemotherapy 27 ± 16; $p < 0.05$).

7. Proliferative Activity of Bone Marrow CFU-C

Using the thymidine-suicide technique, an estimate of the proliferative activity (i.e., of the fraction of cells in S-phase of the cell cycle) can be obtained [3]. Data shown in figure 6 indicate that bone marrow CFU-C increased their proliferative activity almost immediately after chemotherapy. The proliferative activity continued to be increased up to day 10 after chemotherapy, whereupon it returned to pretreatment values around day 14. Figure 6 also shows that the increased proliferative response after the sixth course appeared to be somewhat reduced. However, after discontinuation of chemotherapy, the proliferative activity of marrow CFU-C was in the pretreatment range.

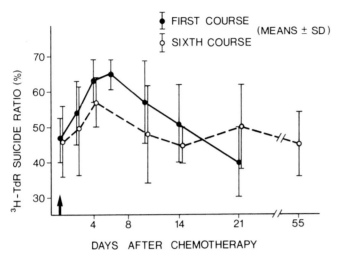

Fig. 6. Proliferative activity (expressed as ^{3}HTdR suicide ratio, ordinate) of bone marrow CFU-C during the first, and during and after the sixth course of adjuvant chemotherapy

Discussion

Data obtained in our patients before treatment are all in the normal ranges of this and of other laboratories; therefore, the conclusion is justified that this study describes the changes of a normal, unperturbed human granulopoiesis in response to the combination chemotherapy administered.

After the first administration of the cytotoxic drugs, there was an almost immediate depletion of the early granulocytic compartments, including the bone marrow CFU-C compartment. This damage, as well as the subsequent granulocytic regeneration proceeded in a wavetype fashion through the successive granulocytic compartments. Granulocytic regeneration was first detected on day 2 from an increased proliferative activity of bone marrow CFU-C, then from increasing relative numbers of bone marrow CFU-C, but only around day 10 granulocytic regeneration was also morphologically identifiable. It is noteworthy that bone marrow CFU-C and the early granulocytic compartments (myeloblasts, promyelocytes) experienced closely related concordant changes, which of course is in agreement with the current concept that the earliest granulocytic cells are the immediate progeny of bone marrow CFU-C.

Around day 14, the proliferative activity of bone marrow CFU-C had returned to pretreatment levels, and the relative number of bone marrow CFU-C was again in the pretreatment range. This indicates that by this time the regeneration at the CFU-C level was completed. The fact that at the same time the segmented neutrophil compartments of peripheral blood and bone marrow were nearly depleted raises some questions as to the regulation of CFU-C proliferation; obviously, a negative feedback control of CFU-C proliferation by segmented neutrophils appears to be of little importance, whereas the possibility of a local CFU-C population size control (by unknown mechanisms) must be considered.

The demonstration of changes at different levels of granulopoiesis persisting after discontinuation of chemotherapy is of considerable interest. In the bone marrow, the relative number of CFU-C was reduced, the M/E ratio was reduced, as were the percentages of cells of the proliferating granulocytic pool and of segmented neutrophils. In the peripheral blood, segmented neutrophils, monocytes and CFU-C were reduced, but lymphocytes showed only a transient decrease and platelets and reticulocytes showed no changes at all. We interpret these late changes as the expression of a production defect of granulopoiesis, i.e. a hypoplasia of the granulocytic system. The hypothesis presented by Schofield during this meeting [10] may explain why so far there has been no sign of repair of this long-lasting defect of granulopoiesis. It remains to be seen from follow-up studies what the granulocytic production defect will ultimately mean for the patients; however, late bone marrow failure appears to be a distinct possibility [2]: Morley and Blake reported late bone marrow failure after short-term administration of busulfan in mice [8]; they later documented that this was caused by a permanent stem cell defect [9]. It is of interest that this marrow aplasia was preceded by a period of a "latent" aplasia; the changes observed during this latent period are almost identical to those observed in our patients after discontinuation of chemotherapy.

Kern et al. [4] have presented evidence for a persistent defect of granulopoiesis in patients in "remission" from aplastic anemia. In Table 2, this defect of granulopoiesis in patients in remission from aplastic anemia is compared to that observed in our patients after discontinuation of adjuvant chemotherapy. It is obvious that qualitatively these changes are quite comparable. It may well be that a common pathomechanism is responsible for these late defects; thus, a granulopoiesis recovering from severe (idiopathic or cytotoxic drug-induced) marrow failure may not be able to recover completely, and Schofield's hypothesis offers an explanation for this phenomenon.

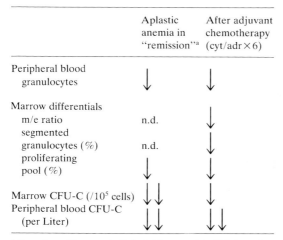

	Aplastic anemia in "remission"[a]	After adjuvant chemotherapy (cyt/adr × 6)
Peripheral blood granulocytes	↓	↓
Marrow differentials		
m/e ratio	n.d.	↓
segmented granulocytes (%)	n.d.	↓
proliferating pool (%)	↓	↓
Marrow CFU-C (/10⁵ cells)	↓↓	↓
Peripheral blood CFU-C (per Liter)	↓↓	↓↓

Table 2. Qualititative comparison of the defects of granulopoiesis in patients with "remission" from aplastic anemia *(left half)*, and in patients after sixth courses of adjuvant chemotherapy *(right half)*. n.d.: not done

[a] Data from Kern et al., Brit. J. Haemat. *35*, 613 (1977)

References

1. Dale, D. C., Fauci, A. S., Guerry, D., Wolff, S. M.: Comparison of agents producting a neutrophilic leukocytosis in man. Hydrocortisone, prednisone, endotoxin, and etiocholanolone. J. Clin. Invest 56, 808–813 (1975)
2. Hellman, S., Botnick, L. E.: Stem cell depletion: An explanation of the late effects of cytotoxins. Int. J. Radiat. Oncol. Biol. Phys. 2, 181–184 (1977)
3. Iscove, N. N., Till, J. E., McCulloch, E. A.: The proliferative states of mouse granulopoietic progenitor cells. Proc. Soc. Exp. Biol. Med. 134, 33–36 (1970)
4. Kern, P., Heimpel, H., Heit, W., Kubanek, B.: Granulocytic progenitor cells in aplastic anemia. Brit. J. Haemat. 35, 613–623 (1977)
5. Lohrmann, H.-P., Schreml, W., Lang, M., Betzler, M., Fliedner, T. M., Heimpel, H.: Changes of granulopoiesis during and after adjuvant chemotherapy of breast cancer. Brit. J. Haemat. 40, 369–381 (1978)
6. Lohrmann, H.-P., Hansi, W., Heimpel, H.: Human placenta-conditioned medium for stimulation of human granulopoietic precursor cell (CFU-C) colony growth in vitro. Blut 36, 81–88 (1978)
7. Lohrmann, H.-P.: Thymidine suicide of human granulocytic progenitor cells (CFU-C). Biomedicine 28, 319–323 (1978)
8. Morley, A., Blake, J.: An animal model of chronic aplastic marrow failure. I. Late marrow failure after busulfan. Blood 44, 49–56 (1974)
9. Morley, A., Trainor, K., Blake, J.: A primary stem cell lesion in experimental chronik hypoplastic marrow failure. Blood 45, 681–688 (1975)
10. Schofield, R.: Mechanism of damage to the stem cell population. Presented at the Intern. Symposium on Aplastic Anemia, Reisensburg, see p. 63

Discussion

Mangalik: I have two questions. One is a general question, the other is a specific question for Dr. Nissen. We have a long-term deficiency of the colony forming unit in these patients and we had a demonstration from Dr. Lohrmann that in these patients at least with chemotherapy that CFU-c were normal. I do think that there are two possibilities to consider. Is it an effect on the soil? And the second one, is it an effect on some kind of hormonal regulator? Is there any evidence that your chemotherapeutic treatment produces effects in the environment in which these stem cells or precursor cells grow? Dr. Nissen, have you studied along with the burst forming units the burst promoting activity on a long-term basis in aplastic anemia patients after recovery?

Lohrmann: There were two reports recently suggesting the possibility of a failure of the microenvironment after chemotherapy[1,2], one[1] pointing to a stem cell defect as a major cause of hemopoietic failure after cytotoxic drug treatment. Otherwise I don't know of any data that chemotherapy induces permanent changes of the microenvironment.

Moore: Yes, there was a stem cell defect and an environmental defect shown.

Wickramasinghe: But the environmental defect was rather trivial. In Morley's model he suggested that there might be an imbalance between differentiation and proliferation of the stem cell compartment. This may account for a lot of stem cell defects in aplastic anemia. Do you have any opinions about that possibility?

---

1 Fried, W., Kedo, A., Barone, J.: Effects of cyclophosphamide and of busulfan on spleen colony-forming units and on hematopoietic stroma. Cancer Res. 37, 1205–1209 (1977)
2 Morley, A., Trainor, K., Blake, J.: A primary stem cell lesion in experimental chronic hypoplastic marrow failure, Blood 45, 681–688 (1975)

Schofield: The fact that we have to consider, when we are talking about drugs, is the very specific effect they have on different species of cells and looking at simple depression of CFU-S in the mouse, we can get similar levels of depression, say by Myleran and Isopropylmethanesulphonate, which are totally different in action. Myleran probably effects primarily the hemopoietic stem cells, whereas if we take another alkylating agent it will effect the more mature types of CFU. Looking at Morley's observation it is most likely to be the case that this is really affecting the primary stem cells. We are getting a permanent damage and possibly also because of damage to the environment. But then, as Dexter pointed out in his cultures, we can begin to look at environmental and stem cell defects in the culture. We are using this model of Myleran depleted bone marrow, using that as both a monolayer feeder and normal monolayer, hopefully to see whether these two effects occur in the Myleran treated and in the normal animal.

Heimpel: There is at least one example which can be used against the effect of high-dose chemotherapy on the microenvironment. Although receiving high doses of cyclophosphamide patients who recovered from bone marrow transplantation had apparently no hemopoietic defect. This would at least argue against an effect on microenvironment, which is relevant for the maintainance of hemopoietic stem cell renewal systems. Dr. Thomas could you comment on this?

Thomas: Yes, that is correct.

Moore: I don't really know if we need to pay a lot of consideration to this environment aspect, at least in the context of aplastic anemia, except that there is one cell type that is part of the microenvironment and that was shown by Dr. Dexter as one of the three cell types in the adherent layer in the continuous marrow culture. This is in fact the macrophage and that will be presumably the only cell that in a conventional bone marrow transplant could conceivably become a component or restore a component of what otherwise might be a defective microenvironment. But I am personally not aware of any information on whether there is any form of macrophage defect associated with aplastic anemia, except the inflammatory pictures we see and they may be in a state of activation, that is incompatible with the function of the regulatory cells.

3.10 General Discussion

Moderators: M. A. S. Moore, N. S. Wickramasinghe

Wickramasinghe: It is the purpose of this particular discussion to try to corelate the fundamental data we heard to the haemopoietic failure of aplastic anemia. I think we should start by re-asking some of the questions we heard right at the beginning today, and try to see if we are any closer to the answers. One of the questions that was raised, "why do patients with aplastic anemia have a late onset of the aplastic anemia?" I wonder if Dr. Schofield in light of this model would like to explain that?

Schofield: Myleran effects essentially the primary stem cells, though the animal will be perfectly fertile for some considerable time after you treated and damaged the primary stem cell by Myleran and functional cells are still produced. The defect will not be observed in functional terms until the supplier cells which would normally be coming through from the stem cells begin to fade.

Wickramasinghe: Some patients with i.e. chloramphenicol-induced aplasia will not get aplastic anemia until 6 months after the course of chloramphenicol and I suppose one has to think of other effects operating at the stem cell level.

Schofield: Can we look at all these aplasias as simply a primary stem cell defect of this sort? Maybe Dr. Moore could comment on the role of suppressor cells in the delayed aplasia?

Moore: I think that it certainly could be a topic of quite a lot of discussion tomorrow. What we are asking is, can you provide an alternative explanation based on your stem cell model?

Schofield: I dont't think so. I think that purely on kinetic models you find difficulty in explaining a great delay on the initial damage.

Lohrmann: The problem is, that you don't know the causative agent. You assume it is Chloramphenicol, but it could be something else affecting hemopoiesis four weeks earlier.

Schofield: On truly kinetic considerations I don't think it is possible to determine that yet.

Moore: I was rather surprised by the presentation of the data on the impairment of CFU-c in recovered aplastics. Some years ago we did some experiments on chloramphenicol-sensitivity of patients with known chloramphenicol-induced aplastic anemias who had recovered. They were certainly within the normal range of CFU-c and they showed no difference in sensitivity to chloramphenicol.

Heit: We did extensive titration curves for normal bone marrow versus aplastic anaemia and bone marrow of patients in remission after chloramphenicol and from others after phenylbutazone for instance. In no marrow specimen so far there was a clearcut hypersensitivity in vitro and even in those where we assumed that chloramphenicol induced the initial failure we do not know if the CFC in these aplastic anemia patients of the CFC we detected with stimulator are still subpopulations of the whole stem cell population. But it is so that they are all depressed and my own opinion is that we don't just know it.

Heimpel: When we first used clonal assays in aplastic anemia, we did so to detect possible hypersusceptibility of committed granulopoietic stem cells to chloramphenicol in patients which went into remission from chloramphenicol-induced aplastic anemia. To our surprise, the colony incidence was so low that we had difficulties to test in vitro inhibition by drugs. As outlined by Dr. Heit, we found such a late defect in the majority of our cases. Truly, blood counts in patients in complete remission were often at the lower margin of the normal range. When we follow these patients for more years, they may become completely normal. We cannot yet decide, whether they are still in a slow phase of recovery, or already on an abnormal plateau of hemopoietic function, which was present already before exposition to the noxious factor.

Gordon-Smith: We have preliminary data on 4 patients who have complete remission as judged by their blood counts and in each of those patients the CFU-c are low compared with the normal range. But we also have 1 patient who, following allogeneic bone marrow transplant lost his marrow after 6 months and regenerated his own marrow. In that patient we were able to follow CFU-c from virtually zero at the time of the second attack of aplasia back entirely to normal. That patient of course had received large doses of Cyclophosphamide at the time of the transplant, whether that had anything to do with it or not.

Nissen: I have seen 9 patients in complete remission and have done CFU-E, CFU-c and BFU-E and in some of them the CFU-E recovered after some time, most remained low and in some the BFU-E's never recovered, although they showed completely normal blood counts.

Singer: I was just going to add that the CFU-C studies on approximately 50 patients' post-grafted marrow, those with aplastic anemia and leukemias essentially CFU-C numbers were normal after day 100. Most have become normal by day 70 or so. So I would say that we don't see the low incidence of CFU-c seen in the spontaneous recovered aplastics.

Shadduck: One comment: We seem concerned that the number of CFU-c in at least some cases of aplastic anemia do not return to normal values after recovery of haemopoiesis. One would wonder what normal values were for these patients before they developed their aplasia, that is to say, have they had sufficient hits or damage to the stem cell compartment, so that it was partly depleted before the development of overtly aplastic anemia.

Dexter: I think that counts of peripheral blood cells could be misleading. One could have an effectively depleted stem cell compartment, but because of an increased amplification in the transit compartments you can end up with normal blood counts, although the stem cell compartment is extremely depleted. There is some evidence from irradiated mice where CFU-c can be decreased, but the BFU-E can be back to almost minimal levels and blood counts back to normal levels at times when the pluripotent stem cells are only 10% of normal.

Kubanek: I fully agree with you Dr. Dexter, that there are black boxes with an unknown kinetic pattern in this system which we are not able to measure yet. Though the fact remains and has to be explained that the CFU-c compartment does not recover to normal size for a long time. This may be reflecting a failure of recovery of the pluripotent stem cell compartment although the production of hemopoietic functional cells is near normal.

Dexter: Yes, you're right, I am just emphasizing there, that the stem cell compartment could be lower than what we are measuring.

Schofield: Dr. Moore, you have given evidence for an age structure in the CFU-c compartment from the colonies produced at different times in culture. Can you give us any idea of the amplification that takes place in this compartment? Could this possibly help to see whether during recovery from aplasia there is an increased amplification?

Moore: We can get an idea of amplification. The reason we can make a statement about an age structure is that when we separate a bone marrow population, we get a fraction that we call "the CFU-d diffusion chamber" which does not form colonies in agar, unless it is incubated for 7 days in suspension culture and then it gives rise to colonies that are maximum when scored at day 14. Taking the cells which form colonies at day 14 from velocity sedimentation fractions and incubating them, they give rise 7 days later to colonies which would be scored at day 7 and then degenerate and don't go on within the restrictions of our suspension culture system. Not knowing facts as plating efficiency, the amplification would be not more than 2-fold. I suspect it is probably just separating these cells by one or at the most two divisions, and as they divide they move into the next compartment. Another matter which should have been pointed out: we are also scoring eosinophil colonies. Quite a large number of colonies that are scored at 14 days are in fact eosinophil colonies, and what their relationship is to the pathology of aplastic anemia I just don't know.

Heimpel: Dr. Iscove, following your concept, BPA determines the rate of cell production by the hemopoietic stem cell pool, but does not regulate the differentiation into one of the single cell lines. Which factors do then determine the balance between the production of erythropoietic and granulopoietic cells, and how can the body respond to the potential need for more red cells or more granulocytes? Could your concept explain the paradoxical observation, that erythropoietic cell production in aplastic anemia is of greater prognostic value than the production of platelets or granulocytes, even though the lack of these latter cells is far more dangerous for survival?

Iscove: I wish we had the answers to those questions. I think they are crucial not only to a basic understanding, but clearly to the disease processes that we are interested in. I think that at present we do not have the experimental tools that allow us to anser or distinguish between the two mechanisms here suggested for stem cell determination. The two mechanisms are: that you have an instructive process that comes from the environment to tell the stem cells to go in the erythroid direction rather than the granulocytic. The alternative model to that is a selective one, in which one has stem cells undergoing determination stochastically, and then the particular environment in which the progeny of them stem cells find themselves then give pathway specific queues turning on one line or the other. So selective incidence of the stem cell throwing out determined progeny, but whether or not they want to proliferate and mature is determined by the environment. We do not have very good tools for looking at that at the moment. We can make some attempts at it. Dr. Gregory in particular has looked for example at individual spleen colonies and asked how many BFU-E and CFU-c are in a spleen colony, trying to answer the question of whether the so-called inductive microenvironment pushes the stem cells one way or another. But that experimental design is inadequate because, as Trentin has argued, by the time a colony is large enough to be dissected out easily, it has encroached on neighbouring microenvironments in the spleen and is going to be mixed anyway, so that is a problem, and I think it's going to be very difficult to solve by the present methods.

Moore: If you don't expect an instructive model of pluripotent stem cell differentiation, if you accept the argument that there is a vast excess of pluripotential stem cells with extensive self renewal capacity, they could withstand a lot of waste in the sense that the purely random commitment to various progenitor lineage committed progenitor cells, that could be justified on the basis of what we know of the impact of humoral regulators in terms of allowing the balance expression of erythroid or myeloid elements. We had a specific example in that, some studies that we have done and a lot of other people have done, looking at competition, if any, between erythropoietin and colony stimulating factors in terms of the balance of ways one can get granulocytic colonies or erythroid colonies. We used pure erythropoietin and we used macrophages as a source of burst promoting activity. We found that, as we increased the concentration of erythropoietin in the cultures, the BFU-c and CFU-e increased and the CFU-c, the granulocytic colonies, decreased in almost perfect balance. It turned out that what happened was that there was not competition on the level of a cell which had receptors for both activities, but because the macrophage happened to have a response to pure erythropoietin and erythropoietin switched on the production of prostaglandin E, and prostaglandin E inhibits CFU-c proliferation at a concentration of 10^{-8} Mol. Prostaglandin does not inhibit CFU-E proliferation, if anything, it enhances it. So, if one could extrapolate from this in vitro system to in vivo, one could say that you have a regulator and a biologically relevant suppressive agent that could determine the exact balance between erythroid and granulocytic. It might be that these sort of factors acting in a short range manner, as Dr. Dexter's system would suggest, are in fact these microenvironments that people talk about may not be cell-to-cell interactions.

Iscove: Then that argument would favour, if I understand it, a selective model rather than an inductive one.

Kubanek: Dr. Iscove, you showed in your model that BPA is acting on the pluripotent stem cell pool, and I wonder what evidence you have on that.

Iscove: It is a theoretical proposal. It is made in the absence of any evidence which says that the factors turing on the pluripotent stem cell and in the absence of any evidence that a specific factor could be separated from the factor necessary for the early stage of burst development. If somebody could separate the two activities chemically, then you could say that there are two factors. One fact I neglected was to draw a distinction between a cell that can form a mixed colony in culture and the question whether or not that is a stem cell. It is pluripotential if we accept that those colonies are clones, but we have no idea if they are stem cells.

Kubanek: Why do you think you get such low incidence of mixed colonies in your adult system, whereas in the fetal system the incidence, at least in CBA mice as shown by Johnson and Metcalf is much higher.

Iscove: It is not much higher actually, but the comparison is difficult to make, because it is a normalization per 10^5 nucleated cells. It is the 10^5 that is determining the figure, not the number of colony forming cells. Adult marrow is filled with almost mature cells and fetal liver is not.

Heimpel: Dr. Nissen, the elevation of BPA in human serum seems to be rather specific for aplastic anemia. There is apparently an inverse relationship between BPA and marrow cellularity, because BPA becomes normal after recovery. Do you believe that the elevation of BPA is secondary to some mechanisms relevant for the pathogenesis of aplastic anemia, or is there just less consumption of this regulator when the marrow is aplastic, as in the case of erythropoietin?

Nissen: There is something that I did not mention. The father of one of our aplastic patients also had elevated BPA, although he was hematologically completely normal, though I don't know what that means. We haven't looked at more relatives of aplastic patients and about what BPA does in aplastic anemia. I would think that maybe one mechanism by which it may act is that BPA maybe primary. My speculation would be that there is some lymphocyte/macrophage cooperation that creates an excess of BPA which then causes an excess of pluripotent stem cells to go into cycle. This again would deplete eventually the pluripotent stem cell pool.

Gordon-Smith: Just one question in relation to that. You said that they responded to immunosuppression. Can you tell us exactly what immunosuppression is?

Nissen: Fourteen patients were treated with ALG, five without marrow and the others with addition of marrow. And the remaining three of these seventeen patients go cyclophosphamide for bone marrow transplantation, but rejected, however lived long enough for evaluation for the immunosuppressive effect of cyclophosphamide. So, that makes the 17 actually immunosuppressed patients.

Gordon-Smith: Patients treated with cyclophosphamide didn't take. They perhaps died, but there are no changes of BPA.

Nissen: No. When they take the graft the BPA level falls, but when they do not take a graft, it normalizes in a few, but not in others. An I don't know why.

Wickramasinghe: Have you looked at BPA production by other cell types, other than lymphoid cells? You said that simulated lymphoid cells were good sources.

Nissen: Well, I just took human lymphocyte conditioned medium as a source of BPA. My standard source of BPA is just plain human leukocyte conditioned medium.

Iscove: It has been published a year ago that human embryos release the same activity into the supernatant.

Moore: So it seems to line up that you have mitogen stimulated T-lymphocytes, you have macrophages, you have kidney cells, you have embryonic fibroblasts and it comes down to the same sort of confusion that we had about the relevance of colony stimulating factor. All of those same types of cells producing granulocyte macrophage colony stimulating factor. Maybe they produce erythropoietin too.

4. The Rationale of Therapeutic Approaches in Aplastic Anemia

4.1 Stimulation of Hemopoiesis in Aplastic Anemia by Hormones and Other Agents

H. Heimpel

Many attempts to treat aplastic anemia have been attempts to stimulate directly or indirectly the production of blood cells by the hemopoietic tissue. The majority of compounds which have been used are hormones or hormone-analoges. This is easy to understand, because for a long time it was assumed that the regulation of hemopoiesis is under hormonal control, and because it was speculated that aplastic anemia might be a disorder of regulation. The intention of this lecture is to review the results of such treatments, including two nonhormonal drugs, Cobalt and Lithium, and to discuss to what extent the clinical results are consistant with the current hypothesis on the pathogenesis of aplastic anemia [27, 7, 33].

Table 1 summarizes the main clinical observations which induced trials of certain drugs.

Leukocytosis and erythrocytosis is present in chronic hypercorticism, as in Cushings disease, and granulocytosis is a well known side effect in patients receiving prednison or related steroids for non-hematological disease. Conversely, anemia is present in Addison's disease and disappears after corticosteroid substitution. Another experience, suggesting the potential benefit from corticosteroids was their effect in auto-immune-hemolytic anemia and idiopathic

Table 1. Clinical observations inducing the trial of different drugs in aplastic anemia

Corticosteroids	Erythrocytosis and leukocytosis in Cushings disease Anemia in Addisons disease Effect in autoimmunhemolytic anemia and ITP
Androgens	Anemia in hypogonadism Erythrocytosis in androgen treated patients with breast cancer
Etiocholanolon	Leukocytosis after i.m. injection and in mediterranian fever
Growth hormone (STH)	Anemia in Sheehans' syndrome and in hypophysectomized animals
Cobalt chloride	Erythrocytosis in normal animals and human subjects
Lithium	Leukocytosis in patients treated for mental disorders

thrombocytopenia as well as in pancytopenia associated with lupus erythemato-
des disseminatus.

Hypogonadism also results in anemia. Androgens as testosteron or methylte-
stosteron normalize the red cell values and result in marked and sustained
erythrocytosis when given in unphysiologically high dosis. This has been shown
most impressively in women with breast cancer [30, 29]. The same is true for
synthetic derivatives, the so-called anabolic hormones, like methenolon or
oxymetholon.

Metabolism of testosteron in vivo leads to some steroids without androgenic
activity. One of these, etiocholanolon, induces fever, local inflammation and
granulocytosis when injected intramuscularely in normal volonteers [18]. High
etiocholanolon serum concentrations are found into some patients with the
so-called mediterranian fever, which is also associated with distinct granulocytosis.

The use of growth hormone (STH) is based only on the anemia in deficiency
states and hypophysectomized animals.

The detection of Cobalt as stimulating erythropoiesis stems from its presence
in the Vitamin B 12-molecule, which was regarded as an erythropoietic
stimulator before its biochemical effects were discovered.

The newest drug to be discussed is Lithium. It has been used by psychiatrists
for about 10 years, mainly in patients with depressive mental disorders. An
incidental observation was a sustained elevation of leukocyte counts, beginning
about one week after administration. Closer investigation of the blood cells
revealed an increase of peripheral granulocytes, eosinophiles and platelets,
a decrease of lymphocytes, no effect on red cells and reticulocytes and no changes
of bone marrow composition and morphology [5].

Our knowledge of the mechanisms responsible for the effects of these
substances on the blood forming tissue is still limited. Some of the data and
interpretations, which may be relevant for the subject of aplastic anemia are
summarized in Table 2.

It has been clearly shown, that glucocorticoids like prednison or dexametha-
son raise granulocyte counts in the peripheral blood by shifts into the circulating
pool from other pools, mainly the marrow storage pool and the marginal
intravascular pool, and by chronic slowing of the output of cells from blood into
the tissues [13, 14]. Attempts to demonstrate stimulating activities of glucocorticoids on hemopoiesis in intact animals or in cell culture systems did not yield
conclusive results. Stimulation of erythropoiesis by dexamethasone or predniso-
lon was reported by Mangor et al. [34] but inhibition by Gordon et al. [21].
Only few data exist on the effect of glucocorticoids on hemopoietic cell cultures.
Golde and coworkers [19, 20] observed enhancement of CFU-E but inhibition of
CFU-C and suggested the presence of specific glucocorticoid receptors on
committed hemopoietic stem cells.

Extensive investigations have been carried out to elucidate the mechanisms of
the action of androgens and their synthetic analoges. Their main target compartment seems to be the committed erythropoietic progenitor cells (CFU-E) or
erythropoietin responsive cells (ERC). Androgens exert their effect on these cells
both indirectly by an increase of the endogenous erythropoietin production [1]

Table 2. Clinical observations inducing the trial of different drugs in aplastic anemia

Corticosteroids	Shift from marrow pool into peripheral blood Inhibition of CFU-C Enhancement of CFU-E
Androgens	Stimulation of EP-production Direct EP-dependent effect on ERC's Enhancement of CSA-production? Induction of proliferation of CFU-S
Etiocholanolon	Shift from marrow into peripheral blood Stimulation of ERC?
Growth hormone (STH)	EP-dependent stimulation of erythropoiesis?
Cobalt chloride	Stimulation of EP-production
Lithium	Enhancement of CSA-production

and directly by an EP-dependent or EP-independent [23] action on the target tissue. The synthetic derivatives oxymetholone and fluoxymetholone were found to enhance erythroid colony formation in rats stronger than the naturally occurring androgens [41].

No effect on granulopoiesis was observed in vivo, but growth of human CFUc can be enhanced by addition of testosteron or methandrostenolon to cultures [37], probably through the enhancement of CSA-production of monocytes [16].

An important observation for the discussion of potential effects of androgens was made by Byron and coworkers from Manchester [9, 11]. In mice, they were able to trigger into cycle CFU-S cells as demonstrated by the effect of cytocidal doses of 3HTdR after an in vivo injection of testostone to normal or polycythemic animals. This property of testostone is shared by non androgenic 5-beta-derivatives as etiocholanolon. However, these data can hardly explain possible effects of androgens in human aplastic anemia. Experimental and clinical data in human aplastic anemia favour a diminution of the number of stem cells going along with an increased rather than a decreased proliferation [27, 7, 33]. Similar changes are observed in late chronic failure after busulfan damage in rats [36]. Experimental use of nondrolon in such animals failed to increase of the numbers of CFU-S, CFU-C and total nucleated cells pro tibia [35]. In contrast, in mice with a decreased hemopoietic stem cell pool from irradiation with 150 rad, CFU-S recovery could be enhanced by giving steroids, as testosterone or ocymetholon in the postirradiation period [10].

Another effect of etiocholanolon and other 5-beta-steroids is the stimulation of erythropoiesis in normal and polycythemic mice as observed by Gordon and coworkers [22]. A rise of leukocyte counts is the only effect of etiocholanolon on the blood which has been observed in normal human subject. This change is mediated merely by a shift from storage into circulating pools, as in the case of glucocorticoids [18].

Only few data are available on the action of STH, suggesting EP-dependent stimulation of erythropoietic bone marrow cells [15]. Cobalt was shown to work through increase of erythropoietin production in the kidneys [17]. Lithium enhances CSA-production from monocytes in mouse and human agar culture [24, 3]. This enhancement is held to be responsible also for the in vivo effects on granulocyte production.

Many observations suggest that in aplastic anemia there is a quantitative and qualitative defect of the early pluripotential hemopoietic stem cells, leading consecutively to a reduction of the committed stem cells and morphologically identifiable bone marrow precursors [27, 7, 33] with an increase of the growth fraction in the CFU-C compartment at least in cases with partial remission [31]. If this is true, one may expect that the majority of mechanisms shown in Table 2 would not be suitable to improve the situation in aplastic anemia. This is consistant with the disappointing results of clinical trials as shown in Table 3.

Many therapeutic effects of glucocorticoids claimed on the basis of pilot studies did not survive more critical investigations. Most clinicians with wide experience on the treatment of aplastic anemia agree that there is no benefit from sustained application of large doses of prednison or related drugs. There is one large series of 55 patients having received prednison for more than one year and showing a complete remission rate of 40% [45], which is more than in the most European and American series [28]. However, prognosis of aplastic anemia is dependent largely on the primary selection of cases [32]. It may well be that the

Table 3. Therapeutic effects of different drugs in aplastic anemia

Corticosteroids	Ineffective in majority of cases Effective within weeks in few cases
Androgens	Effect in the majority of cases controversial No effect detected in randomized trials No effect in busulphan-induced marrow failure Androgen dependency in single cases
Etiocholanolon	Rise of reticulocytes, platelets, PMN's in 2/3 patients
Growth hormone (STH)	Effect doubtfull, no controlled trial
Cobalt chloride	No effect
Lithium	No sustained effect in 1 patient with aplastic anemia Effective in congenital marrow failure?

results reported by Uzuka et al. [45] relate to a group of patients with rather good outlook and may not be significant if controlled by an randomized trial in true severe aplastic anemia.

Some clinicians advocate the use of corticosteroids in high doses (1–2 mg/kg) only for a few weeks, because patients with aplastic anemia showed a rather rapid response to prednison if any [40].

Having adopted this policy of prednison treatment for many years, we have observed 5 quick "responses" amongst approximately 80 patients treated as aplastic anemia. Three of them later manifested acute lymphatic leukemia and were apparently misdiagnosed as aplastic anemia because prednison had already been given before the first biopsy was done. In two others, the early remission may have well been spontaneous. A similar early recovery has been recently observed after we omitted initial high dose prednison from standard treatment (Figure 1).

Improvement of bone marrow failure has been reported after STH [39, 42] and cobalt [43], but these reports are not convincing and could not be confirmed by other investigators [for literature see 26]. Recent successes with etiocholanolon

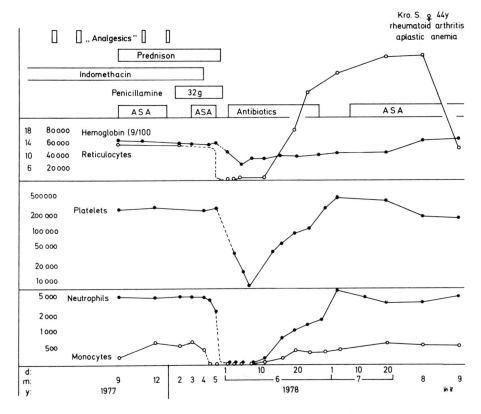

Fig. 1. Early hemopoietic recovery in aplastic anemia, induced by penicillamine and/or indomethacin. Prednison was given for treatment of the underlying disease before the onset of the hemopoietic failure

[4] on single cases await confirmation. Lithium was ineffective in one case of aplastic anemia [2] but improved the granulocyte production in patients with Fanconi's anemia or congenital neutropenia [3]. In the latter conditions the marrow is normal or hypercellular. Granulopoiesis may be therefore stimulated in contrast to aplastic anemia, because the target cells of lithium-enhanced CSA are still present.

The problem of androgens will be reviewed in detail by Dr. Dresch in the next lecture. One recent prospective but non-controlled study [25] and one retrospective study using historical controls [44] seem to support previous positive results [38]. However, this is in contrast to the combined experiment of Camitta et al. [12] who could not demonstrate a prolonged survival of severe aplastic anemia by a treatment with androgens. A few small randomized trials also failed to detect beneficial effects of different androgenic preparations [6, 8]. It seems not worthy to mention that these androgens are clearly effective in conditions in which the stem cell compartments are less severely effected, as in secondary anemia of cancer [30] renal anemia [6] or pancytopenia with hyperplastic bone marrow and smoldering leukemia (own unpublished observations). In our own study on aplastic anemia, control patients did slightly better than patients on mesterolon, but this difference was not statisticly significant (Figure 2). On the other hand we have observed androgen-dependency in a few cases of less severe aplastic anemia, and similar observations have been reported by others. It may well be that androgens are effective in patients with less severly

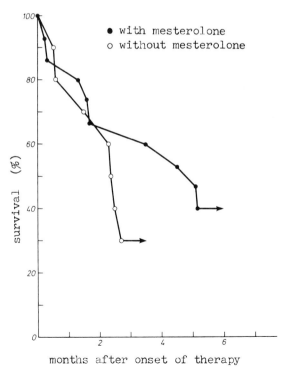

Fig. 2. Survival of patients with severe aplastic anemia, with (n=20) and without (n=11) treatment by androgens

depleted stem cell pools. They may be effective also in a small subgroup of patients with severe aplastic anemia of a different pathophysiology as compared to the majority of non-responders.

In summary, the manifold efforts to stimulate hemopoietic cell production in aplastic anemia by hormones or other pharmacological agents did not alter the outcome in the vast majority of patients. This may be due to the fact, that the agents tested éxert their effect not on the cell population primarely affected, but on progenitor populations with the main function of adaption and regulation but not of longterm cell renewal.

References

1. Alexanian, R.: Erythropoietin excretion in man following androgens. Blood 28, 1007 (1966)
2. Barret, A. J., Griscelli, C., Buriot, D., Faille, A.: Lithium therapy in congenital neutropenia. Lancet II, 1357 (1977)
3. Barret, A. J., Hugh-Jones, K., Newton, K., Watson, J. G.: Lithium therapy in aplastic anaemia. Lancet I, 202 (1977)
4. Besa, E. C., Wolff, S. M., Dale, D. C., Gardner, F. H.: Aetiocholanolone and prednisolone therapy in patients with severe bone-marrow failure. Lancet I, 728 (1977)
5. Bille, P. E., Krogh Jensen, M., Jensen, J. P. and Poulsen, J. C.: Studies on the haematologic and cytogenetic effect of lithium. Acta med. scand. 198, 281 (1975)
6. Bock, E. L., Fülle, H. G., Heimpel, H. und Pribilla, W.: Die Wirkung von Mesterolon bei Panmyelopathien und renalen Anämien. Med. Klinik 71, 539 (1976)
7. Boggs, D. R. and Boggs, S. S.: The pathogenesis of aplastic anemia: A defective pluripotent hematopoietic stem cell with inappropriate balance of differentiation and self-replication. Blood 48, 71 (1976)
8. Branda, R. F., Amsden, T., Jacob, H. S.: Randomized study of Nandrolone therapy for anemias due to bone marrow failure. Arch. intern. Med. 137, 65 (1977)
9. Byron, J. W.: Effect of steroids on the cycling of haemopoietic stem cells. Nature 228, 1204 (1970)
10. Byron, J. W.: Manipulation of the cell cycle of the hemopoietic stem cell. Exp. Hemat. 3, 44 (1975)
11. Byron, J. W., Testa, N.: Stem Cell Kinetics after Treatment with Androgenic-Anabolic Steroids. Proc. 13th International Congr. Haemat. München: Lehmanns Verlag, 1970, p. 199
12. Camitta, B., Thomas, E. D., Nathan, D., Santos, G., Gordon-Smith, E. and Rappeport, E.: Severe aplastic anemia: Effect of androgens on survival. Blood 50, Suppl. I, 313 (1977)
13. Cartwright, G. E., Athens, J. W. and Wintrobe, M. M.: The kinetics of granulopoiesis in normal man. Blood 24, 780 (1964)
14. Cream, J. J.: Prednisolon-induced granulocytosis. Brit. J. Haemat. 15, 259 (1968)
15. Fisher, J. W., Roh, B., Couch, C. and Nightingale, W.: Influence of cobalt, sheep erythropoietin and several hormones on erythropoiesis in bone marrows of isolated perfused hind limbs of dogs. Blood 23, 87 (1964)
16. Francis, G. E., Berney J. J., Bateman, S. M., Hoffbrand, A. V.: The effect of androstanes on granulopoiesis in vitro and in vivo. Br. J. Haematol. 36, (4), 501 (1977)
17. Fried, W. and Gurney, C. W.: The erythropoietic-stimulating effect of androgens. Am. New York Acad. Sc. 149, 356 (1968)
18. Godwin, H. A., Zimmerman, T. S., Kimball, H. R., Wolff, S. M. and Perry, S.: The effects of etiocholanolone on the entry of granulocytes into the blood. Blood 31, 461 (1968)
19. Golde, D. W., Bersch, N. and Cline, M. J.: Potentiation of erythropoiesis in vitro by dexamethasone. J. clin. Invest. 57, 57 (1976)
20. Golde, D. W., Bersch, N., Quan, S. G. and Cline, M. J.: Inhibition of murine granulopoiesis in vitro by dexamethasone. Amer. J. Hemat. 1, 369 (1976)

21. Gordon, A. S., Mirand, E. A. and Zanjani, E. D.: Mechanisme of prednisolone action in erythropoiesis. Endocrinology, *81*, 363 (1967)
22. Gordon, A. S., Zanjani, E., Levere, R. D. and Kappas, A.: Stimulation of mammalian erythropoiesis by 5B-H-steroid metabolites. Proc. Nat. Acad. Sc. *65*, 919 (1970)
23. Gorshein, D., Hait, W. N., Besa, E., Jepson, J. and Gardner, F. H.: Rapid stem cell differentiation induced by 19-N-Nortestosteron decanoate. Brit. J. Haemt. *26*, 215 (1974)
24. Harker, W. G., Rothstein, G., Clarkson, D., Athens, J. W. and Macfarlane, J. L.: Enhancement of colony-stimulating activity production by lithium. Blood *49*, 263 (1977)
25. Hast, R., Skarberg, K., Engstedt, L., Jameson, S., Killander, A., Lundh, B., Reizenstein, P., Uden, A., Wadman, B.: Oxymetholone treatment in aregenerative anaemia. II. Remission and survival – a prospective study. Scand. J. Haemat. *16*, 90 (1976)
26. Heimpel, H.: Die Panmyelopathie und andere Formen der Panzytopenie. Handbuch Innere Medizin, 5. Aufl. Heidelberg: Springer-Verlag, 1974, Vol. II/4
27. Heimpel, H. and Kubanek, B.: Pathophysiology of aplastic anemia. Brit. J. Haemat. *30*, (Suppl.), 57 (1975)
28. Heimpel, H., Rehbock, C., v. Eimeren, W.: Verlauf und Prognose der Panrnyelopathie und der isolierten aplastischen Anämie. Eine retrospektive Studie an 70 Patienten. Blut *30*, 235 (1975)
29. Kennedy, B. J.: Stimulation of erythropoiesis by androgenic hormones. Ann. intern. Med. *57*, 917 (1962)
30. Kennedy, B. J. and Gilbertsen, S.: Increased erythropoiesis induced by androgenic hormone therapy. New Engl. J. Med. *256*, 719 (1957)
31. Kern, P., Heimpel, H., Heit, W. and Kubanek, B.: Granulocytic progenitor cells in aplastic anaemia. Brit. J. Haemat. *35*, 613 (1977)
32. Li, F. P., Alter, B. P. and Nathan, D. G.: The mortality of acquired aplastic anemia in children. Blood *40*, 153 (1972)
33. Mackey, M. C.: Unified hypothesis for the origin of aplastic anemia and periodic hematopoiesis. Blood *51*, 941 (1978)
34. Mangor, L. A., Torales, P. R., Klainer, E., Barrios, L. and Blanc, C. C.: Effects of dexamethasone on bone marrow erythropoiesis. Hormone Research *5*, 269 (1974)
35. Morley, A., Remes, J. and Trainor, K.: A controlled trial of androgen therapy in experimental chronic hypoplastic marrow failure. Brit. J. Haemat. *32*, 533 (1976)
36. Morley, A., Trainor, K. and Blake, J.: A primary stem cell lesion in experimental chronic hypoplastic marrow failure. Blood *45*, 671 (1975)
37. Rosenblum, A. L. and Carbone, P. P.: Androgenic hormones and human granulopoiesis in vitro. Blood *43*, 351 (1974)
38. Sanchez-Medal, L.: The hemopoietic action of androstane. Progr. Hemat. *7*, 111 (1971)
39. Schärer, K. und Baumann, Th.: Die Behandlung der idiopathischen Panmyelopathien mit menschlichem Wachstumshormon. Schweiz. med. Wschr. *94*, 1322 (1964)
40. Scott, J. L., Cartwright, G. E. and Wintrobe, M. M.: Acquired aplastic anemia: An analysis of 39 cases and review of the pertinent literature. Medicine *38*, 119 (1959)
41. Singer, J. W. and Samuels, A.: Steroids and hematopoiesis: III. The response of granulocytic and erythroid colony-forming cells to steroids of different classes. Blood *48*, 855 (1976)
42. Stobbe, H., Neumann, P., Knappe, G. und Bach, G.: Therapieversuche mit menschlichem Wachstumshormon bei hämatologischen Erkrankungen. Ber. Ges. inn. Med. *5*, 254 (1967)
43. Thomas, E. D.: Treatment of refractory anaemia with cobalt. Ann. int. Med. *44*, 412 (1956)
44. Tso, S. C., Chan, T. K., Todd, D.: Aplastic anaemia: a study of prognosis and the effect of androgen therapy. Q. J. Med. *46*, (184), 513 (1977)
45. Uzuka, Y., Miyamori, A., Oka, M., Yamagata, S.: Long-term corticosteroid therapy of idiopathic aplastic anemia. Comtemp. Top. Immunobiol. *5*, 207 (1976)

Discussion

Moore: You made the point that androgens might be effective at the level of pluripotent stem cells by triggering them into cell cycle, particularly following on the work of Dr. Byron. He also showed that pluripotent stem cells have cholinergic receptors and can be triggered by agents that react with cholinergic receptors. There is a group in Italy who have actually attempted to treat aplastic anemia using inhibitors of acetyl cholinesterase. In fact one patient that we were treating had a very dramatic response to acetylcholinesterase inhibitors, demonstrated by a very sharp rise in neutrophils within one week, and quite dramatic changes in CSF excretion in the urine.

Fliedner: Then is does mean that it has nothing to do with the stem cell, because the transit time from the stem cell pool to the mature granulocyte in blood is at least ten days.

Moore: That would be true in steady state situations, but not in an aplastic marrow.

Barrett: Concerning Dr. Heimpel's remarks on lithium: One should not totally abandon to consider this type of agent. Lithium induced leukocytosis occurs in normal individuals who presumably have normal CFU-c levels. It is therefore not unreasonable to assume that it might be the same thing in relation to aplastic anemia. Obviously this will not correct the underlying abnormality in aplastic anemia, but if one can introduce useful stimulation of neutrophils I would very much like to see a trial of this substance in aplastic anemia.

4.2 Evolution of 352 Adult Patients Treated with Androgens: Short and Long-term Results of a Prospective Study

C. Dresch

for the Cooperative Group for the Study of Aplastic and Refractory Anemia. Secretaries: Y. Najean, A. Pecking

The short and long-term evolution of aplastic anemias and the choice of their treatment remain a matter of discussion. Since the techniques of homologous bone marrow transplants have made it possible to offer a new, but hazardous therapy to some patients, with low chances of survival, the need of accurate factors of prognosis, and a better knowledge of the short- and long-term prognosis of androgen-treated cases, are particularly necessary.

For this reason, a joint study was initiated in France and Spain in 1971, involving more than 30 hematological institutions. The present results concern 352 cases, in which enough initial and evolutionary data have been obtained, so that statistical analysis may be done.

Studied Patients

The following criteria were assigned for inclusion into the cooperative prospective study:
- anemia with Hb rate lower than 10 g %, or need for blood transfusion, granulocytopenia lower than $2 \cdot 10^9/1$, and/or thrombocytopenia lower than $100 \cdot 10^9/l$ (in fact, 92% of the included cases had pancytopenia);
- absence of excess of blast cells in the bone marrow;
- no malignant disease, treated by chemo – or radiotherapy;
- in the case of possible toxic etiology, a delay of three months between diagnosis and inclusion into the protocol was required, so that spontaneously remitting patients could be excluded from the study.

Initial evaluation included bone marrow biopsy and radioisotopic study in most of the cases.

All the patients have been allocated by randomization to one of the four following drugs:
- oxymetholone: 2.5 mg/kg/day
- methandrostenolone: 1 mg/kg/day
- metenolone: 2.5 mg/kg/day
- norethandrolone: 1 mg/kg/day

This treatment was given at full dosage for 10 months, and continued for another 10 months period in the case of incomplete recovery; in the case of androgen-induced jaundice, it was suggested to stop the drug temporarily and to resume the treatment after its disappearance.

Prognosis Factors

1. Correlation of Individual Parameters with Survival

Several possible prognostic parameters have been studied, by comparing survival curves and by calculating the correlation of survival rate with these parameters at different times after initial evaluation. It is clear, from these data, that the prognosis significance of the clinical and biological parameters is mainly valuable for the first three months after initial evaluation.

The following parameters have been shown to be of prognosis significance, and are classified according to their statistical association with survival at the 3rd month:
- granulocytopenia: $0.05 > p > 0.02$
- platelet count: $0.1 > p > 0.05$
- non myeloid cells in the bone marrow: $0.1 > p > 0.05$
- reticulocyte count: $0.2 > p > 0.1$
- 59 Fe incorporation into the R B C: $0.2 > p > 0.1$
- reticulinic desorganization on the B.M. biopsy: $0.2 > p > 0.1$

All the other parameters, including sex, age, transfusion requirement, lymphocyte count, apparent density of the bone marrow, plasmocytosis on the B.M. slides, are not statistically correlated with the early survival at $p < 0.2$. We have excluded from this list the prognosis significance of current infections or hemorrhages, which are strictly similar, and redundant, respectively with the number of circulating granulocytes and platelets. As previously observed, evolution of primitive and possible toxic aplastic anemias was similar.

When analyzing these data, two facts are obvious: these clinical and biological parameters are largely redundant, their prognostic value is chiefly clear for the first 3–6 months of the evolution on androgen therapy.

The redundancy of the used parameters is shown by the following table (Table 1), correlating the most important prognostic features.

Table 1. Redundancy of the prognosic parameters in aplastic anemia (352 studied cases)

	Platelet count	Reticulocyte count	Non-myeloid cells in the bone marrow	59 Fe utilization
Granulocyte count	$r = 0.201$ $0.05 > p > 0.02$	$r = 0.164$ $0.1 > p > 0.05$	$r = 0.196$ $0.05 > p > 0.02$	$r = 0.016$ $p > 0.1$
Platelet count		$r = 0.086$ $p > 0.1$	$r = 0.167$ $0.1 > p > 0.05$	$r = 0.121$ $p > 0.1$
Reticulocyte count			$r = 0.193$ $0.05 > p > 0.02$	$r = 0.182$ $0.05 > p > 0.02$
Non myeloid cells in the bone marrow				$r = 0.272$ $0.02 > p > 0.01$

Thus a multiparametric analysis has been done, for attempting a better definition of the short-term prognosis factors; and a study of the significance of other parameters than the initial situation has been attempted.

2. Multiparametric Analysis of the Prognostic Factors

Three methods have been used, to try to define the best way of predicting short-term development in individual patients.

As previously done by Lynch et al. [15] in a lower number of cases, a discriminant linear function using the same criteria has been calculated: 0.544 A−0.027 B+0.416 C+0.209 D−0.024 E−0.134 F−0.055 G+3.652, where A=percentage of the non myeloid cells in the bone marrow, B=reticulocyte count (10^9/l), C=hemorrhagic symptoms (1=absent, 2=moderate, 3=severe), D=sex (1=male, 2=femal), E=granulocyte count (10^6/l), F=platelet count (10^9/l), G=delay between the first clinical symptoms and diagnosis.

It is interesting to notice that, contrary to previously published data [15], such an index only enables a focus on satisfactory progress (85% of the patients with initial index indicating good prognosis survive beyond the 10th month), and does not secure a short-term prevision of death, since almost half the patients classified as the most severe did not die in the first 6 months after the initial evaluation.

Factorial analysis[1], using all the clinical and biological parameters obtained at the initial evaluation, enabled us to individualize 4 groups of patients, arranged in 4 ellipses on the factorial plane, each of them including 95% of the cases; groups 2 and 3, which only differ in their rate of non-myeloid cells in the bone marrow, underwent similar development; group 1 progressed better; the actuarial survival of group 4, including these cases which are on the verge of the definition of the disease, was excellent (Table 2, Fig. 1).

Table 2. Factorial analysis of the prognostic factors in aplastic anemia. Only the most important factors are given in the present figure

	Group 1 (111 cases)	Group 2 (68 cases)	Group 3 (57 cases)	Group 4 (22 cases)
Granulocyte count (10^6/l)	1290±920	596±520	496±426	1560±930
Platelet count (10^9/l)	55± 68	29± 33	25± 27	127±150
Reticulocyte count (10^9/l)	33± 30	17± 25	12± 18	20± 19
B.M. non myeloid cells (%)	19±15	20±19	60± 21	15± 10

The third analysis only consists of a study of the possible complementary information concerning evolution, when summing bad prognosis parameters. Table 3 shows that even when all the most severe clinico-biological parameters are associated, approximately 40% of the patients survive the first critical period of three months after diagnosis; from recent data, it appears that this percentage could be increased if all the participating institutions could use the recent clinical possibilities of medical reanimation, as they are usually used for bone marrow grafted patients.

---

1 These statistical studies have been done by Dr. Rioux, laboratory of clinical statistics, Hôpital de la Pitié, Paris (Director: Pr. F. Gremy)

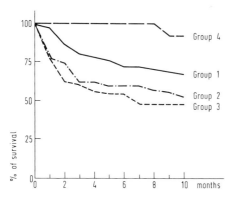

Fig. 1. Survival according to the initial severity of the disease (factorial analysis – see Table 2)

From this analysis, one point is clear: if it is easy to predict satisfactory development in the relatively slight cases, it is difficult to focus the "chance" of death in the severe cases, which would be particularly important before deciding to take the risk of a bone marrow graft. It also appears that the bad prognostic parameters are not additional, in terms of prognosis. At least, the proposed formulae are probably not adequate, since in fact, the prognostic value of the evaluated biological parameters is not a "continuum", for instance there is no difference of evolution between the patients with 0.5 to 1 and 1 to 2–10^9/l granulocytes at the initial study, or between the patients with 30 to 50 and 50 to 100–10^9/l platelets. In fact, it clearly appears that there is a threshold value for each biological parameter, under which the risk of death clearly increases. Such an observation may explain the importance of the maintenance therapy in the first critical months of the disease, and the significance of any pssible, even if modest, improvement during this lapse of time, as it will be shown in the following chapter.

3. Prognostic Significance of the Evolution

The preceding data did underline the difficulty for predicting middle – and long-term evolution from initial biological data. The actuarial curves of survival, when drawn for the only patients surviving at the 4th month after diagnosis, clearly indicate absence of any correlation with the initial prognosis features (Fig. 1).

We have tried to study whether the progress of the cases during this initial period of time could be an adequate prognosis factor. As shown on Table 4,

	% of death at the 3rd month after initial evaluation		
Granulocyte count less than	<0.5	10^9/l	...46%
+ platelet count lower than	<30	10^9/l	...52%
+ reticulocyte count lower than	<20	10^9/l	...56%
+ non-myeloid cells in the bone marrow exceeding 75%			...63%

Table 3. Cumulative significance of the prognostic parameters for predicting short-term survival

Table 4. Survival rate at the 10th and 20th month of the cases still surviving and evaluated at the 3rd month, according to their initial severity (Table 2) and their progress during the first three months on androgen therapy

Initial classification		Deterioration	Stable or slight improvement	Clear improvement
		Development during the first three months of survey		
		21 cases	24 cases	34 cases
Moderate	10th m.	58%	90%	84%
(group 1)	20th m.	43%	68%	73%
		8 cases	7 cases	16 cases
Severe	10th m.	40%	56%	100%
(group 2)	20th m.	37%	56%	85%
		11 cases	8 cases	11 cases
Very severe	10th m.	42%	60%	92%
(group 3)	20th m.	23%	53%	66%

it is clear that improvement or deterioration are very clear prognostic factors: even slight improvement in the most severe cases suggests better prognosis than slight deterioration in relatively mild initial conditions. From an insufficient series (only 86 adequate data), it even appears that as soon as the 6th week after initiating androgen therapy, useful prognostic indications could be obtained from developmental data.

From these data, it appears that the prognosis of aplastic anemia, after the initial critical period, can only be drawn from evolutive data; it appears to the members of our group that pejorative prognosis and the decision of graft, could in some cases be delayed, the patient being submitted to the same maintenance therapy during this time that he would have if grafted and the decision taken in light of objective development on androgen therapy; an argument however against this is the increased risk of graft rejection in multi-transfused patients.

Survival and Improvement on Androgen Therapy

1. Comparative Survival as a Function of the Androgen Used

As shown in Figure 2, survival differs, according to the drug used; at any given time between the 3rd and 20th month, the differences in survival between the four groups are statistically significant at p value lower than 0.1. The analysis of the curves, as a whole, using a log-rank test[1], demonstrates better survival of methandrostenolone treated patients ($p<0.05$); however the pejorative survival of oxymetholone-treated patients only results from excessive early death in this group of patients (first three months).

---

1 Dr. J. Goldberg, Bio-statistical Department. The Mount Sinaï Hospital, New-York

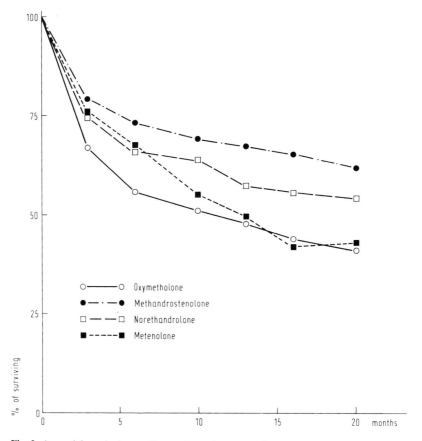

Fig. 2. Actuarial survival according to the androgen used

As in other published series, our curves (Figures 2 and 3) demonstrate two components, which inflect at approximately the 3rd month after the beginning of the androgen therapy: As it was underlined in the first part of this document, the clinico-biological parameters useful for prognostic study are only significant for the prediction of survival during the first trimester of the survey. So, it seems that evaluation of longterm survival on several drugs would exclude the early deaths. In these conditions (Figure 3), the survival of oxymetholone-treated patients does no longer differ from that of other patients, in part due to the early death of the most severe cases in this group; it appears that methandrostenolone-treated cases have the best chance of survival ($0.05 > p > 0.02$) and that metenolone-treated patients have the lowest ($0.02 > p > 0.01$).

In order to exclude possible bias in randomization, which could have affected more severe cases in some therapeutic groups, statistical comparison of these groups has been done. It demonstrated that no statistically significant difference exists between the four groups, as far as the prognostic factors are concerned. As shown in Table 5, excessive initial mortality rate is observed in oxymetholone-

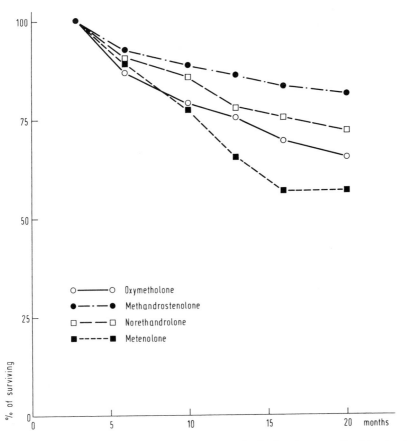

Fig. 3. Actuarial survival excluding the early deaths (in the first three months of treatment)

treated patients with initial factors of prognosis similar to those of other cases. It then appears that the excess of early death in these patients, and the excess of death after the third month in metenolone-treated cases, are not due to a faulty assignment to these drugs.

Table 5. Early (first three months) mortality in paired series, according to the assigned therapy

	Oxymetholone	Methandrostenolone	Metenolone	Norethandrolone
Initial granulocyte count≤500–10⁶/l	61%	38%	30%	42%
Initial platelet count ≦30–10⁹/l	45%	40%	42%	28%
Initial reticulocyte count≤20–10⁹/l	55%	30%	35%	27%
Bone marrow myeloid cells lower than 20%	61%	40%	26%	31%

2. Comparative Improvement as a Function of the Androgen Used

As shown in the preceeding part of this document, it is clear that the degree of improvement is the most valuable factor of prognosis, after the 3rd month. Differences in the degree of improvement, according to the used drug, could then be expected, if the difference of survival between the classes were valid.

Table 6 (A and B) shows that at the 3rd month, and between the 3rd and the 6th month, less improvement was observed in metenolone treated surviving cases than in the other patients.

Table 6A. Improvement during the first three months of androgen therapy, according to the androgen used

	Oxymetholone	Methandrostenolone	Metenolone	Norethandrolone
Hb (g %) (in the cases with initial Hb<10 g %)	1.12	2.37	1	1.87
Granulocytes (10⁶/l) (in the cases with initial count<1000)	270	270	200	290
Platelets (10⁹/l) (in the cases with initial count<100)	9	10	11	15

Table 6B. Improvement during the second trimester

	Oxymetholone	Methandrostenolone	Metenolone	Norethandrolone
Hb (g %) (in the cases with Hb<10 by the 3rdm.)	1.90	1.38	0.62	0.77
Granulocytes (10⁶/l) (in the cases with PMN<1000 by the 3rdm.)	430	310	230	250
Platelets (10⁹/l) (in the cases with platelet count 100 by the 3rdm.)	11	12	4	14

3. Other Factors of Improvement

Long-term glucocorticoid treatment was administered to 57 patients, at an approximate dose of 1 mg/kg/day for 6 months or more. The mortality rate in this group was 33% at the 3rd month as against 25% in the non corticoid-treated series, 44% at the 6th month (as against 32%), and 52% at the 10th month (as against 38%). As glucocorticoids were used in the most thrombocytopenic cases, the same comparison has been done in the only group of cases with very low initial platelet count; once again, lower survival was observed every time in the glucocorticosteroid treated group.

It has been suggested that sex and initial severity of aplastic anemia could influence the possibility of improvement on androgen therapy. From our data, no statistically significant difference was observed according to the sex of the patients; if at the 3rd month 61% of the surviving women did improve their Hb rate as against 45% of the males, 55% of the female patients improved their granulocyte count as against 53% of the males and 39% of the females improved their platelet count as against 27% of the surviving men, the difference between males and females disappears in the patients evaluated at the 6th and the 10th month.

On the other hand, it seems that, if the patients survive beyond the 3rd month, their chance of survival and improvement after this date is similar for all of them, whatever their initial severity was (Figure 1 and Tables 7A and B).

	Survival		
	6th m.	10th m.	20th m.
Initial granulocyte count			
<500–10^6/l	89%	75%	61%
>500–10^6/l	89%	82%	63%
Initial platelet count			
<30–10^9/l	88%	81%	64%
>30–10^9/l	90%	80%	64%

Table 7A. Survival rate of patients surviving the first three months, according to their initial severity

	Improvement of granulocytes by 300×10^6 or more:	
	by the 6th m.	by the 10th m.
Initial granulocyte count		
<500–10^6/l	70%	80%
>500–10^6/l	64%	76%

	% of surviving cases reaching 100–10^9/l	
	by the 6th m.	by the 10th m.
Initial platelet count		
<30–10^9/l	14%	25%
>30–10^9/l	32%	36%

Table 7B. Improvement of the patients surviving the first three months, according to their initial severity

At last a comparison between the possible toxic[1] cases and the probable idiopatic aplastic anemias did not demonstrate statistical differences, neither concerning survival, nor concerning the rate and degree of improvement.

---

1 In this series the most often implicated toxic drugs had been: benzene (22 cases), chloramphenicol (19 cases), phenylbutazone (19 cases), sulfamides (13 cases), noramidopyrine (12 cases), neuroleptics (11 cases)

Long-term Evaluation

Few studies give information on the long-term evolution of patients with aplastic anemia surviving more than the first two years after initial evaluation.

In our prospective study, from 352 initially evaluated cases, 162 died in the first 20 months, 94 can not be evaluated for long-term study (most of them due to loss of follow-up), 96 have been randomized between two therapy choices: slow decrease (i.e. one tenth of the full dose every second month), or fast decrease (i.e. one third of the full dose every month). In only 76 of these cases, enough information enables present analysis.

Relapse:	17 cases (2 patients died)
Patients living, with biological failure of androgen therapy at the 20th month:	12 cases (8 died during the following 12 months)

Table 8. Long-term evaluation of aplastic anemia after progressive cessation of androgen therapy (76 evaluable cases)

Still in remission: 47 cases
- – 1 died from another cause (gastric carcinoma)
- – 9 in remission for less than 6 months
- – 9 in remission for 6 to 12 months
- – 11 in remission for 12 to 24 months
- – 26 in remission for more than 2 years

As shown in Table 8, 17 of the patients followed for 12 months or more after the beginning of the decrease of androgen therapy relapsed (32%). An analysis of these 17 cases shows that:
- relapse is more frequent when androgen therapy at full dose is interrupted at the 10th month (9 relapses in 27 cases) than when it is interrupted at the 20th month (8 cases in 49) $0.1 > p > 0.05$;
- it is observed soon after the complete discontinuance of the drug in the cases randomized for rapid decrease (6 of the 8 relapses appeared in the first six months), or during the slow decrease in the other cases (6 of the 9 relapses in this group have been observed during the first 18 months, i.e. before complete discontinuance of androgen therapy);
- we did not find any difference concerning the quality of the previous remission in the relapsing and non-relapsing cases (previous complete remission, i.e. Hb more than 10 g %, PMN count higher than $1,500-10^6/l$, platelet count higher than $100-10^9/l$, was obtained at the end of the full dose androgen therapy in 61% of the further relapsing patients, and in 46% of the not-relapsing cases);
- no statistically significant difference could be found between the four previously used androgens.

All the 17 relapsing cases have been treated again with the same androgen previously used, at the full dose as in the induction period. As shown in Table 9, similar efficiency to that previously observed was noted in most cases. A further decrease of androgen therapy was attempted in several cases; at the present time,

Not presently evaluable:	1	
Early death:	1	
Not efficient:	2	(1 died; 1 still living)
Efficiency similar to that previously observed:	10	
More complete improvement than previously:	3	

Table 9. Efficiency of the re-initiation of androgen therapy at full dose after relapse (17 cases)

4 cases relapsed again: one of them died and three were put in remission after re-initiation of androgen therapy.

The very long-term development (more than 4 years) of both androgen-dependant and presently non-relapsing cases cannot be drawn in this prospective study.

Discussion

Comparison between the survival series of patients with aplastic anemia is difficult, because of the selection of more or less severe cases included in the published studies. Our group of cases, studied before grafting possibilities could introduce a selection of cases, are assigned to androgen therapy alone in this series. A satisfactory picture of the several clinical and developmental aspects of the disease appears.

Initial prognosis factors, wich have been discussed in a large series of cases, generally correlate survival with few parameters [5, 8, 9, 12, 13, 17, 21]. It is rarely stressed that most of these parameters are largely mutually redundant: however it is clear that infections and granulocytopenia, hemorrhages and thrombocytopenia, are quasi-similar parameters; and in general red cell, granulocyte and platelet count decrease in accordance with the degree of bone marrow myeloid cell depletion. On the other hand, conflicting data concerning the prognostic value of bone marrow biopsy [7, 28, 30], and isotope studies [16, 18], could be due to technical problems. A multiparametric analysis of our cases could enable us to define the value of the different clinical and biological parameters for the prognostic evaluation and the decision of treatment.

From our prospective study, including all patients treated with high-dose androgens, the following conclusions can be drawn.

The risk of death is high during the first three months, and it is possible to predict such development, at least from a statistical point of view, if using a multiparametric analysis [15, 18]. However, if the correct prediction of survival is good in the cases classified as moderate, the risk of false prediction is extreme in the most severe cases, in which, perhaps due to adequate maintenance therapy, the chance of survival is not as low as statistically expected. So it appears extreme to assert that the risk of any therapy (for instance bone marrow graft) could be easily taken in a given patient according to his low expectancy of survival, if founded only on initial clinico-biological investigation; even the comparison of survival curves of grafted and androgen-treated patients could not have any significance, if the maintenance were not similar in the two groups of patients.

The clinical evolution after the 3rd month does not apparently depend on the initial clinico-biological situation. In contrast with other papers [1, 6, 8, 14, 15], our data clearly indicate that the most severe cases can improve when the first critical period of the disease is over. Thus it appears necessary to provide the same careful maintenance therapy to all cases, whatever the protocol used. On the other hand it appears necessary to obtain accurate biological data two or three months after beginning androgen therapy, since an improvement suggests probable long-term survival, and deterioration indicates a poor prognosis and would be an argument for the decision of bone marrow graft.

The effectiveness of androgens on survival in aplastic anemia is still diversely accepted, twenty years after the first, enthusiastic, publication [23]. According to some authors, their effectiveness is absent [2], or only episodic [8, 15], or could be the prerogative of the less severe cases [1, 10, 20], but other series suggest a very high level of effectiveness [4, 6, 21].

The present study brings some data, which can be useful for interpreting the possible efficiency of these drugs.

The observed differences in survival, as well as the parallel differences in the degree of improvement, are strong arguments favouring an objective efficiency of these drugs. These results agree with some published data suggesting different efficiency of drugs, either in aplastic anemia in man [29], or on normal colony forming cells studied by in vitro methods [25]. Another positive argument would be the difference, as far as the improvement rate is concerned, between cases treated with low – and high-dose androgens [20, 21].

Our present study gives a useful complement to a soon to be published series, with regard to long-term survival. However, our survey is too short to enable conclusions other than indicative. From a previous, retrospective, study [19] it seemed that about 50% of the improved patients did relapse after cessation of androgens, but in this study the duration of the initial androgen cure was generally shorter than in the present series, and we observed in this study that the shorter the initial full-dose androgen course, the higher the frequence of relapses. It appears however that relapses occur as frequently in patients in complete remission as in incompletely improved cases.

The most interesting observation is that most of the relapsing cases did improve again when androgen therapy was resumed, as was already noted in a shorter, retrospective, analysis [19].

These data demonstrate that at least one third of the patients surviving in the second year after the beginning of aplastic anemia are really androgen responsive, and androgen dependent.

Conclusion

From a prospective, multi-center study of aplastic anemia in adults, the following conclusions can be drawn:

1. Short-term (survival to the 3rd month) prognosis is difficult to establish in individual cases, chiefly in the severe cases, which makes the decision of bone marrow graft difficult.
2. Middle-term prognosis (i.e. survival more than 3 months) depends on the initial improvement, and not on the initial severity of the disease.
3. Gluco-corticoid treatment, not only does not improve, but worsens the prognosis, in terms of survival. It does not decrease the frequency of complications due to androgen therapy.
4. Survival was higher, at any period, in the 20 months of full dose therapy, in patients treated with methandrostenolone and norethandrolone, than in the patients treated with oxymetholone and metenolone. The difference could be due to an abnormal early death rate in oxymetholone treated cases, but the excessive mortality of metenolone treated patients after the 3rd month parallels with less improvement of these cases.
5. Long-term (20 months) androgen therapy could be useful instead of only short (10 months) treatment: platelet count increases in the patients still abnormal at the 10th month; and the rate of relapse is higher in cases in which the induction therapy was stopped early.
6. At least one third of the surviving patients relapsed during the first 12–18 months after decreasing androgen therapy, and demonstrated androgen-sensitivity and androgen-dependance.

Acknowledgement

The study was supported by the Institut National de la Santé et de la Recherche Médicale, grants no. 76–4071 3

References

1. Alexanian, R., Nadell, J., Alfrey, C.: Oxymetholone treatment for the anemia of bone marrow failure. Blood 40, 353 (1972)
2. Camitta, B. M., Thomas, E. D., Nathan, D. G., Santos, G., Gordon-Smith, E. C., Gale, R. P., Rappeport, S. M. and Storb, R.: Severe aplastic anemia: a prospective study of the effect of early marrow transplantation on acute mortality. Blood 48, 63 (1976)
3. Corneo, G., Cartellaro, M. and Maiolo, A. T.: Some clinical and laboratory aspects and an evaluation of the androgen-glucocorticoid treatment. Acta Haemat. 46, 50–59 (1971)
4. Daiber, A., Herve, L., Coni, I. and Donosco, A.: Treatment of aplastic anemia with nandolone decancate. Blood 36, 748–753 (1971)
5. Davis, S. and Rubin, A. D.: Treatment and prognosis in aplastic anemia. Lancet 1, 871–873 (1972)
6. Duarte, L., Lopez Sandoual, R., Esquivel, F. and Sanchez Medal, L.: Androstane therapy on aplastic anemia. Acta Haemat. 47, 140–145 (1972)
7. Frich, B., Lewis, S. M.: The bone marrow in aplastic anemia. Diagnostic and prognostic feature. J. Clin. Pathol. 27, 231 (1974)
8. Heimpel, H., Reebock, C., Van Eimeren, W.: Verlauf und Prognose der Panmyelopathie und der isolierten aplastischen Anämie. Eine retrospektive Studie zu 78 Patienten. Blut 30, 235 (1975)

9. Hellriegel, K. P., Zuger, M., Gross, R.: Prognosis in acquired aplastic anaemia – An approach in the selection of patients for allogenesic bone marrow transplantation. Lancet *II*, 647 (1976)
10. Hirota, Y. and Hibino, S.: Effect of Androstanes on Aplastic Anemia in Japanese. In: Aplastic Anemia. Japan Med. Res. Found (ed.). Tokyo 1978, 377–383
11. Hast, R., Skarberg, K. O., Engstedt, L., Jameson, S., Killander, A., Lundh, B., Reizenstein, P., Uden, A. M. and Wadman, B.: Oxymetholone treatment in agenerative anemia. II: Remission and survival – a prospective study. Scand. J. Haematol. *16*, 90–100 (1976)
12. Lewis, S. M.: Course and prognosis in aplastic anaemia. Brit. Med. J. *1*, 1027 (1965)
13. Lohrmann, H. P., Kern, P., Niethammer, D., Heimpel, H.: Identification of high risk patients with aplastic anemia in selection for allogeneic bone marrow transplantation. Lancet *11*, 647 (1976)
14. Lepson, J. H., Gardner, F. H., Gorshein, D., Hait, W. M.: Current concepts of the action of androgenic steroids on zrythropoiese. J. Pediatrics *83*, 703 (1973)
15. Lynch, R. E., Williams, D. M., Reading, J. C., Cartwright, G. E.: The prognosis in aplastic anemia. Blood *45*, 517–528 (1975)
16. Mac Neil, B. J., Rappeport, J. M., Nathan, D. G.: Indium chloride scintigraphy: an index of severity in patients with aplastic anaemia. Brit. J. Haematol. *34*, 599 (1976)
17. Najean, Y., Bernard, J., Wainberger, M., Dresch, C., Boiron, M., Seligmann, M.: Evolution et pronostic des pancytopénies idiopathiques. Etudes de 116 observations. Nouv. Rev. Fr. Hémat. *5*, 635 (1965)
18. Najean, Y., Pecking, A.: The prognosis factors in acquired aplastic anemia – A study of 352 cases. To be published
19. Najean, Y., Laprevotte, Y., Schaison, G., Dresch, C. et Ardaillou, N.: Evolution tardive des pancytopénies chroniques traitées avec succès par les androgenes. Nouv. Rev. Fr. Hémat. *9*, 23–32 (1969)
20. Perugini, S., Lusvarghi, E. and Vaccari, G.: Androgenes et corticosteroides associés dans le traitement des aplasies medullaires acquises. Schweiz Med. Wschr. *100*, 1982–1984 (1970)
21. Sanchez-Medal, L., Gomez-Leal, A., Duarte, L. and Rico, M.: Anabolic androgenic steroids in the treatment of acquired aplastic anemia. Blood *34*, 283–300 (1969)
22. Scott, J. L., Cartwright, G. E., Wintrobe, M. M.: Acquired aplastic anemia: an analysis of 39 cases and review of the pertinent literature. Medicine *38*, 119–172 (1959)
23. Shahidi, N. T. and Diamond, L. K.: Testosterone-induced remission in aplastic anemia of both acquired and congenital type: further observations in 24 cases. New Engl. J. Med. *264*, 953–967 (1961)
24. Shinohara, K., Matsumoto, N., Tajiri, M., Nakashima, K., Ariyoshi, K., Miwa, S. and Miyaji: Oxymetholone treatment in aplastic anemia. Acta Haemat. Jap. *37*, 255–265 (1974)
25. Singer, J. W. and Adamson, J. W.: Steroids and hematopoiesis. III: The response of granulocytic and erythroid colony-forming cells to steroid of different classes. Blood *48*, 855–854 (1976)
26. Thong, K. L., Mant, M. J. and Grace, M. G.: Lack of effect of prednisone administration on bleeding time and platelet function of normal subjects. Brit. J. Haemat. *38*, 373–380 (1978)
27. Van Der Weyden, M. and Firkin, B. G.: The management of aplastic anaemia in adults. Brit. J. Haemat. *22*, 1–7 (1972)
28. Velde, J. T., Haak, H. L.: Aplastic anaemia. Histological investigation (methacrylate embidded bone marrow biopsy specimens: correlation with survival after conventional treatment in 15 adult patients). Brit. J. Haemat. *35*, 61 (1977)
29. Whang, K. S.: Aplastic Anemia in Korea. A Clinical Study of 309 Cases. In: Aplastic Anemia. Japan Med. Res. Found. (ed.). Tokyo, 1978, 225–240
30. Williams, D. M., Lynch, R. C., Cartwright, G. E.: Drug induced aplastic anemia. Seminars Hemat. *10*, 195–223 (1977)
31. Yamagishi, M., Kariyone, S. and Wakisaka, G.: Hematological effect of corticosteroids and anabolic steroids on aplastic anemia patients. Acta Hemat. Jap. *36*, 9–24 (1973)
32. Zittoun, R., Bernadou, A., Blanc, C. M., Bilski-Pasquier, G. et Bousser, J.: La metenolone dans le traitement des aplasies médullaires. Presse Méd. *76*, 445–449 (1968)

Discussion

Thomas: Dr. Dresch, the criteria you used for aplastic anemia were not very strict, and include many cases that we have not included. Our protocol criteria require a granulocyte count of less than 500, a platelet count of less than 20,000, anemia as you had and a hypoplastic marrow. I wonder if you have any figures on the fraction of your patients which would need these criteria for severe aplastic anemia.

Dresch: If you take patients with a very low counts of granulocytes and of platelets, most of these patients would come in the first slope of the survival curve and die before 3 months, but not all of them. Some of these patients recover right afterwards.

Fliedner: It is very important to keep this argument in mind, because we have to discuss it later in this meeting, whether you are discussing the same disease, what Dr. Thomas called "severe aplastic anemia", and what you called "aplastic anemia". Is this just a spectrum of severity of one disease or are they pathogenetically or etiologically different diseases with the same result as far as the peripheral blood is concerned?

Lohrmann: Aplastic anemia is a disease with a very unpredictable course. I wonder how meaningful your data are, if you do not have an untreated control group. It could well be that the course of your untreated control group is in the range of your best curve and what you then prove is not that androgens are helpful, but rahter that some of them are harmful in the initial treatment of aplastic patients. Do you see what I mean?

Fliedner: Dr. Lohrmann, what do you mean by untreated controls in the patients with this diagnosis, because untreated means that you do nothing.

Lohrmann: Not androgen-treated, of course. I assume, that these patients were treated with supportive care in addition to androgens, and that is why I mean you need a control group.

Camitta: Dr. Dresch, in your survival curves after 20 months the curves had all come down to about 50% and there was no plateau. This suggests that the patients were still dying, approximately at the same rate, except for the initial few months. What has happened to those survival curves after the first 20 months? If the patients are still dying, what evidence do you have that androgens have altered the course of the disease? You have to plot the data on a semilog survival versus time, because only if there is a plateau on that plot, you can say there is truly a decrease in the death rate.

Dresch: There have been a few deaths after the 20th month, but they occur less and less.

Camitta: Another question to Dr. Dresch. From your 130 patients, many have relapsed on withdrawal of androgens. What are your criteria for relapse, if you put any one of us in this room on androgen for a couple of months and took us off, we would have a "relapse" also.

Dresch: They are the criteria of the initial evaluation which means falling below 10 g Hb or below 1000 granulocytes/μl.

Camitta: If I understand you correctly it would be a relapse if the hemoglobin rose to 10.1 g and then after the therapy was discontinued, the hemoglobin dropped to 9.9 g %?

Dresch: Yes, but usually – as I showed you – it came back roughly to the initial value. We usually restored treatment again before it fell too low.

Lohrmann: In your study, have you been able to define at the onset of desease any hematological parameters that predicted a response or no response to androgens? E.g., do patients with high reticulocyte counts respond better?

Dresch: No, not exactly. We defined some sort of prognostic factor on the patient which will die in 3 months. The problem is not that those dying in the 3 months are not responding to androgens, but that they have no time to respond.

Singer: Was there any sex difference in the response rate?

Dresch: No.

Gordon-Smith: I am afraid that we cannot draw any conclusions from such a trial about the therapeutic effects of androgens, because there is no control. We can say nothing about whether androgens work or not. But there is an interesting point about the difference between various androgens in the survival curves. But how did you choose the dosage of the different preparations? Did you use the anabolic effects, the androgenic effects, or the virilizing effects? 2.5 mg/kg oxymetholone, as conventionally used, is guesswork, some people use much more. How were the doses chosen and could that influence the different effects of these drugs, related perhaps to side effects?

Dresch: It was just estimated as an equivalent dose.

Gross: May I repeat my question, that I addressed to Dr. Heimpel previously. Did you see any different side effects with these different drugs, especially on liver function?

Dresch: No, there was not. In all branches there were few cases of jaundice, but usually you can stop and take androgens again afterwards.

Weber: There were no hepatomas?

Dresch: No.

Nissen: I want to make another comment. There is a paper by Frey-Wettstein and Craddock[1], which describes the immunosuppressive effects of androgens, and we were wondering if this is exactly what the androgens do, and all the rest is only a by-product.

Fliedner: That comes back to the question, what this therapy does and what we learn from it with respect to pathogenesis.

Heimpel: As I mentioned this morning, glucocorticoids were tried in aplastic anemia in different dosages and for different types without clear-cut effect. This is an argument against a possible immunosuppressive effect of the androgens, because glucocorticoids provide a much stronger immunosuppressive effect.

---

1 Frey-Wettstein, M., Craddock, C. G.: Testosterone-induced depletion of thymus and marrow lymphocytes as related to lymphopoiesis and hematopoiesis. Blood *35*, 257–271 (1970)

4.3 Insight into the Pathophysiology of Aplastic Anemia Provided Through the Results of Marrow Transplantation

E. D. Thomas

Syngeneic Transplants

The patient with aplastic anemia who has a normal identical twin to serve as marrow donor provides a unique opportunity for therapy by marrow transplantation which in turn offers insight into the basic nature of the disease. In 1964, in describing the second successful such case transplanted in Seattle [16], we argued that restoration of marrow function by a simple infusion of normal marrow, without any conditioning regimen, argued strongly against the disease being due to lack of a nutrient or hormone or a persisting toxic factor. Rather, correction of the defect by an infusion of "stem cells" indicated that the probable mechanism involved an acquired intrinsic stem cell defect. In addition, the etiological agent responsible for the original marrow failure must have been transient in nature.

The Seattle marrow transplant team has now performed a total of 6 marrow transplants from normal identical twins into recipients with aplastic anemia [2, 6, 7, two cases unpublished]. One of these recipients had paroxysmal nocturnal hemoglobinuria which evolved into severe aplasia. None of the recipients were prepared for engraftment with immunosuppressive drugs. All 6 transplants were successful as evidenced by rising peripheral blood counts beginning within 10–20 days after transplantation and eventual complete recovery of peripheral counts and marrow cellularity. The recipients required from 1–3 months for full recovery. Since, with an identical twin donor, no blood genetic marker is available to confirm engraftment, the recovery might have reflected recovery of the patient's own marrow. However, our results with 6 consecutive patients strongly suggest that the recovery was due to the stem cell infusion. Two other similar successful cases have been reported [4, 5] and some failures can be attributed to the terminal state of the patient at the time of attempted marrow transplantation [6].

Two recent case reports, however, indicate that some cases of aplastic anemia may involve a different mechanism. One patient at the Royal Marsden [15] did not recover after a simple infusion of syngeneic marrow but recovery did occur after the administration of cyclophosphamide and syngeneic marrow. One patient at UCLA did not recover after an infusion of syngeneic marrow nor after cyclophosphamide and syngeneic marrow [3]. These differing results may simply be an indication of multiple pathophysiologic mechanisms involved in aplastic anemia.

Allogeneic Marrow Grafts

The results of the Seattle experience in patients with severe aplastic anemia transplanted from HLA identical donors have been presented in detail elsewhere [17, 10, 11, 12, 14]. Table 1 summarizes the patients conditioned for engraftment by the administration of cyclophosphamide, 50 mg/kg/day on each of 4 days followed 36 hours later by the marrow infusion.

Table 1. Patients conditioned with cyclophosphamide

Group		Incidence	
		Rejection[a]	Survival
With prior transfusions	before Oct. 1975	20/56 (36%)	26/60 (43%)
	after Oct. 1975	8/40 (20%)	27/41 (66%)
Without prior transfusions		0/17 (0%)	17/18 (94%)

[a] Excludes 6 patients who died within 1 week

Marrow Graft Rejection

Excluding patients who died too early to be evaluated, 20 of 56 multitransfused patients (36%) rejected the marrow graft. Only 2 patients survived marrow graft rejection, 1 with regeneration of his own marrow and the other successfully transplanted from a second HLA matched sibling. Thus, marrow graft rejection is a major cause of death.

In a logistic regression analysis two factors were found to be important in predicting marrow graft rejection [13]. The first was in vitro evidence of an immune reaction of the recipient's lymphocytes against the lymphocytes of the HLA matched sibling donor. The in vitro tests utilized were the relative response index in the mixed leukocyte culture and the chromium release test. The second factor was the marrow cell dose with the probability of graft rejection diminishing as the dose of marrow cells increased. Since most of the patients had been multiply transfused before coming to transplantation, it was not possible to tell whether the immunologic reaction of the recipient against the donor was a manifestation of the basic disease mechanism or reflected sensitization to "minor" transplantation antigens induced by transfused blood.

Results of Transplantation in Patients Without Prior Transfusion

We have now had the opportunity to carry out marrow transplants in 18 patients who had not received blood transfusions before coming to transplantation (Table 1). They were prepared with the cyclophosphamide regimen. One patient died too early to be evaluated but the other 17 patients are living, and there has been no instance of graft rejection. We conclude, therefore, that in most patients the immunologic mechanisms that result in graft rejection are iatrogenic (that is, induced by previous blood transfusions) and not a manifestation of a pathogenetic mechanism of aplastic anemia.

The Role of Donor Buffy-Coat Cell Infusions Following Marrow Infusion

In procuring marrow from the donor, we have for a number of years made a maximal effort to secure the largest possible number of marrow cells through multiple aspirations of the anterior and posterior iliac crest. Being unable to get more marrow cells, and having demonstrated that graft rejection was less likely with a larger number of marrow cells [13], we turned to the donor's peripheral blood as an added source of donor cells [8, 14]. These cells could be of interest for several reasons: (1) additional donor cell antigen might facilitate engraftment through "paralysis" of residual host lymphoid cells, (2) donor buffy-coat cells might contribute additional lymphoid cells which could facilitate engraftment and/or result in an increased incidence in severity of graft-versus-host disease, and (3) donor buffy-coat cells might contribute additional hematopoietic stem cells.

For all these reasons we decided to give additional donor buffy-coat cells, leukapheresed from 4 units of peripheral blood, on each of 5 days following the marrow infusion, for patients who were shown by in vitro tests to be sensitized and thus at great risk of graft rejection. Thirteen of 16 (81%) such sensitized recipients given marrow only rejected the graft while of those given marrow plus buffy-coat only 3 of 20 (15%) rejected the graft. Survival in the two groups was 25% and 70% respectively. The incidence of acute graft-versus-host disease was not different in the two groups. However, more patients in the group given marrow plus buffy-coat developed chronic graft-versus-host disease, perhaps because more patients survived.

Studies in the dog laboratory have shed further light on these observations [1]. First, buffy-coat cells killed by irradiation are not effective in facilitating engraftment. Secondly, living thoracic duct cells, known not to contain hematopoietic stem cells [9], are effective in facilitating engraftment.

From all these observations we conclude that some type of lymphoid cell present in the peripheral blood and the thoracic duct is of value in preventing graft rejection by the sensitized recipient and in overcoming natural resistance to engraftment when donor and recipient are not identical at the major histocompatibility complex.

General Inferences

1. Successful engraftment of syngeneic or allogeneic marrow in patients with aplastic anemia argues strongly against a pathogenetic mechanism that involves a defect in the structural microenvironment of the marrow, a persistence of a "toxin" or a lack of an essential nutrient or hormone.
2. Our consistent success with syngeneic grafts has lead us to believe that an

immunologic mechanism to explain the marrow aplasia is unlikely since the syngeneic marrow of the donor should be equally susceptible to the hypothesized immunologic attack. However, the one patient described by others suggests that some cases might involve such an immunologic mechanism.

3. In vitro evidence of an immune reactivity of the patient's residual lymphocytes against lymphoid cells or marrow cells of unrelated individuals and even an HLA identical sibling is, in the majority of instances, related to sensitization by prior blood transfusions. No untransfused recipients have, as yet, rejected a marrow graft. A small fraction of such recipients have shown some in vitro evidence of immunologic reactivity (see data presented by Dr. Singer), but the significance of this finding is unknown since these patients did accept a marrow graft after conditioning with cyclophosphamide.

4. One of our patients, and now 4 similar cases reported by others, recovered autogenous marrow function after being given cyclophosphamide and marrow from an HLA identical sibling. The role of cyclophosphamide in these cases of "spontaneous" recovery is unknown. These cases indicate that even in a severely hypoplastic marrow there are stem cells capable of recovery and regeneration of the entire marrow, presumably after termination of the original pathogenetic mechanism.

Conclusions

1. Most cases of severe aplastic anemia involve an acquired stem cell defect. The nature of the insult that appears to involve virtually all of the stem cells in the body is, of course, unknown. The leading candidates as etiologic agents appear to be a virus or a chemical agent. Perhaps, since aplastic anemia is a rare disease, a coincidental occurrence of chemical exposure and viral infection may be involved.

2. An immunologic mechanism for severe aplastic anemia may occur but only in less than one-fifth of the patients. Identification of these patients by in vitro immunologic studies would be of great importance but these studies must be carried out before transfusions are given.

3. A major difficulty in studying patients with aplastic anemia is the fact that patients do not become symptomatic until almost complete organ failure has occurred. Therefore, the clinical investigator almost never has the opportunity to study these patients until the end stage of the disease when the original pathogenetic mechanism may no longer be present.

This investigation was supported by Grant Number CA 18029, awarded by the National Cancer Institute, DHEW.

Dr. Thomas is a recipient of a Research Career Award AI 02425 from the National Institute of Allergy and Infectious Diseases.

References

1. Deeg, H. J., Storb, R., Weiden, P. L., Shulman, H. M., Graham, T. C., Torok-Storb, B. J. and Thomas, E. D.: Abrogation of resistance to and enhancement of DLA-nonidentical unrelated marrow grafts in lethally irradiated dogs by thoracic duct lymphocytes. Blood 53, 552–557 (1979)
2. Fefer, A., Freeman, H., Storb, R., Hill, J., Singer, J., Edwards, A. and Thomas, E. D.: Paroxysmal nocturnal hemoglobinuria and marrow failure treated by infusion of marrow from an identical twin. Ann. Intern. Med. 84, 692–695 (1976)
3. Gale, R. P., Falk, P., Feig, S. A. and Cline, M. J. for the UCLA Bone Marrow Transplant Team: Failure of recovery following syngeneic marrow grafting in aplastic anemia. Exp. Hematol. 5, (Suppl. 2), 103 (1977) (abstract)
4. Melvin, K. E. W. and Davidson, J. N. G.: Aplastic anaemia treated by transplantation of isologous bone marrow. N. Z. Med. J. 63, 93–95 (1964)
5. Mills, S. D., Kyle, R. A., Hallenbeck, G. A., Pease, G. L. and Cree, I. C.: Bone-marrow transplant in an identical twin. J. Am. Med. Assoc. 188, 1037–1040 (1964)
6. Pillow, R. P., Epstein, R. B., Buckner, C. D., Giblett, E. R. and Thomas, E. D.: Treatment of bone-marrow failure by isogeneic marrow infusion. N. Engl. J. Med. 275, 94–97 (1966)
7. Robins, M. M. and Noyes, W. D.: Aplastic anemia treated with bone-marrow transfusion from an identical twin. N. Engl. J. Med. 265, 974–979 (1961)
8. Storb, R., Epstein, R. B., Bryant, J., Ragde, H. and Thomas, E. D.: Marrow grafts by combined marrow and leukocyte infusions in unrelated dogs selected by histocompatibility typing. Transplantation 6, 587–593 (1968)
9. Storb, R., Epstein, R. B. and Thomas, E. D.: Marrow repopulating ability of peripheral blood cells compared to thoracic duct cells. Blood 32, 662–667 (1968)
10. Storb, R., Thomas, E. D., Buckner, C. D., Clift, R. A., Johnson, F. L., Fefer, A., Glucksberg, H., Giblett, E. R., Lerner, K. G. and Neiman, P.: Allogeneic marrow grafting for treatment of aplastic anemia. Blood 43, 157–180 (1974)
11. Storb, R., Thomas, E. D., Buckner, C. D., Clift, R. A., Fefer, A., Fernando, L. P., Giblett, E. R., Johnson, F. L. and Neiman, P. E.: Allogeneic marrow grafting for treatment of aplastic anemia: A follow-up on long-term survivors. Blood 48, 485–490 (1976)
12. Storb, R., Thomas, E. D., Weiden, P. L., Buckner, C. D., Clift, R. A., Fefer, A., Fernando, L. P., Giblett, E. R., Goodell, B. W., Johnson, F. L., Lerner, K. G., Neiman, P. E. and Sanders, J. E.: Aplastic anemia treated by allogeneic bone marrow transplantation: A report on 49 new cases from Seattle. Blood 48, 817–841 (1976)
13. Storb, R., Prentice, R. L. and Thomas, E. D.: Marrow transplantation for treatment of aplastic anemia. An analysis of factors associated with graft rejection. N. Engl. J. Med. 296, 61–66 (1977)
14. Storb, R. and Thomas, E. D. for the Seattle Marrow Transplant Team: Marrow transplantation for treatment of aplastic anaemia. Clinics in Haematology. Thomas, E. D. (ed.), pp. 597–609, London: Saunders 1978
15. The Royal Marsden Hospital Bone-Marrow Transplantation Team: Failure of syngeneic bone-marrow graft without preconditioning in post-hepatitis marrow aplasia. Lancet II, 742–744 (1977)
16. Thomas, E. D., Phillips, J. H. and Finch, C. A.: Recovery from marrow failure following isogenic marrow infusion. J. Am. Med. Assoc. 188, 1041–1043 (1964)
17. Thomas, E. D., Buckner, C. D., Storb, R., Neiman, P. E., Fefer, A., Clift, R. A., Slichter, S. J., Funk, D. D., Bryant, J. I. and Lerner, K. G.: Aplastic anaemia treated by marrow transplantation. Lancet I, 284–289 (1972)

4.4 Pathogenesis of Severe Aplastic Anemia: Inferences from Therapeutic Trials

B. M. Camitta

Introduction

The pathogenesis of severe aplastic anemia is not known. Therapeutic approaches have usually involved uncontrolled empiric trials of various stimulants of hematopoiesis. Recently, a prospective controlled trial evaluated the efficacy of androgens and bone marrow transplantation for treatment of severe marrow aplasia [3, 4]. Results of this study may permit limited inferences as to the pathogenesis of bone marrow failure.

Cooperative Aplastic Anemia Study Results

The therapeutic plan of the International Cooperative Aplastic Anemia Study is summarized in Figure 1. Oral androgen was oxymetholone (3–5 mg/kg/day). Intramuscular androgen was nandrolone decanoate (3–5 mg/kg/week). Transplanted and nontransplanted patient groups were similar with regard to disease severity, duration, etiology and other factors possibly affecting prognosis. Only newly diagnosed patients were eligible for study. Further details of patient selection, evaluation and care have been presented elsewhere [3, 4].

Cooperative Study results as of November 1, 1977 are shown in Figures 2–4. As utilized in this study, androgens did not influence survival. They also did not alter the percent of patients improving, time to initial response, rate of response or completeness of response. In contrast, early histocompatible bone marrow transplantation significantly improved survival and hematologic recovery.

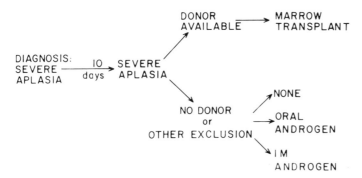

Fig. 1. Therapeutic strategy for Cooperative Aplastic Anemia Study

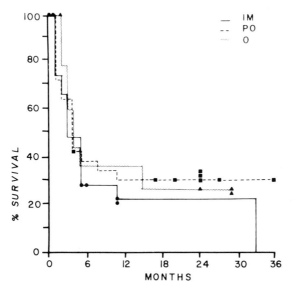

Fig. 2. Kaplan-Meyer plot of survival of nontransplanted patients in the Cooperative Study according to initial therapeutic regimen. IM=intramuscular androgen, PO=oral androgen, O=no androgen. All three groups received otherwise identical supportive care. Data as of 11. 1. 77

RESPONSE	ANDROGEN		
	NONE	PO	IM
CR	2	2	0
PR	1	5	2
NI	1	1	4
DIED	9	19	18
TOTAL	13	27	24

Fig. 3. Hematologic status of nontransplanted Cooperative Study patients. Data as of 11. 1. 77. CR=complete remission, PR=partial response, NI=no improvement

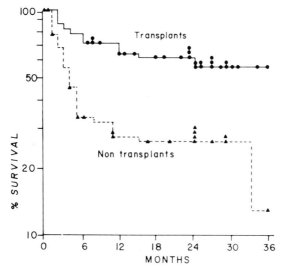

Fig. 4. Kaplan-Meyer plot of survival of transplanted and nontransplanted patients in the Cooperative Study. Data as of 11. 1. 77

Pathogenesis of Aplastic Anemia: Inferences from Therapeutic Trials

Prognosis in aplastic anemia depends upon severity of disease. Severely affected patients do poorly whether or not they receive androgens. HLA types do not appear to influence development of aplasia or subsequent prognosis; however, numbers of nontransplanted survivors are small and further data is needed. In the Cooperative Study, nontransplanted patients with idiopathic disease had a better prognosis than patients with aplasia of known etiology. However, in previous studies there has been no consistant relationship between prognosis and etiology of aplasia. Etiology of disease did not affect prognosis in transplanted patients in the Cooperative Study.

Androgenic hormones primarily influence erythropoiesis. Effects on granulocyte and platelet production are less consistent and less prominent. Since lack of the latter elements in blood may lead to fatal complications, it is not surprising that androgens do not influence the course of severe aplastic anemia.

Androgens with 5-unsaturated and 5α-H configurations (e.g., oxymetholone) affect erythropoiesis by increasing erythropoietin production [5]; steroids with a 5β-H configuration (e.g., metabolites of nandrolone decanoate) have a direct action on erythroid committed stem cells [7]. Since erythropoietin levels are usually increased in patients with marrow aplasia [1], failure of 5α-H compounds to alter recovery is not unexpected. However, failure of 5β-H compounds implies that target stem cells are absent or are defective in their ability to respond to the hormone.

In a broader context, if residual stem cells exist in severe aplastic anemia, the aplasia should be repaired at a rate dependent upon the rate of stem cell replication. Failure of recovery implies that: (a) stem cells are absent (b) the patient dies before recovery is apparent (c) residual stem cells do not recognize need to self replicate [2] or (d) residual stem cells have a decreased ability to self replicate due to chromosomal or other damage [6, 8]. Histologic and in vitro colony forming data as well as autologous marrow recovery (with or without transplantation) suggest that most patients with aplastic anemia have residual stem cells. In the Cooperative Study half of responding patients began recovery within 2 months. The other initial responses were noted as late as 18 months from diagnosis. This skewed pattern of recovery has been seen in other series and implies a mechansim such as d (or c). In this model, as the number of stem cells decreased, it would be less likely that any normal stem cells remained and the distributions of survival times would be more skewed. Other interpretations of this data are of course possible.

Only 30% of severely aplastic individuals reject histocompatible marrow grafts. Recent data suggests that rejection is almost always attributable to transfusion induced sensitization against minor transplantation antigens [12]. As presented by Drs. Singer and Thomas at this conference, graft rejection has not been seen in eighteen untransfused patients with severe aplastic anemia. This data implies that at the time of diagnosis, persistence of marrow aplasia is due to a stem cell defect (and/or deficit) rather than to a microenvironmental defect.

Another interpretation of the transplantation data should be considered. Persistent aplastic anemia could be due to immune suppression of hematopoiesis.

Transplantation success could then be due to pretransplant immune suppression. However as presented by Dr. Thomas, identical twin transplants for aplastic anemia are usually (but not always) successful without use of immune suppression [10, 14]. Furthermore, coculture evidence for immune suppression of hematopoiesis is seen in only a small minority of aplastic patients if they are untransfused and histocompatible sibling marrow is used as a target [11]. In these patients suppression did not correlate with graft rejection. Thus, although immune suppression of hematopoiesis may be demonstrated in a small percentage of patients, its role in causing persistence of marrow aplasia must still be shown. Another interpretation of marrow graft failure in these individuals is the persistence of the initial process causing aplasia. Thus, future studies of immunosuppressive therapy in aplastic anemia must be rigorously controlled both in regard to patient care and correlation of outcome with pretherapy in vitro studies.

Transplanted bone marrow is a source of macrophages and lymphocytes in addition to hematopoietic cells. Traditionally, microenvironmental cells such as macrophages are not supposed to circulate [13]. However, their role in repair of human marrow aplasia requires further definition. Recently anemia in W/W^v mice has been attributed to lack of a theta-sensitive cell [15]. This observation raises additional possiblities as to the pathogenesis of aplastic anemia. Finally, leukemic cells may suppress normal hematopoiesis via diffusable (? nonimmunologic) inhibitory substances [9]. Could transplanted marrow abrogate such an effect by providing suppressor cells?

In conclusion, well planned clinical studies may help to elucidate the pathogenesis of aplastic anemia. These trials will be most useful when integrated with concurrent in vitro assays of stem cell and immunologic function.

Supported by grants CA 17700, CA 17851 and CA 17997 from the National Cancer Institute; the International Aplastic Anemia Study was supported in part by a grant from Organon, Inc.

References

1. Alexanian, R.: Erythropoietin excretion in bone marrow failure and hemolytic anemia. J. Lab. Clin. Med. *82*, 438–445 (1973)
2. Boggs, D. R., Boggs, S. S.: The pathogenesis of aplastic anemia: A defective pluripotent hematopoietic stem cell with inappropriate balance of differentiation and selfreplication. Blood *48*, 71–76 (1976)
3. Camitta, B. M., Thomas, E. D., Nathan, D. G., Santos, G., Gordon-Smith, E. C., Gale, R. P., Rappeport, J. M., Storb, R.: Severe aplastic anemia: A prospective study of the effect of early marrow transplantation on acute mortality. Blood *48*, 63–70 (1976)
4. Camitta, B. M., Thomas, E. D.: Severe aplastic anemia: A prospective study of the effect of androgens on haematologic recovery and survival. Clin. Haematol. *7*, 587–595 (1978)
5. Fisher, J. W., Samuels, A. I., Malgor, L. A.: Androgens and erythropoiesis. Israel. J. Med. Sci. *7*, 892–900 (1971)
6. Hashimoto, Y., Takaku, F., Kosaka, K.: Damaged DNA in lymphocytes of aplastic anemia. Blood *46*, 735–742 (1975)
7. Levere, R. D., Gidari, A. S.: Steroid metabolites and the control of hemoglobin synthesis. Bull. N. Y. Acad. Sci. *50*, 563–575 (1974)

8. Morley, A., Trainor, K., Seshadri, R., Sorrell, J.: Is aplastic anaemia due to abnormality of D.N.A.? Lancet 2, 9–12 (1978)
9. Quesenberry, P. J., Rappeport, J. M., Fountebuoni, A., Sullivan, R., Zuckerman, K., Ryan, M.: Inhibition of normal murine hematopoiesis by leukemic cells. N. Engl. J. Med. 299, 71–75 (1978)
10. Royal Marsden Hospital Bone-Marrow Transplantation Team: Failure of syngeneic bone-marrow graft without preconditioning in post-hepatitis marrow aplasia. Lancet 2, 742–744 (1977)
11. Singer, J. W., Brown, J. E., James, M. C., Doney, K., Warren, R. P., Storb, R., Thomas E. D.: Effect of peripheral blood lymphocytes from patients with aplastic anemia on granulocytic colony growth from HLA-matched and mismatched marrows: Effect of transfusion sensitization. Blood 52, 37–46 (1978)
12. Storb, R., Prentice, R. L., Thomas, E. D.: Marrow transplantation for treatment of aplastic anemia: An analysis of factors associated with graft rejection. N. Engl. J. Med. 296, 61–66 (1977)
13. Tavassoli, M.: Studies on hemopoietic microenvironments. Exp. Hemat. 3, 213–226 (1975)
14. Thomas, E. D., Storb, R., Clift, R. A., Fefer, A., Johnson, F. L., Neiman, P. E., Lerner, K. G., Glucksberg, H., Buckner, C. D.: Bone marrow transplantation. N. Engl. J. Med. 292, 832–843, 895–902 (1975)
15. Wiktor-Jedrzejczak, W., Sharkis, S., Ahmed, A., Sells, K. W., Santos, G. W.: Theta-sensitive cell and erythropoiesis: Identification of a defect in W/Wᵛ anemic mice. Science 196, 313–315 (1977)

4.5 Infusion of Fetal Liver Cells in Aplastic Anemia

G. Lucarelli, T. Izzi, A. Porcellini, C. Delfini

Several experimental data support the hypothesis that fetal liver of early embryonation represents a good source of stem cells for transplantation purposes with a low degree of immunocompetence [1, 2, 6]. For these reasons fetal liver transplantation in bone marrow failure in humans lacking an histocompatible donor has been considered. There are few reports in the literature on fetal liver transplantation in aplastic anemia [3, 4, 5, 7].

We have performed fetal liver transplantation (FLT) in 12 cases of bone marrow failure five of whom had aplastic anemias (AA) (Table 1) and are the subject at this report. The infusion of fetal liver cells in aplastic hematological malignancies has been performed to evaluate hematologic recovery, engraftment and possible GVHD using varied doses of fetal liver cells of different embryonation age. All lymphoma and leukemia patients were aplastic and profoundly immunosuppressed as a consequence of previous intensive chemo-therapy. The five cases of A.A. did not receive immunosuppression prior to FLT. All cases were serially histologically documented.

Case 1. A 49 year old female with 300 granulocytes per cu.mm, Hb 8 gr/100 ml, 9,000 reticulocytes per cu.mm, 120,000 platelets per cu.mm had a hypoplastic marrow. She did not responde to conventional treatment and was given i.v. 1×10^8 fetal liver cells from a single 10 week old female fetus obtained shortly after hysterectomy. Seven days after FLT, erythroid and myeloid precursors were present in bone marrow smears and sections. Twenty eight days after FLT, bone marrow and peripheral blood returned to normal values. Normal hematopoietic function lasted 8 months. The bone marrow again became hypoplastic with peripheral pancytopenia. The patient received a second FLT with 1×10^9 cells from a male fetus 14 weeks old. She died of bilateral pneumonia 12 days following FLT. Post mortem examination revealed a hypercellular bone marrow with extramedullary hemopoiesis in liver and spleen. There were no markers to identify engraftment.

Case 2. A 33 year old male, had marrow hypoplasia with 2,100 reticulocytes per cu.mm, Hb of 8 g/100 ml, 300 granulocytes per cu.mm, 30,000 platelets per cu.mm. He received a total dose of 5.1×10^7/Kg recipient body weight of fetal liver cells from two twin fetuses 14 weeks of age.

Erythroid and myeloid precursors were present in marrow smears and sections on day 7 following FLT at which time the granulocytes level in the peripheral blood was 1,700 per cu.mm. Complete marrow reconstitution was reached at 21 days with morphological signs of dyserythropoiesis which disappeared within three months. Ineffective erythropoiesis continued. The patient is in good, clinical and hematological condition 12 months following FLT without evidence of engraftment.

Table 1. AML.: Acute myelogenous leukemia; A.A.: Aplastic anemia; L.L.: Lymphocytic lymphoma; ALL.: Acute lymphoblastic leukemia; F.L. Age: Embryonation age of fetal liver in weeks; T.N. Cells: Total number of cells obtained from the fetal liver; Cells × Kg: Cells per Kg of recipient body weight; Bone marrow reconstitution: Appearance in the bone marrow sections and smears of the recipient of myeloid and erythroid precursors and of megacaryocytes; Peripheral blood effect: Rise in granulocytes, reticulocytes and platelets per cu.mm; GVHD: Graft versus host disease according to the Seattle criteria (Grade 1 in case 12)

| Case | Age | F.L. Age | T.N° Cells | Cells × Kg | 8 Days after fetal liver infusion | | | |
					Bone marrow reconstitution	Peripheral blood effect	GVHD	Duration of hematopoietic bone marrow reconstitution
1 O.F.	17-AML 3rd relapse	10w	1×10^8	1.4×10^6	no	no	no	
2 P.C.	34-AML 2nd relapse	12w	9×10^8	1.8×10^7	no	no	no	
3 N.V.	58-AML 2nd relapse	12w	1×10^9	1.7×10^7	no	no	no	
4 P.F.	34-AML 1st relapse	14w	3×10^9	6.0×10^7	yes	yes	no	15 days (relapse)
5 P.I.	49-A.A.	10w	1×10^8	2.0×10^6	yes	yes	no	8 months
6 M.F.	56-A.A.	12w	5×10^8	2.0×10^6	yes	yes	no	>1 year
7 C.E.	33-A.A.	16w	4×10^9	5.0×10^7	yes	yes	no	>1 year
8 V.A.	17-A.A.	16w	5×10^9	6.0×10^7	no	no	no	
9 F.I.	59-L.L. 4th relapse	14w	3×10^9	6.0×10^7	yes	yes	no	25 days (relapse)
10 M.T.	3-ALL chemioresistent	16w	4×10^9	4.0×10^8	yes	yes	no	C.R. 90 days (relapse)
11 B.E.	9-ALL 4th relapse	2×16w	8×10^9	2.7×10^8	yes	yes	no	C.R. 45 days (relapse)
12 R.F.	3-A.A.	2×16w	8×10^9	7.7×10^8	yes	yes	yes	>90 days

Case 3. A 56 year old male with A.A. resistant to conventional treatment received 5×10^8 fetal liver cells from a male 12 weeks old fetus. Complete hemopoietic reconstitution with normal peripheral blood values was achieved 21 days after FLT and he is still hematologically compensated at one year. Engraftment could not be documented.

Case 4. A 17 year old male with an acellular bone marrow and profound peripheral pancytopenia received three consecutive FLT from 14–16 weeks old fetuses with a total amount of 5×10^9 cells, equal to 0.6×10^8 xKg. The patient died 70 days after the onset of aplasia, 21 days after the first and 10 days after the last FLT with sepsis and bilateral pneumonia without showing any hemopoietic recovery.

Case 5. A 2½ year old female was treated with chloro-amphenicol and pirazolone for fever and pneumonia. One month after termination of treatment she developed pancytopenia and had a hypocellular bone marrow. She did not respond to conventional treatment. She received, without immunosuppression, a FLT using a total dose of cryopreserved cells of 4×10^9 from a 16 weeks old female fetus. Two days after a second FLT using a total amount of 3.7×10^9 fresh cells from a 16 weeks old male fetus was carried out. Complete marrow reconstitution and normal peripheral blood count were seen within 15 days following FLT. A skin rush, a macropapular eruption involving 25–50% of the body surface, occurred. Pathological GVHD grade 1 according to the Seattle criteria, was first seen 10 days after FLT and is still present 60 days following FLT. She is in good clinical and hematological condition. HLA typing of circulating lymphocytes indicated reconstitution of host cells. Male chromosomes in perypheral blood and marrow were not seen. These studies indicate that fresh or cryopreserved fetal liver cells can be administered safely. Four of 5 patients with aplastic anemia had prompt hematological reconstitution and 3/5 are currently survivors with adequate hematopoiesis. There has been no documentation of engraftment except for possible GVHD in case N. 5. No patient was immunosuppressed. The possibility of a stimulatory effect on residual stem cells from a humoral factor contained in fetal liver suspensions cannot be ruled out at the moment. Possible factors resulting in recovery must wait longer clinical trials with and without immunosuppression in patients with A.A. without HLA matched donors.

Supported by C.N.R., Roma/Italy, and Stiftung Volkswagenwerk

References

1. Bortin, M., Rimm, A. A., Rosp, W. C., Truitt, L., Saltzstein, C.: Transplantation of hematopoietic and lymphoid cells in mice. Transplantation 21, 4 (1976)
2. Lowenberg, B., Dicke, K. A., Van Bekkum, D. A., Dooren, L.: Quantitative aspects of fetal liver cell transplantation in animal and man. Transpl. Proc. 8, 4 (1976)
3. Kelemen, E.: Recovery from chronic idiopathic bone marrow aplasia of a young mother after intravenous injection of unprocessed cells from the liver (and yolk sac) of her 22 mm CR-lenght embryo. Scand J. Haemat. 10, 305 (1975)

4. Lucarelli, G., Izzi, T., Delfini, C., Grilli, G.: Fetal liver transplantation in severe aplastic anemia. Haematologica *63*, 1 (1978)
5. O'Reilly, R. J., Pawa, R., Kagan, W., Kapoor, N., Sorell, M., Meyers, P., Good, R. A.: Reconstitution of hematopoietic function in post-hepatic aplasia following high-dose cyclophosphamide and allogeneic fetal liver transplantation. Experim. Hemat. *5*, Suppl. 2, 46 (1977)
6. Polchi, P., Moretti, L., Manna, M., Benetti, P., Lucarelli, G.: Transplantation of cryopreserved fetal liver in the rat. Haematologica *63*, 3 (1978)
7. Scott, B. R., Matthias, J. Q., Constandonlakis, M., Kay, H. E. M., Lucas, P. F.: Hypoplastic anaemia treated by transfusion of foetal haemopoietic cells. Brit. Med. J. *II.*, 1368 (1961)

4.6 Antilymphocyte Globulin Treatment in Severe Aplastic Anemia – Comparison with Bone Marrow Transplantation. Report of 60 Cases

E. Gluckman, A. Devergie, A. Faille, A. Bussel, M. Benbunan, J. Bernard

Introduction

The treatment of severe aplastic anemia (S.A.A.) is still controversial because the physiopathogeny is still uncertain. Direct or indirect damage to stem cells is discussed.

Experimental models in mice show that these two types of lesion can be demonstrated by bone marrow graft experiments.

Clinical experience in man shows that the disease has heterogeneous origin. The success of bone marrow transplant (B.M.T.) strongly supports the hypothesis of direct stem cell damage. In contrast, the observation of autologous marrow reconstitution after cyclophosphamide [1, 7, 9, 10, 11] or antilymphocyte globulin (ALG) with or without marrow infusion suggests the possibility of a different mechanism [3, 4, 5, 8].

In our bone marrow transplant unit, we have studied 60 consecutive patients with S.A.A. Thirty seven were treated with B.M.T. from an HLA identical sibling, twenty eight were treated with A.L.G. and androgens. This paper reports the results of our study.

Patients – Data at Admission

Sixty patients with S.A.A. were admitted to our unit for consideration for bone marrow transplantation. Patients with an HLA identical sibling were grafted except in the last year of the study where they were randomized to receive A.L.G. or B.M.T. Patients without an HLA identical donor were treated with A.L.G. and androgens.

I. Twenty eight patients received A.L.G. There were 15 males and 13 females. Their age ranged from 4 to 56 years (median 21). Nineteen had aplastic anemia of unknown etiology. In four patients, the disease was associated with drugs, in two with hepatitis, in two it occured after acquired paroxysmal nocturnal hemoglobinuria (PNH), and in one after a pregnancy.

The duration of aplastic anemia ranged from 0,5 to 96 months (median 4 months). Twenty six patients had received therapy with androgenic steroids without effect during the preceding 14 days to 96 months (median 2 months).

Granulocytes ranged from 0 to 1,500/mm³ (median 250/mm³), reticulocytes ranged from 0 to 100,000/mm³ (median 10,000/mm³). All patients had less than 20,000/mm³ platelets.

Nine patients were severely infected at admission, 25 had hemorrhagic problems. All had received multiple random red blood cell and platelet transfusions.

II. Thirty seven patients were treated by bone marrow transplantation. There were 23 males and 14 females. Their age ranged from 3,5 to 31 years (median 19). Twenty seven had aplastic anemia of unknown origin. In two patients, the disease was associated with drugs, in six with hepatitis, and two patients had Fanconi anemia. The duration of the marrow aplasia ranged from 8 days to 7 years (median 3 months). Granulocyte counts ranged from 0 to 1,000/mm^3 (median 80/mm^3), reticulocyte counts ranged from 0 to 15,000 (median 0). All patients had platelet counts of less than 20,000/mm^3. All had received multiple random red blood cells and platelets transfusions. One had received leucocyte transfusions from her mother.

Five patients received A.L.G. treatment without effect before being proposed for marrow transplantation.

Methods

I. A.L.G. Treatment

A description of the preparation and utilisation of A.L.G. has been published elsewhere [3]. Two consecutive batches of horse A.L.G. were used during this study. A.L.G. was diluted in 250 ml isotonic dextrose saline and 15 mg/kg/day of IgG was given by intraveinous infusion over six hours on five consecutive days.

All patients were treated in a conventional ward without bacteriological decontamination. They received prophylactic irradiated blood products.

Fever and chills were routinely observed during the first dose and platelet requirements were increased during treatment. Serum sickness occured on day 9 in 21 patients. It was transient and easily controlled with steroids.

All patients continued androgen treatment after A.L.G. at the same dose (1 mg/kg norethandrolone).

II. Bone Marrow Cultures

Bone marrow and blood CFU-C were carried out using the method of Pike and Robinson [6]. In vitro studies with A.L.G. were performed before treatment: 6×10^5 bone marrow mononuclear cells were incubated with 8,5 ng/ml of A.L.G. and 10% fresh autologous serum for 1 hour at 37°C before being washed and put to culture.

Control incubations without A.L.G. were treated in the same manner. Similar experiments with normal bone marrows from patients without hematological disorders were simultaneously carried out. Patient's serum was assayed for its inhibitory effect on normal colony growth, by incubating 6×10^5 bone marrow cells in the presence of 10% patient's serum with 10% rabbit complement for 1 hour at 37°C. The marrow cells were then washed, resuspended in agar and plated in triplicate at a final concentration of 2×10^5 cells per plate. As control sera from normal adults subjects were treated in the same manner. The ratio of colony growth after exposure to normal sera versus colony growth after exposure to test serum was used to calculate an inhibition index. To assess the in vivo action of A.L.G., CFU-C were assayed in bone marrow and blood before and at various points after treatment.

III. Bone Marrow Graft [2]

In all instances, the marrow donor was an HLA identical sibling. Patients were isolated and received bacteriological decontamination. Nine patients were conditioned with a combination of procarbazine, A.L.G. and cyclophosphamide. Seventeen patients were conditioned with cyclophosphamide 50 mg/kg i.v. on each of 4 successive days. Ten patients were conditioned with cyclophosphamide 60 mg/kg i.v. on two successive days followed by 1,000–800 rads total body irradiation with 400–500 rads to the lungs. After grafting, methotrexate was administrated as described by the Seattle group [12].

Results

I. A.T.G. Group

1. Response During the First Month

Twenty patients had no modification of their hematological status during and after A.L.G. treatment. The failure of any improvement of the blood count was associated with a failure of improvement of the bone marrow aspirate and biopsy. No major improvement in the clinical condition occured. But hemorrhagic problems were less frequent because platelet increaments were better after immunosuppression with A.L.G.

These patients formed a group of "non responder" to A.L.G. which contrasted strongly with a second group of eight patients, called "responder", who had a prompt improvement of their blood counts during and after treatment (Table 1). In this group, daily blood counts showed a granulocyte rise within 2–3 days after beginning A.L.G. treatment. A peak of granulocyte counts from 1,500–4,350/mm^3 was reached between day 3 to 30. The median granulocyte counts during the first month was 1,200 contrasting with a pretreatment median of 650/mm^3 (p<0.02). Reticulocytes showed a similar rise between day 8 to day 30 with a peak ranging from 32,000/mm^3 to 200,000/mm^3. The median reticulocyte counts during the first month was 35,000/mm^3 contrasting with a pretreatment median of 12,500/mm^3. Platelets remained low during the first month.

CFU-C studies showed that before treatment colony growth was poor. In 5 out of 6 responder patients, an increase of total groups was observed after

Table 1. Blood counts in 8 responder patients before ALG treatment and during first month after treatment

Patient		Granulocytes/mm^3				Reticulocytes/mm^3			
		Before ALG (median)	After ALG median	After ALG maximum (day)		Before ALG (median)	After ALG median	After ALG maximum (day)	
1		600	1,000	2,600	(day 3)	10,000	40,000	60,000	(day 9)
2		715	1,200	1,550	(day 19)	10,000	50,000	100,000	(day 16)
3		550	2,000	4,350	(day 10)	15,000	25,000	40,000	(day 30)
4	1st	200	1,050	1,600	(day 3)	6,000	15,000	70,000	(day 23)
	2nd	650	1,500	2,000	(day 5)	40,000	40,000	100,000	(day 30)
5		600	1,000	1,700	(day 3)	15,000	50,000	60,000	(day 16)
6		725	2,200	3,550	(day 4)	15,000	30,000	60,000	(day 8)
7		0	1,000	2,500	(day 21)	0	150,000	200,000	(day 21)
8	1st	1,000	1,650	2,950	(day 10)	15,000	20,000	32,000	(day 8)
	2nd	250	1,200	1,600	(day 30)	10,000	15,000	15,000	

incubation with A.L.G. In contrast such an effect was observed in 3 out of 14 non responder patients (p=0.036). The presence of a serum inhibitor of normal colony growth was observed in 6 out of 7 responder patients and in none out of 6 non responder patients (p=0.05). After A.L.G. treatment, 5 out of 7 responder patients had a rise of CFU-C, 4 out of 10 non responder patients had a rise of CFU-C despite the absence of hematological reconstitution. The difference between this two groups was not significant (p=0.2).

2. Long-term Follow-up

The overall survival curve shows that 46% of the patients survived after 6 months. After this period a plateau was obtained (Fig. 1).

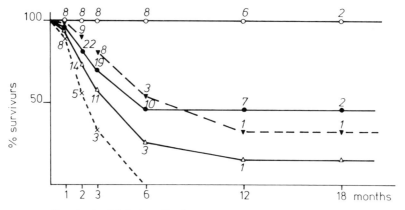

● Survival of total patients.
○ Survival of 8 responder patients.
△ Survival of 20 non responder patients.
× Survival of 10 non responder patients treated with ALG and androgens
▼ Survival of 10 non responder patients treated with ALG androgens then
     another treatment (immunosuppression or BMT)

Fig. 1. Actuarial survival curve of 28 patients treated with ALG and androgens

a) Non-Responder Group

Only six patients out of 20 are still alive, the actuarial survival curve shows a 25% survival at 6 months. 10 patients were treated with A.L.G., androgens and supportive care. None survived after six months. 10 patients received additional therapy after failure of A.L.G. Two patients were treated a second time with a different batch of A.L.G. One patient did not improve and died within two months after the second attempt, one patient with PNH is improving but is too early to evaluate fully. 5 patients received a bone marrow transplant from an HLA identical sibling. Three patients survived, one had an autologous marrow reconstitution after bone marrow graft rejection, two are chimeras with total hematological reconstitution with a follow up of more than 83 days and 63 days. Two patients died after bone marrow transplantation one from acute hepatitis on day 128, and one from CMV interstitial pneumonitis on day 63. Three patients were treated with cyclophosphamide 50 mg/kg intravenously on each of 4 successive days. One patient received fetal liver cells in addition. Two patients died within one month from sepsis, one patient is surviving for more than 2 months without any sign of hematological reconstitution.

b) Responder Group

None of these patients have died. The follow up currently ranges from 9 to over 24 months. In 6 patients granulocytes and sometimes reticulocytes showed a transient fall during the second month after treatment which recovered spontaneously.

Table 2. 8 responder patients – long-term follow up after ALG treatment blood counts

	Granulocytes/mm³			Reticulocytes/mm³			Platelets		
	3 months	6 months	Last counts (date)	3 months	6 months	Last counts (date)	3 months	6 months	Last counts (date)
1	1,200	1,300	1,350 (15M)	80,000	80,000	120,000 (15M)	20,000	25,000	54,000 (15M)
2	1,500	1,400	1,850 (18M)	100,000	300,000	90,000 (18M)	35,000	25,000	280,000 (18M)
3	1,600	2,000	2,950 (24M)	16,000	67,000	72,000 (21M)	15,000	40,000	37,000 (24M)
4 1st	500	/	/	40,000	/	/	8,000	/	/
2nd	800	/	/	50,000	/	/	10,000	/	/
5	1,100	1,400	3,050 (13M)	70,000	40,000	40,000 (13M)	40,000	190,000	160,000 (13M)
6	1,300	3,000	1,120 (24M)	48,000	30,000	29,000 (24M)	35,000	80,000	45,000 (24M)
7	800	4,000	3,600 (12M)	30,000			10,000	60,000	160,000 (12M)
8 1st	1,000	300	/	14,000	3,000	/	10,000	< 5,000	/
2nd	840	1,800	/	15,000	25,000	/	25,000	25,000	/

Table 2 shows hematological data at the third month, the sixth month and at the most recent examination. The median granulocyte count at 3 months was 1,050/mm³, at 6 months it was 1,600/mm³. Latest counts range between 1,120 to 3,050/mm³ between 12 to 24 months.

The reticulocyte count rose progressively. Red cell transfusion were discontinued in 7 patients between 1 to 7 months after A.L.G., only one patient (case No. 4) required occasional transfusion (Table 3). Platelet counts rose progressively. In 7 patients, it was above 20,000/mm³ at 6 months. Platelet transfusions were discontinued in all patients between one and 6 months after A.L.G.

Two patients had a relapse after 7 months and 5 months.

They received a second course of A.L.G. using a different batch. They both had a response similar to the first one. The first one (patient No. 8) has markedly improved 9 months after the second course, the second one (patient No. 4) requires occasional R.B.C. transfusions but has no hemorrhages despite a low platelet count four months after the second course.

Three patients have normal blood counts, in two androgens are being progressively reduced.

The other patients are clinically normal but they continue to require androgens at the initial dose (1 mg/kg) (patients 8, 7, 4). The other received

Table 3. 8 responder patients. Long-term survival after ALG treatments supportive care and survival

Patient	Last transfusion RBC	After ALG platelets	Androgenotherapy	Survival
1	7 months	1 month	1 mg/kg → 9 months slow discontinuation	> 15 months
2	2 months	6 months	1 mg/kg → 12 months slow discontinuation	> 20 months
3	14 months	1,5 months	1 mg/kg → 10 months slow discontinuation	> 24 months
4	Occasional	1 month after second course	1 mg/kg	> 9 months after 1st course > 4 months after 2nd course
5	1 month	1 month	1 mg/kg → 8 months slow discontinuation	> 13 months
6	1,5 month	1,5 month	1 mg/kg → 10 months slow discontinuation	> 24 months
7	2,5 months	1,5 month	2 mg/kg	> 15 months
8	3,5 months	1 month	1 mg/kg	> 16 months after 1st course > 9 months after 2nd course

1 mg/kg for 8 to 12 months then the dose was progressively reduced. One patient (No. 6) relapsed on a very low dose of androgens (10 mg) and the dose was increased to ½ mg/kg. One patient relapsed during a viral hepatitis, but seems to be recovering.

II. Bone Marrow Transplant Group

Thirty seven patients with severe aplastic anemia received a bone marrow graft from an HLA identical sibling. Seventeen patients (46%) survived with a follow up range from 53 days to 4 years (Fig. 2). Twenty seven patients were conditioned with cyclophosphamide (200 mg/kg) associated or not with antilymphocyte globulin and procarbazine. The main problem was marrow rejection which was

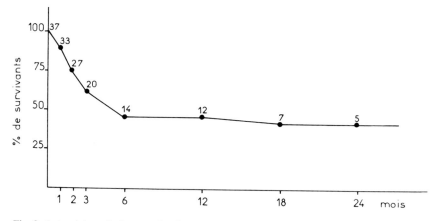

Fig. 2. Actuarial survival curve after bone marrow transplantation

found in 16 patients (59%). Among this rejection group, only two patients survived, one after autologous marrow reconstitution, the other after a second graft with another HLA identical sibling. Ten patients received cyclophosphamide 120 mg/kg associated with 800 rads total body irradiation with lung shielding (400 rads). No rejection was observed (Table 4).

Table 4. A. A. conditionning regimen and results

	No. patients	Take	Rejection	Severe GVH	No. patients Alive
Chemotherapy	27	11 40,7%	16 59%	4 15%	10[a] 37%
Cytoxan + TBI	10	10 100%	0	4 40%	7 70%

[a] 1 autologous reconstitution 1 second BMT with a take

The second problems is graft versus host disease, it was observed in 8 cases. 4 patients died, 2 had a complete recovery, one had chronic graft versus host disease, one is still under A.L.G.

Eleven patients had an uneventful recovery after grafting. Two patients died on day 68 and 130 of viral complications. This study shows the interest of a conditioning regimen including total body irradiation.

Discussion

This study confirms the effect of A.L.G. and androgens on myelopoiesis in some cases of severe aplastic anemia. The mechanism of action of A.L.G. is still unknown. The main hypothesis are (1) that A.L.G. eliminates a population of suppressor lymphocytes that inhibit the growth of bone marrow hemopoietic stem cells or alter the microenvironment, (2) A.L.G. could stimulate stem cells directly by a non immunological process, (3) A.L.G. might potentiate the effect of androgens by permitting the differentiation of stem cells to a point where they become accessible to the stimulation of androgens.

This treatment differentiates between two categories of patients: the responder and the non responder patients.

The comparison of pretreatment blood counts, granulocytes, reticulocytes and platelets did not show any significant difference between both groups ($p < 0.08$). Marrow smears and biopsies were also similar. None of the patients with drug induced or posthepatitis severe aplastic anemia entered in the responder group.

There was also no difference with age, sex, ABO blood group and HLA antigens. The only difference observed was the study of CFU-C in vitro. The responder group had significantly more often a stimulation of marrow CFU-C by A.L.G. and the presence of a serum inhibition of normal colony growth.

The comparison of the actuarial survival curve of patients treated with A.L.G. and with bone marrow transplant shows no difference. Nevertheless patients who were non responders to A.L.G. had a poor prognosis and died within 6 months, except the patients who received a B.M.T. after A.L.G. treatment failure.

From a practical point of view, a strategy of treatment for S.A.A. could be defined from this study. A.L.G. treatment is indicated in S.A.A. especially if an in vitro efficacity can be demonstrated. The absence of major toxicity to A.L.G. and the rapidity of response carries a great advantage for this form of treatment. The negative points are the possible androgen dependance, and the lack of knowledge of the long term follow up of these patients.

Bone marrow transplantation is indicated in post hepatitis aplastic anemia, in Fanconi anemia and after failure of A.L.G. treatment. This treatment is limited by the necessity of having an HLA identical sibling. Hematological reconstitution is usually rapid and complete with establishment of a complete chimerism. Mortality is related to bone marrow graft rejection, graft versus host disease (GVHD) and persistant immune deficiency which causes late deaths from opportunistic infections. The absence of in vitro tests able to predict bone marrow graft rejection with accuracy led us to give TBI to all our patients. This opportunity has the advantage of decreasing the number of bone marrow graft rejections but the disadvantage of precluding the possibility of an autologous marrow reconstitution. The late potential hazards of TBI are still unknown.

The problems of G.V.H. and opportunistic infections are still unsolved. On long term follow-up, morbidity and mortality are high during the first six months; after this period chronic G.V.H. is the main concern.

Further work is needed to explain the mechanism of this heterogeneous disease. The use of other immunosuppressive drugs such as cyclophosphamide requires study, as sometimes autologous marrow reconstitution has been observed after bone marrow graft rejection [7, 9, 10, 11].

The addition of incompatible stem cells has been claimed by other authors [4] to improve the rate of autologous marrow reconstitution. Controlled studies comparing these various regimens are mandatory.

References

1. Baran, D. T., Griner, P. F., Klemperer, M. R.: Recovery from aplastic anemia after treatment with cyclophosphamide. N.E.J.M. 295, 1522–1523 (1976)
2. Gluckman, E., Devergie, A., Marty, M., Bussel, A., Rottembourg, J., Dausset, J., Bernard, J.: Allogeneic bone marrow transplantation in aplastic anemia. Report of 25 cases. Transpl. Proc. 10, 141–145 (1978)
3. Gluckman, E., Devergie, A., Faille, A., Barrett, A. J., Bonneau, M., Boiron, M., Bernard, J.: Treatment of severe aplastic anemia with antilymphocyte globulin and androgens. Exp. Hematol. (in press)
4. Jeannet, M., Speck, B., Rubinstein, A., Pelet, B., Wyss, M., Kummer, H.: Autologous marrow reconstitution in severe aplastic anemia after ALG pretreatment and HLA semi compatible bone marrow cell transfusion. Acta. Hematol. 55, 129–139 (1976)
5. Mathe, G., Amiel, J. L., Schwarzenberg, L., Choay, J., Trolard, P., Schneider, M., Hayat. M., Schlumberger, J. R., Jasmin, C.: Bone marrow graft in man after conditioning by antilymphocyte serum. Brit. M. J. 2, 131–136 (1970)
6. Pike, B. L., Robinson, W. A.: Human bone marrow colony growth in agar gel. Journal of Cell. Phys. 76, 77–84 (1970)

7. Sensenbrenner, L. L., Steele, A. A., Santos, G. W.: Recovery of hematologic competce without engraftment following attempted bone marrow transplantation for aplastic anemia: a report of a case with diffusion chamber studies. Exp. Hemat. *5*, 51–58 (1977)
8. Speck, B., Gluckman, E., Haak, H. L. and Van Rood, J. J.: Treatment of aplastic anemia by antilymphocyte globulin with and without allogenic bone marrow infusion. The Lancet *2*, 1145–1148 (1977)
9. Speck, B., Cornu, P., Jeannet, M., Wissen, C., Burri, H. P., Groff, P., Nagel, G. A., Buckner, C. D.: Autologous marrow recovery following allogenic marrow transplantation in a patient with severe aplastic anemia. Exp. Hemat. *4*, 131–137 (1976)
10. Territo, M. C., for the UCLA B.M.T. Team: Autologous bone marrow repopulation following high dose cyclophosphamide and allogenic marrow transplantation in aplastic anemia. Brit. J. of Hemat. *36*, 305–312 (1977)
11. Thomas, E. D., Storb, R., Giblett, E. R., Longpre, B., Weiden, P. L., Fefer, A., Witherspoon, R., Clift, R. A., Buckner, C. D.: Recovery from aplastic anemia following attempted marrow transplantation. Exp. Hemat. *4*, 97–102 (1976)
12. Thomas, E. D., Storb, R., Clift, R. A., Fefer, A., Johnson, F. L., Neiman, P. E., Lerner, K. G., Glucksberg, H., Buckner, C. D.: Bone marrow transplantation. New Engl. J. Med. *292*, 832–843, 895–902 (1975)

4.7 Aplastic Anemia: Evidence of Immuno-Pathomechanisms and Successful Treatment with Absorbed ATG

B. Netzel, R. J. Haas, M. Helmig, H. Rodt, E. Thiel, B. Belohradsky,
E. Kleihauer, S. Thierfelder

Introduction

In 1976 and 1977 several reports tried to elucidate by in vitro studies the pathogenesis of aplastic anemia (AA) in man by suggesting that a subpopulation of lymphocytes is able to inhibit stem cell proliferation [2, 6, 7, 10]. From these data it was concluded that some forms of aplastic anemia can be considered a disease representing a suppression of hemopoietic stem cells by cellular immunomechanisms.

Further evidence of immunomechanisms in AA has been obtained by clinical studies, showing that immunosuppressive therapy in AA led to autologous marrow recovery [3, 18, 19, 23]. It has also been found that treatment with ATG had a beneficial effect in some patients with severe AA [1, 12, 20, 21].

This report describes our in vitro studies performed with various techniques in order to detect immunomediated mechanisms in aplastic anemia. In addition, it presents data on the various effects of crude and absorbed ATG on hemopoietic stem cells and other hemopoietic cells.

Our results show that ATG purified from crossreacting antibodies against hemopoietic stem cells can be applied successfully in the management of severe aplastic anemia. In contrast, the clinical application of crude ATG, which is highly toxic for hemopoietic stem cells [13], failed to induce hemopoietic reconstitution in all of our aplastic patients challenged to this regimen.

Materials and Methods

Cell Preparation

Bone marrow and blood samples were collected using preservativefree heparin as an anticoagulant and separated by Ficoll-Isopaque density gradient sedimentation.

In some of the experiments the mononuclear cells were made monocyte deficient by use of the iron carbonyl technique [4].

Antisera

ATG was prepared from rabbits, specifically absorbed, and purified as described previously by Rodt et al. [15].

For clinical application of ATG, the following criteria had to be fulfilled: sterility, absence of pyrogens (rabbit test, limulus test), no hemoagglutination titer, no anti-glomerulum basal membrane crossreaction, no toxicity on stem cells, no crossreaction with plasma proteins, no activation of complement, absence of general toxicity.

Test Systems

Granulocyte-Monocyte Colonies (CFU-C)

A double-layer agar technique as described by Pike and Robinson [14] was used with minor modifications [13]. After gelation at room temperature, the agar dishes were incubated for 12 to 14 days at 37° C in a fully humidified atmosphere continuously flushed with 5% CO_2. Colonies defined as groups of 50 or more cells were counted using a Leitz-Diavert microscope.

Erythropoietic Colony Formation (CFU-E)

The clonal growth of erythrocytic precursors in methyl cellulose medium [8], containing 30% selected fetal calf serum, 1% deionized bovine serum albumin and 2 units of sheep plasma erythropoietin step III (Connaught Labs, Canada) as a source of erythrocytic colony-stimulating activity was performed as described previously [13].

Lymphocyte Colonies (CFU-L)

Mononuclear cells were stimulated with ATG in McCoy medium + 20% fresh human plasma for 18 hrs. and seeded into 0.33% agar medium, using a feeder layer system with ATG as a source of mitogen. Cultures were incubated at 37° C in a fully humidified atmosphere of 5% CO_2 for 5–7 days.

Groups of more than fifty cells were scored as colonies. Colonies were mass-harvested by transferring intact groups of proliferating cells with a fine Pasteur pipette to TC 199 culture medium, dispersed and washed twice.

After an overnight incubation, cells were washed and tested for rosette formation and immunoautoradiography with F(ab)₂ anti T and anti IgM.

Rosette Formation

Assays for lymphocytes forming spontaneous rosettes with sheep erythrocytes (SRBC) were performed at 4° C [9] and after 1 h incubation at 37° C [5].

Immunoautoradiography

Immunoautoradiography of lymphocytes with ^{125}I-labeled F(ab')₂ anti T and F(ab')₂ anti IgM was performed by a quantitative photometric method described in detail by Thiel et al. [22].

Complement Fixation

Complement fixation using human lymphoid cells as antigens was performed by a micromethod as described previously [15]. The 50 percent lysis of sheep red cells was defined as the titer of the antiserum.

Co-cultures

Equal amounts of bone marrow mononuclear cells of patients with AA and of healthy individuals were co-cultured in TC 199 medium and 10% FCS. After a preincubation period of 18 hrs. the cells were counted and added to an agar culture mixture and plated on feeder layers. Marrow cells from the patient and the healthy donor were similarly treated as a control.

The plates were incubated for 12–14 days and colonies were counted as described.

Inhibition was defined as positive, when less than 80% of control CFU-C were present in the co-culture dishes.

ATG + C Treatment

Bone marrow mononuclear cells from patients with AA were incubated with absorbed ATG (10 mg/ml) in a final volume of 0.5 ml (final dilution of ATG = 1:16) for 30 min. at 4° C. Then 25% selected rabbit complement was added and the suspension being further incubated for 60 min. at 37° C. Then the cells were washed, counted and tested in the CFU-C assay as described.

Results

1. Specificity of Absorbed ATG

The specific activity of ATG absorbed in different ways against several lymphoid populations is shown by the complement fixation test (Fig. 1).

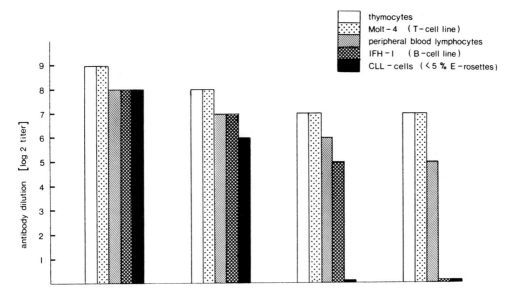

Fig. 1. Antibody titers of differently absorbed ATG in the complement fixation test against various lymphatic cell populations. All ATG preparations were reconcentrated to 10 mg/ml before testing. ATG (LK), absorbed with liver-kidney homogenate. ATG (LK, CLL), absorbed additionally with pooled B-cells from chronic lymphatic leukemias. ATG (LK, CLL, LCL), absorbed additionally with a pool of B-type lymphoblastoid cell lines

Crude ATG cannot discriminate between the various cell populations, whereas stepwise absorption of ATG caused an increasing loss of activity against B cells (IFH-I, CLL) and increasing specificity for T lymphocyte populations (Molt 4, thymocytes). Marrow cells were incubated with ATG for 30 min. at 4°C and thereafter with complement for 60 min. at 37° C. Incubation with crude ATG and complement completely inhibited colony formation in the CFU-C assay. When marrow cells were incubated with ATG absorbed with LK, only some of the antibodies crossreacting against granulocytic precursor cells were removed. No inhibitory activity against CFU-C was detected when ATG was absorbed with LK and pooled B-cells from chronic lymphatic leukemias or with ATG additionally absorbed with a pool of B-type lymphoblastoid cell lines (Fig. 2).

Crude ATG completely inhibited hemopoietic precursor cells from forming colonies of erythrocytic cells (CFU-E) in a methylcellulose medium in the presence of erythropoietin (Table 1). In contrast, no detectable inhibitory activity was caused by ATG absorbed with LK and CLL.

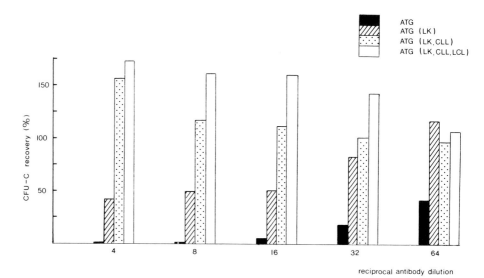

Fig. 2. Recovery of CFU-C after incubation of marrow cells with enhancing dilutions of crude and differently absorbed ATG in the presence of complement. 100% = control value (marrow cells incubated with normal rabbit globulin in appropriate dilutions in the presence of complement). The protein concentration of ATG (LK, CLL, LCL) was 10 mg/ml. All other ATG preparations were adjusted to the same cytotoxic activity against T-cells using the complement fixation test for determination of activity

		[a]CFU-C/ 2×10^5 nucl. marrow cells	CFU-E/ 10^5 nucl. marrow cells
ATG not absorbed	1:16+C	0	0
ATG absorbed	1: 8+C	84±3	52±6
	1:16+C	68±6	59±4
Normal rabbit globulin	1:16+C	72±5	48±4

Table 1. Recovery of CFU-C and CFU-E after treatment of marrow cells with absorbed ATG (crude and absorbed ATG were adjusted to the same cytotoxic activity against T lymphocytes before testing)

[a] CFU-C and CFU-E ± s.e.m. from triplicates

2. In Vitro Side Effects of ATG

2.1 Enhancement of Stem Cells

Short-term incubation of normal marrow cells with crude ATG in the absence of complement led to an increase in the number of CFU-C (Fig. 3).

This effect was also found with absorbed ATG in the presence of complement (Fig. 2), and when crude or absorbed ATG was added directly into the agar dishes and was present for the whole culture period. Previous culture studies indicate that the colony enhancement results from an interaction of ATG on the cell surface of colony-forming cells or earlier stem cells, which thus become more responsive to the colony-stimulating factor [13].

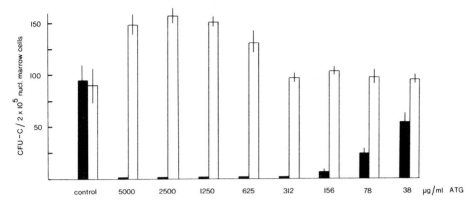

Fig. 3. Survival of CFU-C after treatment of normal marrow cells with various concentrations of crude ATG alone (white columns) or together with complement (black columns). Bars represent mean number of colonies ± s.e.m. from triplicates. Control: mean value ± s.e.m. using normal rabbit globulin in appropriate protein concentrations

2.2 Induction of Lymphocyte Colonies

Under certain conditions lymphoid colonies (CFU-L) develop in agar cultures when ATG is added to bone marrow or peripheral blood cell suspensions.

Figure 4 summarises the cell dose relationship and CFU-L growth using two different batches of ATG (No. 411, No. 502). More than 4000 CFU-L were obtained by plating 0.5×10^6 peripheral blood mononuclear cells, the colonies consisting of 150–2000 cells. Interestingly, colonies did not develop, when monocytes had been removed from the cell suspensions by use of the iron carbonyl technique. Cytochemical time monitoring of colony growth showed that single lymphocytes need close cell contact with monocytes in order to form a cluster and to finally grow to colony size after 4–6 days. For the purpose of demonstrating optimal lymphoid colony growth, the cell suspension had to be preincubated in a liquid phase for some hours in the presence of ATG, otherwise cluster and colony growth in the culture dishes was only sporadical.

Table 2 summarises the immunological cell surface characteristics of ATG-induced lymphoid colony cells. Colonies were mass-harvested on day 6 of the culture period and analysed.

Table 2. Analysis of ATG-induced colony cells (data from 12 separate experiments)

Technique[a]	%
E$_4$ rosettes	60–91
^{125}I-F(ab')$_2$ anti IgM	9–62
^{125}I-F(ab')$_2$ anti T	83–97

[a] Colonies were mass-harvested after 6 days of cultivation

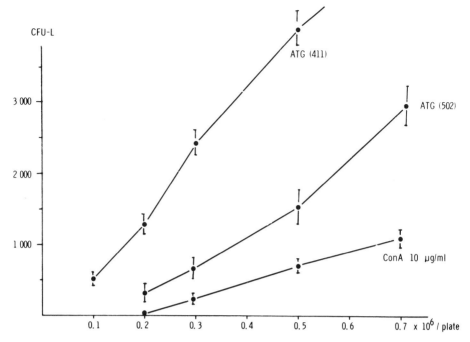

Fig. 4. Growth of lymphoid colonies in agar cultures after stimulation of peripheral blood mononuclear cells with ATG (batch No. 411, 502) or ConA (10 µg/ml). Each point represents the mean ± s.e.m. of three culture dishes

97% of lymphoid cells carry the T antigen, as was shown by photometric immunoautoradiography with ^{125}I-F(ab')$_2$ anti T. A considerable number of the lymphocytes were positive with ^{125}I-F (ab')$_2$ anti IgM. In double-labelling experiments we demonstrated both T antigen and IgM surface globulin on the same cell.

3. In Vitro Culture Studies in Severe Aplastic Anemia

3.1 Suppressor Cells in AA

Bone marrow cells from 22 patients with severe aplastic anemia, mostly of unknown aetiology, were investigated in order to find suppressor mechanisms which could be the cause of ineffective stem cell differentiation. The bone marrow mononuclear cells of patients with AA were co-cultured with normal bone marrow cells in agar dishes after a preincubation period of 18 hrs in a liquid phase. The liquid phase was chosen because in the agar dishes the cells were immobilised and interactions due to cell-to-cell contact would therefore have been missed. At the same time some of the bone marrow cells were incubated with highly absorbed ATG and rabbit complement in order to inactivate the T lymphocytes.

	Increase in CFU-C after ATG+C treatment[a]	
	+	−
Inhibition of normal marrow CFU-C in co-cultures +	1	3
−	0	18

Table 3. Evidence of suppressor cells in marrow cells from 22 patients with severe aplastic anemia

[a] ATG+C=absorbed ATG and rabbit complement

Before and after the incubation procedure the cells were plated for CFU-C growth. The results are summarised in Table 3.

Only 1 patient showed a significant increase in colony growth (3 versus 12 colonies per 10^5 mononuclear cells plated), whereas all other cases studied showed low colony growth (0–5 colonies per 10^5 mononuclear cells plated) without an increase after ATG treatment. So far no significant correlation has been found between patients' cells which induced suppression of normal CFU-C in co-cultures and the increased colony growth after inactivation of the same patients' T lymphocytes.

3.2 Serum Colony Stimulating Activity (CSA) and Inhibitors in AA

Sera from 27 patients with AA were investigated in order to find inhibitory or stimulating activities for the growth of granulocytic precursor cells.

Serial dilutions of the patients' sera were added to normal bone marrow cells and cultured in the CFU-C assay. The serum from one patient inhibited the colony growth (70% inhibition) of unrelated normal bone marrow cells as well as peripheral blood CFU-C from all family members. This patient had received multiple blood transfusions before testing.

10 patients had elevated CSA levels in their sera (140–250% CFU-C), compared with normal serum donors (100%). Most of the AA sera caused colony growth even in the absence of exogenous CSA.

3.3 E_{37} Rosettes

Bone marrow mononulear cells and peripheral blood lymphocytes were evaluated for spontaneous rosetteformation with SRBC. Bone marrow cells from patients with AA invariably had elevated levels of E_4 rosetteforming lymphocytes (29–92%), corresponding to the large number of lymphocytes demonstrated morphologically in smears.

In contrast to normal bone marrow mononuclear cells, some of the patients with AA showed stable rosette-formation (E_{37} rosettes). E_{37} rosettes were usually found in bone marrow lymphocytes, and not in peripheral blood lymphocytes of the same donor, this indicating a distinct population (or at least stage of activation) of T lymphocytes.

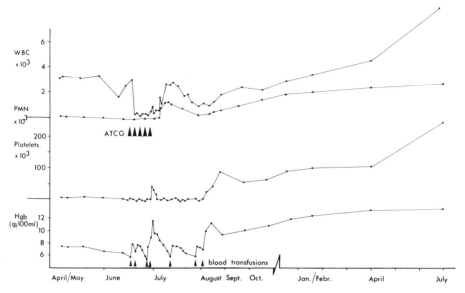

Fig. 5. Course of patient B. M. with severe aplastic anemia. Response of WBC, PMN, platelets and Hgb following treatment with absorbed ATG (15 mg/kg) for 5 days

4. Clinical Application of Absorbed ATG

ATG, absorbed with LK and CLL, had been given to a 14-year-old boy who, for 8 months, had been suffering from severe aplastic anemia resistant to androgens and steroids. The patient had suffered several episodes of bacterial infections and had received 10 blood transfusions before ATG treatment. In vitro data of the patient's bone marrow cells are summarised in Table 4. The patient was transferred to a laminar flow unit and treated with 15 mg/kg ATG intravenously for 5 days. The treatment was tolerated without severe side-effects but was followed by serum sickness 10 days after ATG treatment. ATG caused a profound lymphopenia, followed by a transient rise in platelets and granulocytes (Fig. 5). Three weeks after ATG treatment there was a constant rise in platelets, granulocytes and red cells up to normal values and the patient is now in complete remission for 12 months.

Table 4. Evidence of bone marrow suppressor cells in patient B.M.

% rosettes in patients marrow cells		CFU-C/10^5patients mononucl. marrow cells		CFU-C/10^5 normal marrow cells	CFU-C/ 10^5normal + 10^5patients marrow cells[a]
E_4	E_{37}	before	after ATG+C		
71	40	3 ± 0.7[b]	5 ± 0.7	43 ± 2	45 ± 2

[a] Normal and patients marrow cells were cocultured for 18 hrs in a liquid phase before plating in agar dishes

[b] CFU-C$\pm$s.e.m. from triplicates

Discussion

The specifity of rabbit anti-T cell globulin (ATG) absorbed in different ways has been extensively investigated by various methods such as the cytotoxic test, electron microscopy, photometric immunoautoradiography, and quantitative complement fixation, and is described in detail [15]. Crude ATG is highly toxic for stem cells and the absorption of crude ATG with liverkidney homogenate and pooled cells from patients with chronic lymphatic leukemias has been found to remove antibodies crossreacting with antigens present on the surface of granulocytic and erythrocytic precursor cells, while preserving a high degree of activity against thymocytes and other T-cells [13]. In addition, no inhibitory activity of absorbed ATG on early stem cells was found in diffusion chamber studies [16].

Further evidence of lacking stem cell toxicity of absorbed ATG was obtained recently, when donor bone marrow cells, in order to prevent graft-versus-host reactions [17], were pretreated with ATG before transplantation into lethally irradiated leukemic patients. A complete hemopoietic reconstitution of donor type cells was demonstrated in these patients.

Various in vitro side effects of ATG were observed. Marrow cells from aplastic patients were incubated with ATG in liquid cultures. In two patients a numerical increase in CFU-C was found in short-term cultures (unpublished observations). These data support recent findings that ATG acts directly on the cell membrane of hemopoietic progenitor cells and makes them more responsive to humoral activators [13].

Under certain in vitro conditions, ATG induces lymphoid colony growth. Its mitogenic effect has been demonstrated both with crude and absorbed ATG. It still remains unknown, however, whether this is an in vitro "artefact" or whether there is an in vivo correlation after clinical application of ATG.

Interestingly enough, the analysis of surface membrane markers of the lymphoid cells grown in vitro showed both T antigen and surface immunoglobulin on the same cell, indicating the differentiation potential of early lymphoid cells. The important role of monocytes in lymphocyte transformation is well known [11].

ATG-induced lymphoid cluster or colony formation only occurred when monocytes were present in the cell suspensions and there was close cell to cell contact between monocytes and lymphocytes in the liquid precultures. The removal of monocytes by carbonyl iron thus led to a loss of lymphocyte colony-formation.

Various reports indicate the presence of suppressor cells in the bone marrow of at least some patients with aplastic anemia [2, 6, 7, 10]. Removal of these suppressor T lymphocytes from aplastic bone marrow cells by rosette techniques or their inactivation by ATG treatment resulted in an increase in colony formation.

In addition, bone marrow cells from these aplastic patients were able to inhibit proliferation of CFU-C from normal marrow donors.

In our studies we have so far not found any significant correlation between the described test systems, which would be indicative of suppressor cell activities.

A question still outstanding is to identify, by other in vitro techniques, the immunomechanisms which led to bone marrow failure in some of the patients with severe aplastic anemia. Recent progress in the characterisation of a subpopulation of bone marrow lymphocytes in aplastic patients has been made possible by other techniques. We have thus been able to identify a subpopulation of T lymphocytes in the marrow of some aplastic patients forming stable rosettes ($E_{37}o$) with SRBC. These T lymphocytes were found in bone marrow cell suspensions (0–48%), but usually not in the peripheral blood of the same patients. An interesting result is that we have been able to detect E_{37} lymphocytes in the patient who responded to the treatment with absorbed ATG. The patient's bone marrow cells did not provide any evidence of suppressor cells either in co-culture studies or after in vitro treatment of his marrow cells with ATG + C. A 5-day ATG course in this patient led to a reconstitution of his own hemopoietic cells and he is now in complete remission for 12 months. In our hands crude ATG was unable to achieve hemopoietic remission in patients with aplastic anemia. 7 patients were treated with ATG absorbed with red cells, which showed high stem cell toxicity in the CFU-C assay. 6 patients lived long enough to demonstrate a lack of hemopoietic reconstitution.

These findings imply that the application of ATG, purified from cross-reacting antibodies against hemopoietic stem cells, may be an adequate therapeutical regimen for treating patients with severe aplastic anemia, and that some of them will benefit from this treatment.

The immunological background of the disease, and the response to ATG treatment, cannot be predicted reliably by the culture assays currently available.

Supported in part by the Deutsche Forschungsgemeinschaft, SFB 37, B 8.

References

1. Amare, M., Abdou, N. L., Cook, J. D., Abdou, N. J.: Aplastic anemia (AA) associated with bone marrow suppressor T (BM-ST) cell hyperactivity. Successful treatment with antihymocyte globulin (ATG). Clin. Res. 25, 333A (1977) (Abstr)
2. Ascensao, J., Pahwa, R., Kagan, W., Hansen, J., Moore, M., Good, R.: Aplastic anemia: Evidence for an immunological mechanism. Lancet I, 669 (1976)
3. Baran, D. T., Griner, P. F., Klemperer, M. R.: Recovery from aplastic anemia after treatment with cyclophosphamide. N. Engl. J. Med. 295, 1522 (1976)
4. Bloom, B. R., Glade, P. (eds.): In vitro Methods in Cell-Mediated Immunity. New York: Academic Press Inc., 1971
5. Borella, L., Sen, L.: E receptors on blasts from untreated acute lymphocytic leukemia: Comparison of temperature dependence of E rosettes formed by normal and leukemic lymphoid cells. J. Immunol. 114, 187 (1975)
6. Haak, H. L., Goselink, H. M.: Mechanisms in aplastic anaemia. Lancet I, 194 (1977)
7. Hoffmann, R., Zanjani, E. D., Lutton, J. D., Zalusky, R., Wasserman, L. R.: Suppression of erythroid-colony formation by lymphocytes from patients with aplastic anemia. N. Engl. J. Med. 296, 10 (1977)
8. Iscove, N. N., Sieber, F. and Winterhalter, K. H.: Erythroid colony formation in cultures of mouse and human bone marrow: Analysis of the requirement for erythropoietin by gel filtration and affinity chromatography on agarose-concanavalin A. J. cell. Physiol. 83, 309–320 (1974)

9. Jondal, M., Hohn, G., Wigzell, H.: Surface markers on human T and B lymphocytes. I. A large population of lymphocytes forming non immune rosettes with sheep red blood cells. J. Exp. Med. *136*, 207 (1972)

10. Kagan, W. A., Ascensao, J. A., Pahwa, R. N., Hansen, J. A., Goldstein, G., Valera, E. B., Incefy, G. S., Moore, M. A. S., Good, R. A.: Aplastic anemia: Presence in human bone marrow of cells that suppress myelopoiesis. Proc. Natl. Acad. Sci. USA *73*, 2890 (1976)

11. Lewis, W. R. and Robbins, J. H.: Effect of glass adherent cells on the blastogenic response of "purified" lymphocytes to phytohemagglutinin. Exp. cell. Res. *61*, 153 (1970)

12. Mathe, G., Schwarzenberg, L.: Treatment of bone marrow aplasia by bone marrow graft after conditioning with antilymphocyte globulin. Long-term results. Exp. Hematol. *4*, 256 (1976)

13. Netzel, B., Rodt, H., Hoffmann-Fezer, G., Thiel, E. and Thierfelder, S.: The effect of crude and differently absorbed anti-human T-cell globulin on granulocytic and erythropoietic colony formation. Exp. Hematol. *6*, 410–420 (1978)

14. Pike, B. L. and Robinson, W. A.: Human bone marrow colony growth in agar-gel. J. cell. Physiol. *76*, 77–84 (1970)

15. Rodt, H., Thierfelder, S., Thiel, E., Götze, D., Netzel, B., Huhn, D., Eulitz, M.: Identification and quantitation of human T-cell antigen by antisera purified from antibodies cross-reacting with hemopoietic progenitors and other blood cells. Immunogenetics. *2*, 411 (1975)

16. Rodt, H., Netzel, B., Niethammer, D., Körbling, M., Götze, D., Kolb, H. J., Thiel, E., Haas, R. J., Fliedner, T. M., Thierfelder, S.: Specific absorbed antithymocyte globulin for incubation treatment in human marrow transplantation. Transpl. Proc. *9*, 187 (1977)

17. Rodt, H., Netzel, B., Kolb, H. J., Janka, G., Rieder, I., Belohradsky, B., Haas, R. J., Thierfelder, S.: Antibody Treatment of Marrow Grafts in vitro: A Principle for Prevention of GvH Disease. (ed. S. J. Baum et. al.) Exp. Hemat. Today, 197. New York: Springer 1979

18. Sensenbrenner, L. L., Steele, A. A., Santos, G. W.: Recovery of hematologic competence without engraftment following attempted bone marrow transplantation for aplastic anemia: Report of a case with diffusion chamber studies. Exp. Hematol. *5*, 51 (1977)

19. Speck, B., Cornu, P., Jeannet, M., Nissen, C., Burri, H. P., Groff, P., Nagel, G. A., Buckner, C. D.: Autologous marrow recovery following allogeneic marrow transplantation in a patient with severe aplastic anemia. Exp. Hematol. *4*, 131 (1976)

20. Speck, B., Gluckman, E., Haak, H. L., van Rood, J. J.: Treatment of aplastic anemia by antilymphocyte globulin with and without allogeneic bone marrow infusions. Lancet *II*, 1145–1148 (1977)

21. Speck, B., Cornu, P., Sartorius, J., Nissen, C., Groff, P., Burri, H. P., Jeannet, M.: Immunologic aspects of aplasia. Transplant. Proc. *10*, 1, 131 (1978)

22. Thiel, E., Dörmer, P., Rodt, H., Thierfelder, S.: Quantitative immunoautoradiography at the cellular level. I. Design of a microphotometric method to quantitate membrane antigens on single cells using ^{125}J-labeled antibodies. J. Immunol. Methods *2*, 317 (1975)

23. Thomas, E. D., Storb, R., Giblett, E. R., Longpre, B., Weiden, P. L., Fefer, A., Witherspoon, R., Clift, R. A., Buckner, C. D.: Recovery from aplastic anemia following attempted marrow transplantation. Exp. Hematol. *4*, 97 (1976)

Discussion

Moore: Dr. Netzel, does your absorbed ATG has any activity against null-cells or only on T or B?

Netzel: Absorbed ATG shows a specific cytotoxic activity against T cells and not on B cells or other hemopoietic cells. The mitogenic activity of ATG on peripheral blood lymphocytes results in T cell proliferation, as could be shown in the analysis of ATG-induced lymphoid colony formation and finally, enhancing activities of ATG on other than T cells were demonstrated in our CFU-C experiments, indicating that committed or earlier stem cells were triggered into proliferation and self-renoval.

4.8 The Role of Haplo-Identical Bone Marrow Transfusion in ALG Treated Patients with Severe Aplastic Anemia (SAA)

W. Weber, B. Speck, P. Cornu, C. Nissen, M. Jeannet

Immunosuppression with ALG has become a new alternative in the treatment of SAA [19]. In the original reports ALG was followed by a BM infusion from a family donor [12, 15]. In subsequent experimental and clinical trials ALG has been effective with and without BM infusion [16, 17, 19, 20, 21]. This modality is applicable in the HLA-nonidentical setting; it is not followed by GvHD and leads to autologous BM recovery in more than 50% of patients with SAA. It is still an unsolved problem whether ALG followed by BM is more or less effective than ALG alone.

Basel Experience

20 patients with androgen refractory SAA requiring regular blood component support were treated the following way:
1. 15 patients received 4×40 mg/kg commercially available equine antihuman thoracic duct ALG i.v. which was then followed by HLA haplo-identical, MLC positive family BM.
2. 5 patients received the same dose of the same ALG without BM.

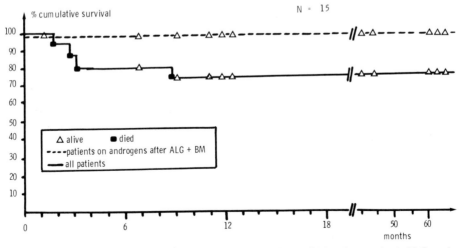

Fig. 1. Survival curve of 15 patients with androgen resistent SAA who received ALG and haplo-identical BM. The overall survival is 73%. All the patients who got androgens after ALG and BM survived (Basel 1973 – June 1978)

Results of group 1: The overall survival is 73% (Fig. 1). 9 patients are in complete remission (CR) at over 6 months to over 5 years. 1 patient still needs platelet transfusions 8 months after ALG + BM. 1 patient is off transfusions 2 months after the procedure. 4 patients died of hemopoietic failure; 2 of them after transient improvement. All the survivors received androgens after ALG and BM. The 4 patients who died were not put on androgens. GvHD was never seen. The reconstitutions were autologous in all cases as documented by genetic markers.

Results of group 2: 1 patient has a CR at 1½ years. 1 has minimal thrombopenia at 1 year. 2 patients showed minimal improvement and require transfusions at over 1 year. 1 patient died from intracranial hemorrhage. All the patients received androgens after ALG and BM.

International Experience

Speck et al. [19] reported results of the centers Basel, Paris and Leiden. In the Paris series of 15 patients with SAA who were treated with ALG alone 60% survived from 2 to more than 13 months.

The European Cooperative Group for Bone Marrow Transplantation (E.B.M.T.) started a randomized trial 1½ years ago comparing ALG alone with ALG + haploidentical BM. At the moment there are 30 patients in the trial [23]. Results are not yet available.

Discussion

Treating SAA in patients without HLA-identical family donor we achieved good results with ALG and haplo-identical BM. The overall survival rate of 73% is superior to the results achieved so far with ALG alone [19]. The patient groups are too small and the observation time too short so that firm conclusions cannot be drawn at the moment.

Androgen therapy after ALG and BM correlates with success. The reason for this is unclear. Oxymetholone and Nor-Ethandrolone were used. It is known that androgens are immunosuppressive [3]. More stem cells might have become responsive to androgens after ALG and BM.

Our clinical results encourage us to postulate a therapeutic effect of the infused BM. This hypothesis has been advanced by Mathé [12] and Speck [18]. Based on experimental evidence the following mechanisms might be operative (Table 1):

1. A transient split hemopoietic chimerism is established by synergistic actions of ALG and infused BM. This stimulates endoreduplication and differentiation of stem cells.

Evidence: 80% of rabbits with benzene or ^3P induced aplasia survive after ALG conditioning and transfusion of mismatched allogeneic marrow as compared to 20% in the control groups which had BM alone, ALG alone or no treatment [16, 17, 20]. Split hemopoietic chimerism was documented for a period of 6 weeks to 6 months and was ultimately followed by complete autologous

Table 1. Hypotheses on the action of transfused bone marrow derived from experimental evidence

Postulated mechanisms	References
Transient split hemopoietic chimerism	Mathé et al. 1973, Speck et al. 1973
Microchimerism	Liégois et al. 1977
Reversal of an ongoing disease process	Kaplan et al. 1953, Metcalf 1978
Provision of an inducer substance	Metcalf 1977
Supply of lymphocytes interacting with bone marrow cells of the host	Rajewsky et al. 1972, Blaese et al. 1974, Kanamaru et al. 1974
Supply of microenvironmental cells	Friedenstein et al. 1974
Synergism between ALG and transfused bone marrow	Chertkov et al. 1972, Liégois et al. 1977, Speck et al. 1971, 1973, 1976, 1978

ALG = Antilymphocyte globulin

reconstitutions. GvHD does not occur in this setting because the take is limited to hemopoietic cells and to B lymphocytes. T lymphocytes remain of recipient origin [13, 16, 17].

A state of microchimerism without GvHD can be established in normal mice with ALS and allogeneic BM grafting [9].

Alloimmunized rabbits to donor can only survive 1200 rad total body irradiation (TBI) if they are given a marrow transfusion. The reconstitutions are all autologous [18, 20, 21].

2. An ongoing disease process is reversed.

Evidence: Kaplan et al. [6, 7] have shown that the injection of normal BM can prevent or reverse the development of lymphoid leukemia in mice. AA can be induced in mice with 600 rad TBI only if allogeneic lymph node cells are infused [8]. Active lymphoid marrow infiltrations have been found in aplastic patients [22].

3. Inducer substances are provided by a marrow infusion.

Evidence: Regulatory substances such as the granulocyte-macrophage-colony-stimulating factor habe been found in animal and human BM [10].

4. Lymphocytes are supplied which interact with the host marrow.

Evidence: In the murine system lymphocyte-dependent allogeneic effects were shown to be operative in antibody and granulocyte production [5, 14]. Additionally humoral immunodeficiency of chicken can be transmitted with bone grafts of affected animals to sublethally irradiated normal animals [1].

5. Microenvironmental cells are supplied.

Evidence: Fibroblast-like colonies were detected in monolayer cultures of animal and human bone marrow [4]. After transplantation of these colonies under the kidney capsule they transformed into bone populated with hemopoietic cells.

The experimental evidence favours the use of haploidentical BM transplantation in ALG treated aplastic anemia patients. Cooperative clinical trials will have to assess its real value.

Supported by the Swiss National Science Foundation Grant 3.890.0.77 and by the Swiss Cancer League Grants FOR 080.AK.75 and FOR 101.AK.77(2)

References

1. Blaese, R. M., Weiden, P. I., Koski, I., Dovley, N.: Infectious agammaglobulinemia: transmission of immunodeficiency with grafts of agammaglobulinemic cells. J. Exp. Med. *140*, 1097–1101 (1974)
2. Chertkov, J. L., Lemeneva, L. N., Mendelvitch, O. A., Udaov, G. A.: Stimulation of hemopoietic colony formation by antilymphocyte serum. Cell Tiss. Kinet. *5*, 387–400 (1972)
3. Frey-Wettstein, M., Craddock, C. G.: Testosterone – induced depletion of thymus and marrow lymphocytes as related to lymphopoiesis and hematopiesis. Blood *35*, 257–271 (1970)
4. Friedenstein, A. J., Chailakhyan, R. K., Latsinik, N. V., Panasyuk, A. F., Keiliss-Borok, I. V.: Stromal cells responsible for transferring the microenvironment of the hemopoietic tissues. Transplantation *17*, 331–340 (1974)
5. Kanamura, A., Kitamura, Y., Kawata, T., Kanamura, A., Naggi, K.: Synergism between lymph node and bone marrow cells for production of granulocytes. 1. Requirement for immunocompetent cells. Exp. Hematol. *2*, 35 (1974)
6. Kaplan, H. S., Brown, M. B., Paull, J.: Influence of bone marrow injections on involution and neoplasia of mouse thymus after systemic irradiation. J. Natl. Cancer Inst. *14*, 303–316 (1953)
7. Kaplan, H. S., Moses, L. E., Brown, M. B., Nagareda, C. S., Hirsch, B.: The time factor in inhibition of lymphoid tumor development by injection of marrow cell suspensions into irradiated C 57 BL mice. J. Natl. Cancer Inst. *15*, 975–979 (1955)
8. Kubota, K., Mizoguchi, H., Miura, Y., Kano, S., Takaku, F.: Experimental hypoplastic marrow failure in the mouse. Exp. Hematol. in Press (1978)
9. Liégeois, A., Escourrou, J., Ouvré, E., Charreire, J.: Microchimerism: a stable state of low-ratio proliferation of allogeneic bone marrow. Transpl. Proc. *9*, 273–276 (1977)
10. Metcalf, D.: Hemopoietic Colonies: In vitro Cloning of Normal and Leukemic Cells. Berlin–Heidelberg–New York: Springer-Verlag, 1977
11. Metcalf, D.: Approaches to some unanswered problems concerning aplastic anemia. Transpl. Proc. *10*, 151–153 (1978)
12. Mathé, G., Amiel, J. L., Schwarzenberg, L., Choay, J., Trolard, P., Schneider, M., Hayat, M., Schlumberger, J. R., Jasmin, C.: Bone marrow grafts in man after conditioning with antilymphocyte serum. Brit. Med. J. *II*, 131–136 (1970)
13. Mathé, G., Kiger, M., Florentin, I., Garcia-Giralt, E., Martyre, M. C., Halle-Pannenko, O., Schwarzenberg, L.: Progress in the prevention of GvH: bone marrow grafts after ALG conditioning with lymphocyte split chimerism, use of lymphocyte "chalone T" and soluble histocompatibility antigens. Transpl. Proc. *5*, 933–939 (1973)
14. Rajewsky, K., Roelants, G. E., Askonas, B. A.: Carrier specificity and allogeneic effect in mice. Eur. J. Immunol. *2*, 592–598 (1972)
15. Schwarzenberg, L., Mathé, G.: Bone Marrow Transplantation after Antilymphocyte Globulin Conditioning: Split Lymphocyte Chimerism. In: A.L.G. Therapy and Standardization. Behring Res. Commun., 1972, pp. 163–175
16. Speck, B., Kissling, M.: Succesful bone marrow grafts in experimental aplastic anemia using antilymphocyte serum for conditioning. Eur. J. biol. Res. *16*, 1047–1051 (1971)
17. Speck, B., Kissling, M.: Studies on bone marrow transplantation in experimental ^{32}P – induced aplastic anaemia after conditioning with antilymphocyte serum. Acta haemat. *50*, 193–199 (1973)
18. Speck, B., Buckner, C. D., Cornu, P., Jeannet, M.: Rationale for the use of ALG as sole immunosuppressant in allogeneic bone marrow transplantation for aplastic anemia. Transpl. Proc. *8*, 617–622 (1976)
19. Speck, B., Gluckman, E., Haak, H. L., van Rood, J. J.: Treatment of aplastic anemia by antilymphocyte globulin with and without allogeneic bone-marrow infusions. Lancet *II*, 1145–1148 (1977)
20. Speck, B., Cornu, P., Nissen, C., Groff, P., Weber, W., Jeannet, M.: On the pathogenesis and treatment of aplastic anemia. Exp. Hemat. To-day. in press (1978a)
21. Speck, B., Gluckman, E., Haak, H. L., van Rood, J. J.: Treatment of aplastic anemia by antilymphocyte globulin with or without marrow infusion. Clinics of Hematology. in press (1978b)

22. te Velde, J., Haak, H. L.: Aplastic anaemia. Histological investigation of methacrylate embedded bone marrow biopsy specimens: correlation with survival after conventional treatment. Br. J. Haematol. *35*, 61–69 (1977)
23. Zwaan, F.: Personal communication (1978)

4.9 The Effect of Antithymocyte Globulin on Abnormal Lymphocyte Transformation in Patients with Aplastic Anemia

L. J. M. Sabbe, H. L. Haak, B. A. Bradley, J. J. van Rood

Introduction

Together with Dr. E. Glückman (Hôpital St. Louis, Paris) and Dr. B. Speck (Kanton Hospital, Basel) we have recently published a study, in which the effect of ATG on the clinical course of severe aplastic anemia was assessed [13]. Our conclusions were that:
a) about half of the patients suffering from severe aplastic anemia improved significantly after 4 daily doses of ATG;
b) it was unlikely that this apparent beneficial effect of ATG was due to patient selection in the sense that we had selected patients which without ATG treatment would have survived anyhow;
c) it was possible that the ATG removed from the bone marrow T cells which interfered with the differentiation of the stem cells. A possibility which is strongly supported by in vitro studies by one of us [8] and others [1].
To gain further insight into the role of abnormal T cells in aplastic anemia the in vitro mitogen and antigen response as measured by the lymphocyte transformation test (LTT) was determined. It could be shown that in some of the patients the mitogen and antigen induced LTT was abnormal, and that ATG treatment in the majority of the cases studied corrected these abnormalities. One patient was also studied using the cell mediated lympholysis (CML) test. An abnormally high Cr^{51} release was found before ATG, which was normalized after ATG.

Materials and Methods

All patients were treated in the Department of Hematology (head Prof. Dr. E. A. Loeliger). The relevant hematological data of these patients are given in Table 1. Numbers 1 to 5 correspond with the numbers 1 to 5 of our previous publication. As is usual after ATG pretreatment, none of the patients receiving bone marrow had a take. Cells were collected before and at several times after ATG treatment.

Case Histories (see Table 1)

Pt. 1 He. The pancytopenia was diagnosed in this female at the age of 50, after she had used oxphenbutason. After treatment with 150 mg/d oxymethalon for 3 months she was treated in an isolation unit using horse ATG 40 mgr/kg for 4 days followed by infusion of her sisters haplo-identical bone marrow. She did not show any improvement and died 3,5 months after therapy.
Pt. 2 Ji: This 14 year old female was also treated with ATG and haploidentical bone marrow after a trial with oxymethalon and prednison. Already during the ATG infusion there was a rise in the

Table 1. Hematological values of the patients described at the time of admission

Case	Age	Sex	Reti's %0	Neutro's ×10^{-9}/L	Ly ×10^{-9}/L	Platelets ×10^{-9}/L	Sx	Therapy	Improved
1 (He)	50	F	0	0.15	1.5	10	–	ATG+BM	–
2 (Ji)	14	F	6	0.27	0.1	22	–	ATG+BM	+
3 (Bo)	24	M	120	0.24	1.0	7	40 d	ATG+BM	+
4 (Re)	42	F	0	0.03	0.5	10	50 d	ATG+BM	+
5 (To)	14	M	4	0.04	1.0	3	50 d	ATG+BM	–
6 (KD)	38	F	14	0.16	0.8	7.5	–	ATG	±
7 (GN)	30	F	0	0.02	1.3	5	–	ATG	+

reticulocyte and granulocyte counts. These were stabilized at a low to normal level. Only the thrombocytes remained in the low range.

Pt. 3 Bo: In this 23 year old teacher a pancytopenia was found together with a positive Acid Ham test, a low pyruvate kinase content of the erythrocytes, and a trisomy of chromosome 6. Methenolon treatment was unsuccessful. 10^{10} B.M. cells from his father were infused after 4 days treatment with horse ATG. 1,5 Months thereafter he was splenectomized. Only the red and white cell counts rose slightly, but the platelets remained at critical levels.

Pt. 4 Re: This 43 year old female was admitted to the hospital with a pancytopenia of unknown origin. She was treated with horse-ATG and HLA-identical bone marrow from a brother. 20 Days after transplantation all donor bone marrow had disappeared. On day 50 she was splenectomized. Shortly thereafter she recovered almost completely.

Pt. 5 To: A 14 year old boy with an idiopathic aplastic anemia. He was treated with horse-ATG and bone marrow from his father. During the ATG treatment the granulocytes rose, red cell count and platelets remained low both before and after splenectomy. After 3 months a second trial, now with rabbit-ATG in a dose of 220 mg/d and bone marrow was performed. There was no improvement. He died 2 months later of a cerebrovascular accident.

Pt. 6 KD: This patient had had recurrent pyogenic infections since birth. At the age of 22 she developed an aplastic anemia after treatment with chloramphenicol. This was treated with 20 mg of prednison and bloodtransfusions. At the age of 38 she gradually deteriorated without known etiology. She was treated with rabbit-ATG only. After the serumsickness, there was a partial remission.

Pt. 7 GN: The pancytopenia in this 30 year old female patient was probably caused by hexachlorobenzene. All blood cell counts rose shortly after rabbit-ATG treatment.

Cell Preparation

Peripheral blood leukocytes were obtained by venipuncture or leukopheresis. Mononuclear cells were separated from the heparinized blood samples by density centrifugation. Cells were washed three times with HBBS and resuspended in RPMI 1640 supplemented with 40% inactivated human AB serum plus 1,5% glutamine (20 mM), 100 U/ml penicillin and 100 µg/ml streptomycin, at a concentration of 3×10^6 cells/ml for mitogen and antigen stimulation and 1×10^6 cells/ml for mixed lymphocyte reaction (MLR).

Cell Culture Conditions

0,1 Ml. aliquots of the cell suspension were mixed with 0,1 ml of mitogen or antigen dilution in flatbottomed microculture plates (Linbro) and cultured for two (mitogen) or five (antigen) days respectively at 37° C in a humidified, 5%

CO_2 atmosphere. For MLR culture, 0.1 ml aliquots of responder cells were mixed in a roundbottomed plate with 0.05 ml of a suspension of 2×10^6 cells/ml of a lymphocyte pool (containing at least three different donor cells, which had been irradiated with a dose of 2000 RAD), and cultured for 4 days under the same conditions. Pulse labelling with 2 μCi of tritiated thymidine in 50 μl of medium and harvesting of cultures onto glass-fiber filters was done at the specified times. Incorporated tritiated thymidine was measured in a Packard scintillation counter. The data are presented as the mean of triplicate cultures.

Stimulants

Mitogens used were, Phytohemagglutinin (PHA) (Wellcome) at a final dilution of 1/100 and 1/400, Pokeweed mitogen (PWM, Grand Island biological company) at the same dilution, and Concanavalin A (Con A, Calbiochem. San Diego) at a final dilution of 5 and 20 μg/ml. As antigens we used Varidase R (Lederle), containing 100,000 IU Streptokinase and 100,000 IU Streptodanase, at a final dilution of 1/200 and 1/2000, Candida albicans allergenic extract (Hollister-Stier) diluted 100 and 400 times, PPD (Statens Seruminstitut Copenhagen) at final concentrations of 0.05 and 0.01 mg/ml, Staphylococcus aureus vaccine at 12.5 and 50 mg/ml, tetanus toxoid at 0.75 and 0.075 U/ml, Cytomegalo virus and varicella 'complement fixing antigen' and medium control, both diluted 200 times. The latter antigens were obtained from the RIV Bilthoven, The Netherlands.

Cell Mediated Lympholysis Test (CML)

This test was performed by Els Goulmy from our laboratory using techniques as described elsewhere [6]. The cytolytic capacity was tested against three random donors.

Results

Figure 1 shows the evolution of the relevant hematological parameters. In all patients except two the granulocyte counts rose above pretreatment values. The same two patients retained low reticulocyte values although in the others they reached supranormal numbers within 3 months after treatment.

Figure 2 gives the mitogen responses of the individual patients against time in counts per minute. Although the PHA LTT was within the normal range before and after ATG treatment (data not shown), this was not the case for the majority of the patients' lymphocytes tested against PWM and Con A. Five of the 7 patients had a lower than normal response against PWM while 3 reacted weakly to Con A before ATG. These are consistent findings in 20 other A.A. patients (unpublished results).

Patient number 3 reacted strongly to all mitogens, but only the Con A response was out of the normal limits. After treatment, all the transformation tests came within the normal range except in patient 1 whose Con A response

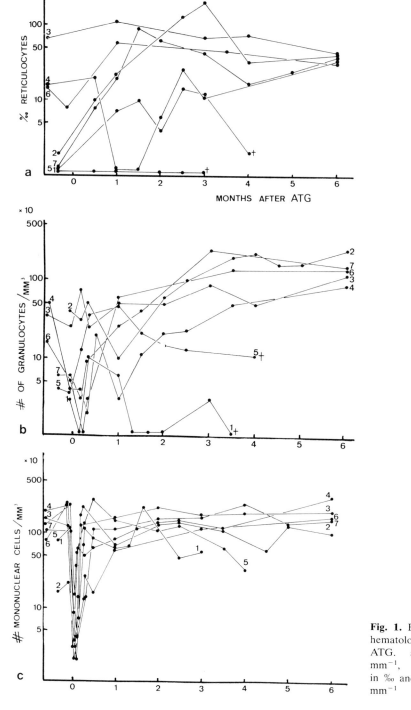

Fig. 1. Evolution of the hematological values after ATG. **a** granulocytes mm^{-1}, **b** reticulocytes in ‰ and **c** lymphocytes mm^{-1}

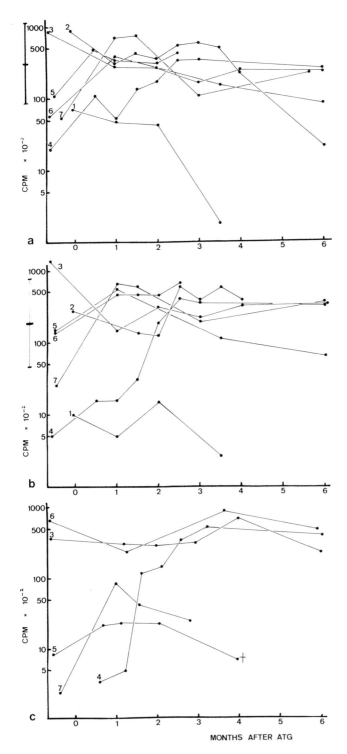

Fig. 2. Evolution of the immunological parameters; **a** response to Concanavalin A, **b** response to Pokeweed mitogen and **c** response to Varidase all in counts per minute

remained almost negative. This patient died 5 months after therapy. Patient number 4 was the only one whose restoration of the clinical and LTT values was delayed. Figure 2 summarizes the results with the antigen LTT (only varidase is shown). A follow-up study was done in 5 patients. Two of them had normal responses before ATG treatment, two showed negative responses for all antigens tested while one was not tested before but was still unresponsive within 1 month after therapy. Two of the non-responders recovered completely, one remained almost unresponsive until he died.

One patient (number 7) was studied in somewhat more detail. Figure I shows the granulocyte numbers before and after ATG treatment. A sharp increase in the numbers of all blood elements was noticed shortly after ATG treatment and makes it unlikely that the amelioration of the aplastic anemia is a coincidence. Figure 3 details the mitogen and antigen LTT, and the MLC and CML results obtained with the lymphocytes before and 1 month after treatment with ATG. The data show that the lymphocyte transformation to PHA was within normal limits before ATG, but PWM, and Con A gave weak responses. No antigen could induce any proliferation. This reaction pattern could be reproduced in exactly the same way with frozen cells of the same bleeding date. One month after ATG the reaction against all mitogens and antigens became high normal, even the

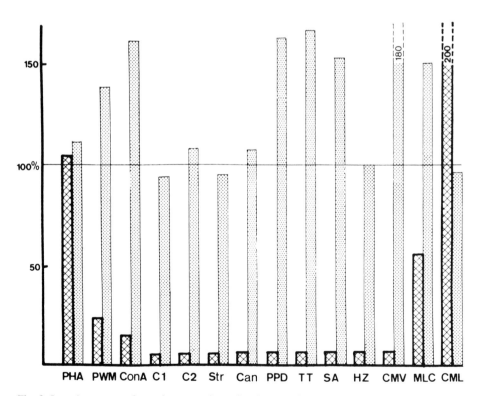

Fig. 3. Lymphocyte transformation test values of patient number 7 and cytotoxic capacity (CML) before (■) and after ATG treatment (□) in percentage of the modal response of a pool of controls

reactions against tetanus toxoid and PPD, although the patient had not been vaccinated nor infected. The MLC against a pool of stimulator cells remained within normal limits. In contrast, the CML reaction, which showed a high cytotoxic capacity before, went down to more acceptable levels after ATG.

Discussion

The results reported in this study show that ATG is not only capable of restoring the bone marrow aplasia in a number of patients, but also the in vitro immunological abnormalities which were shown to be present. These variations were not due to alterations in cell pools as the absolute numbers of lymphocytes and monocytes were not significantly different before and after ATG nor was there an increase in percentage of B cells. These findings lent further credibility to our working hypothesis that an immunosuppressive agent like ATG can, if not cure, at least improve some patients suffering from aplastic anemia. Many clinicians involved in the treatment of aplastic anemia remain highly sceptical on this point. The arguments which led us to think that ATG could be effective have been reviewed before [13]. In brief, they include the fact that severely ill patients show autologous B.M. reconstitution after ATG treatment (see also Figure 1) and that the bone marrow of aplastic anemia patients contains T cells which interfere with the formation of CFU_c [1, 8] and probably the differentiation of stem cells.

Of basic interest is the question how we should interpret these findings and whether they can help us unravel the pathogenesis of aplastic anemia. At least two mechanisms could be envisaged. The first is that the low LTT's are due to a suppressor cell which also has the capacity to block in an aspecific way the differentiation of committed stem cells for the hemopoietic series [1]. Another possibility which we favour is that the primary pathogenic factor is the presence of a clone of specific autocytotoxic cells which react against differentiation antigens which are transiently expressed on committed stem cells and lymphoblasts. There is ample evidence for an immunopathogenic factor in aplastic anemia. Infusion of bone marrow from an identical twin gives reconstitution in only a limited number of cases [5, 12] when administered without immunosuppression. Bone marrow aplasia can be induced experimentally by specific antibodies by infusion of allogeneic cells in immunodeficient patients [9, 10, 11]. That some patients are sensitizied to their own bone marrow has been shown by Chudomel [3] in a LTT and MIF assay using homogenates of the own bone marrow as antigen.

A real cell mediated autoimmune character of the disease has never been proven. A few other diseases have been shown to be caused by autoreactive lymphocytes, like active chronic hepatitis [4, 14] and diabetes mellitus [2].

The high CML killing upon non-identical targets fits with the assumption of a dysregulation of the cytotoxic function. A.A. patients as a group also demonstrate a higher cytolytic capacity than other patients (Goulmy, personal communication). It is of interest that Goulmy found 3 instances of killing on the male H-Y antigen restricted to HLA-A or -B antigens in patients with A.A. and not in multitranfused renal patients with leukocyte antibodies [7]. This could be interpreted as indicating that an A.A. patient is more prone to develop auto-immune cytolytic cells with relatively little priming by non MHC (Major Histocompatibility Complex) antigens.

As already pointed out in the introduction, many patients improve but only a few are really cured from aplastic anemia. If our assumptions are correct they indicate that ATG, although only partly effective, is a useful tool in the removal of the suppressive or cytotoxic lymphocytes causing the aplastic anemia in the majority of the cases.

The immunological studies presented here might help to define protocols to improve the treatment of aplastic anemia by adequate immunosuppression.

In part supported by the National Institute of Health (contract NOl-AI-4-2508), the Dutch Organization for Health Research (TNO), the Dutch Foundation for Medical Research (FUNGO), which is subsidized by the Dutch Foundation for the Advenacement of Pure Research (ZWO), and the J. A. Cohen Institute for Radiopathology and Radiation Protection.

Acknowledgements

We are indebted to A. Thompson and E. Goulmy for performing the LTT and CML test respectively. We are very grateful to J. van Nassau for her patience in typing the manuscript.

References

1. Ascensao, J., Kagan, W., Moore, M., Paheam, R., Hansen, J. and Good, R.: Aplastic anaemia: evidence for an immunological mechanism. Lancet *I*, 669 (1976)
2. Buschard, K., Madsbad, S. and Rygaard, J.: Passive transfer of diabetes mellitus from man to mouse. Lancet *I*, 908 (1978)
3. Chudomel, V. L., Novák, J. T., Pekárek, J., Svejcar, J. and Laznicka, M.: Autoimmunity as possible evoking factor of some so-called idiopathic Bone Marrow Hypoplasias. Z. Immun. Forsch. Bd. *150*, 379–383 (1975)
4. Eddleston, A. L. W. F. and Williams, T.: Inadequate antibody response to HBAg or suppressor T-cell defect in development of active chronic hepatitis. Lancet *II*, 1543 (1974)
5. Gale, R. P. and UCLA Bone Marrow Transplant Team: Bone marrow transplantation in identical twins with aplastic anaemia. Report of a case and review of the literature. In press
6. Goulmy, E., Termijtelen, A., Bradley, B. A. and van Rood, J. J.: HLA restriction of non-HLA-A, -B, -C, and -D Cell Mediated Lympholysis (CML). Tissue Antigens 8, 317–326 (1976)
7. Goulmy, E.: The importance of H-Y incompatibility in human organ transplantation. Transplantation 25, 6, 315–319 (1978)
8. Haak, H. L.: (1978) Acquired aplastic Anaemia in Adults. Thesis.
9. Hathaway, W. E., Githens, J. H., Blackburn, W. R., Pulginiti, V. and Kempe, C. H.: Aplastic anemia, histiocytosis and erythrodermia in immunologically deficient children. New Engl. J. Med. *273*, 953 (1965)
10. Kumar, S. and Saraya, A. K.: Experimental production of bone marrow aplasia by immunological means. Acta Haemat. 27, 306 (1962)
11. Nettleship, A.: Bone marrow changes produced by specific antibodies. Am. J. Path. *18*, 689 (1942)
12. Royal Marsden Hospital Bone Marrow Transplantation: Failure of syngenic bone marrow graft without preconditioning in post-hepatitis marrow aplasia. Lancet *II*, 742 (1977)
13. Speck, B., Gluckmann, E., Haak, H. L. and van Rood, J. J.: Treatment of aplastic anaemia by antilymphocyte globulin with and without allogeneic bone marrow infusions. Lancet *II*, 1145 (1977)
14. Thomson, A. D., Cochrane, M. A. G., McFarlane, I. G., Eddleston, A. L. W. F. and Williams, R.: Lymphocyte cytotoxicity to isolated hepatocytes in chronic active hepatitis. Nature 252, 721 (1974)

Discussion

Camitta: Dr. Sabbe, I think it is important to know whether or not your patients tested have had steroids and androgens. We tested the immune response of patients at the time of diagnosis of aplastic anemia. The immune responsiveness was intact and we did not see any deficit. How long have your patients been ill and what was their pretreatment before you tested them?

Sabbe: All patients were previously treated with androgens and with transfusions, the time of pretreatment varying between one month and several years.

Camitta: That may be an explanation of the findings you had rather than the findings being important in terms of origin of the aplasia.

Haak: All androgens have an immunosuppressive effect in various systems. It should be stressed that the patients tested by Dr. Sabbe were continued on androgens. Before they got ALG, the number of granulocytes stayed the same. We do not actually know the contribution of the different agents, but we know that the reactions were altered very soon after the ALG treatment.

Moore: Just a very specific question. Is it conceivable that the reason for the impaired Pokeweed responses is the adherent cell requirement which is not met by the peripheral blood of patients with aplastic anemia? Could you restore a normal response if you added to the cultures adherent cells as monocytes?

Sabbe: That is what we are doing at this moment. There are several explanations and one possible explanation is a lack of monocytes, because Pokeweed or Con A mitogens need monocytes, while PHA doesn't.

4.10 Recovery from Aplastic Anemia Following Therapy with Antithymocyte Globulin

R. K. Shadduck, A. Winkelstein, Z. Zeigler, J. Lichter, M. Goldstein, M. Michaels, B. Rabin

Introduction

It is generally believed that aplastic anemia results from an injury to or quantitative reduction in the number of hemopoietic stem cells [24]. In part, this notion derives from clinical observations in which the marrows of such patients usually show little or no cellularity with a paucity of erythroid, granulocytic and megakaryocytic differentiation. Recent studies of stem cell content using clonal techniques for the growth of myeloid precursor cells are in accord with this concept as they show a marked reduction in colony formation [6, 8, 14].

The acute aplasia induced in experimental animals by irradiation or administration of cytotoxic drugs is readily reversed by the transplantation of compatible marrow cells [26]. In recent years bone marrow transplantation has also proven successful in the management of human aplasia in those cases in which HLA compatible donors are available [27]. These studies would appear to support the concept that aplastic anemia is due to a reduction in hemopoietic stem cells.

An alternative pathogenetic mechanism for the aplasia has been suggested in a limited number of case studies [2, 9, 10]. In these reports it is proposed that aplasia is secondary to a cellular immune-mediated suppression or destruction of hemopoietic stem cells. In part, this concept derives from observations in which patients with aplastic anemia have shown recovery of hemopoiesis following cytotoxic (conditioning) therapy [20, 22, 25, 28]. Although marrow transplants were employed in most cases, graft rejection or failure of engraftment was followed by autorepopulation by the patient's marrow. Assuming the ability to detect such patients with immune-mediated bone marrow aplasia, therapy might be effectively directed towards eliminating the aberrant cell population rather than attempting transplantation with sources of hemopoietic stem cells.

The present report describes a case of acute aplastic anemia which appeared following atypical infectious mononucleosis. The results of in vitro culture studies and the prompt recovery of the patient after a course of antithymocyte globulin are highly suggestive of an immune-mediated mechanism for this patient's aplastic anemia.

Case Report

A 17 year old white female presented on 9/18/77 with a six day history of bruising and menorrhagia. She had been in good health until one month prior to admission when she developed generalized myalgia, headache and a temperature of 102° F. Physical examination showed no pharyngitis, adenopathy nor splenomegaly. The hematocrit was 43%, leukocyte count 7600 with 64% neutrophils and 36% lymphocytes. Throat culture was negative for pathogens. The illness subsided in four days with no specific therapy.

Physical examination at the time of admission showed only diffuse petechiae and ecchymoses of the lower extremities. The initial hemoglobin was 10.8 gm%, hematocrit 32% and reticulocytes 0.2%. The leukocyte count was 3500/µl with 44% neutrophils and 52% lymphocytes with 8–10% atypical forms. The platelet count was 11,000µl. A heterophile test was positive to a titer of 1:128 with absorption by beef cell but not guinea pig antigens. Initial anti-EB virus titers were positive; the anti-IgG titer was 1:8 and anti-IgM, 1:32. Sucrose hemolysis test and serum hepatitis associated antigen and antibody determinations were negative. A bone marrow aspiration and biopsy (Figure 1a) showed severe aplasia with only scattered late neutrophilic forms. No erythroid nor megakaryocytic cells were identified.

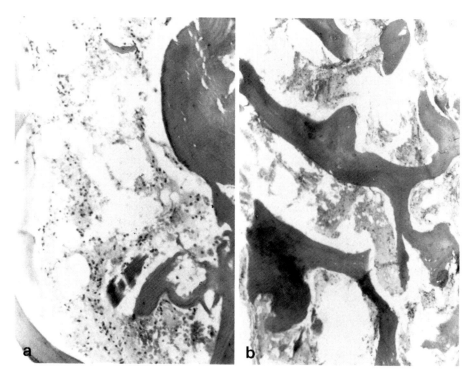

Fig. 1. a *(left panel)* The bone marrow biopsy obtained on 9/20/77 showed only scattered late neutrophilic cells, **b** *(right panel)* repeat biopsy 8 days later was markedly hypocellular with rare lymphocytes and plasma cells

By the fourth hospital day the neutrophil count had fallen to 50/µl and platelets to 7000/µl. Therapy included corticosteroids, androgens and progestational agents; the latter for attempted control of excessive menstrual bleeding. Over the next two weeks the patient became acutely ill with severe pancytopenia, fever and diffuse bleeding. Therapy included multiple red cell, white cell and platelet transfusions and combination antibiotic regimens. Repeat anti-EB virus studies revealed an IgG titer of 1:16 and IgM titer of 1:64. A repeat bone marrow aspiration and biopsy (Figure 1b) showed near total aplasia with only scattered lymphocytes and plasma cells. No erythroid, granulocytic nor megakaryocytic precursors could be identified. Bone marrow transplantation was considered, however there were no HLA compatible donors.

On the 18th hospital day the patient was started on a seven day course of intravenous antithymocyte globulin (ATG). The antiserum (which was kindly provided by Upjohn Company, Kalamazoo, Michigan) was administered in a dose of 15 mg/kg/day over a six hour intravenous infusion. This was tolerated well except for an episode of severe shaking chills during the first infusion. Ten days after starting this therapy, the patient developed serum sickness which was characterized by severe generalized myalgia, joint pains, spiking fevers and splenomegaly. This was associated with a rise in the SGPT to 500 u/dl and depression in serum complement values. The prednisone therapy was increased

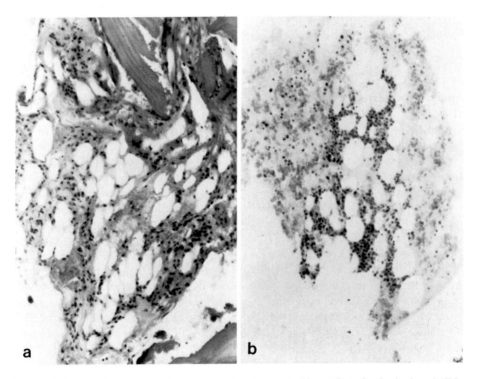

Fig. 2 a *(left panel)* Bone marrow biopsy obtained on 10/19/77, thirteen days after institution of ATG therapy, **b** *(right panel)* bone marrow cellularity on 11/16/77, one month after therapy

to 80 mg daily and within the next 24 hours there was prompt improvement in the patient's symptomatology. Over the next several weeks peripheral blood granulocytes rose to normal levels. As shown in Figure 2a, there was a marked improvement in marrow cellularity. Platelet and red cell recovery was delayed, however, by December 1977 all hematologic values approached normal levels. A repeat bone marrow showed normal cellularity and maturation of all three major cell lines (Fig. 2b). In February 1978, the hematocrit was 45%, white count 4,800/µl and platelets 170,000/µl. Serial blood counts are shown in Table 1.

Table 1. Peripheral blood counts

Date	Hematocrit	Reticulocytes $\times 10^{-3}/\mu l$	Leukocytes /µl	Neutrophils /µl	Platelets /µl
9/19/77	31%	7.1	4000	920	11,000
9/26	28%	–	1000	0	3,000
10/3	21%	–	1500	0	8,000
10/7	39%[a]	4.5	600	0	9,000
10/13	30%[a]	7.0	700	0	49,000[a]
10/19	32%[a]	7.7	800	260	9,000
10/27	43%[a]	5.2	3100	620	7,000
11/2	40%[a]	4.8	3200	1120	21,000
11/15	32%	37.2	4200	2230	21,000
12/19	35%	163.8	6500	4420	143,000
2/27/78	45%	48.6	4800	3265	170,000

[a] Antithymocyte globulin was administered from 10/6 through 10/12/77. Asterisks denote hematocrit and platelet values which were supported by transfusion

Special Studies

Granulocytic stem cells were evaluated by the in vitro granulocyte colony forming technique [18]. Bone marrow from both the patient and normal controls was obtained from the poster iliac crest into heparinized syringes. After sedimentation at room temperature, the buffy coat was aspirated and washed twice in McCoy's medium. Each culture was prepared such that 2×10^5 marrow cells were immobilized in 1 ml of 0.3% McCoy's-agar. Human placental condtiioned medium (0.1 ml) served as a source of CSF [5]. Cultures were incubated for 14 days in a 37° C humidified 7.5% CO_2 atmosphere. Colonies (greater than 50 cells) were scored with the aid of a dissecting microscope.

The effect of serum on colony growth was evaluated by incorporating either 10% patient or normal serum into the normal marrow cultures. Possible cellular-mediated colony suppression was investigated by coculture of normal marrow cells with either patient's marrow or peripheral blood mononuclear cells. The latter were obtained by ficoll-hypaque density gradient separation [4] and glass adherence.

The removal of T lymphocytes from marrow cells was attempted by incubating 2×10^6 marrow cells with 0.5 ml of a 1:100 dilution of antithymocyte globulin [2] for one hour prior to plating. The ATG proved toxic to normal control cultures and as such was not useful for studying the effect on the patient's marrow cells. HLA typing and determination of serum antibodies to HLA antigens was done by the lymphocyte cytotoxicity test [1].

Results

Bone marrow aspiration and biopsy were obtained on 9/28/77 at which time the patient was severely pancytopenic. The marrow was markedly hypocellular with only scattered lymphocytes and plasma cells. In vitro culture of this specimen showed no colony formation, however when the patient's marrow was mixed with a normal control, colonies were reduced to 38% of expected values, Table 2. Repeat bone marrow aspirate on 10/6/77 prior to the institution of antithymocyte globulin also revealed no colony growth. Again, admixture with a normal control marrow caused reduction in colony formation. In conjunction with these studies, the patient's peripheral blood lymphocytes and serum were tested for possible inhibition of colony formation. Neither the patient's lymphocytes nor the mononuclear cells from the normal control caused a consistent inhibition of colony formation in three studies. Likewise the patient's initial serum sample showed no inhibition in three culture studies. Serum used in these experiments showed no evidence of HLA antibodies by lymphocytotoxicity tests.

Table 2. Pre-treatment marrow cultures

		Colonies as % of normal control	
9/28/77	Normal marrow	100	±11%
	Patient	0	
	Mixing study	38%	±5%
10/6/77	Normal marrow	100	±3%
	Patient	0	
	Mixing study	72	±3%

Values are expressed as percent of control colonies ± 1 S.E. The normal marrows yielded 37 ± 4 and 62 ± 2 colonies/ 2×10^5 marrow cells respectively. The two studies were done 3 and 11 days after the initial blood transfusion. Serum samples showed no evidence of lymphocytotoxic reactions

One week after completion of the course of antithymocyte globulin, a repeat bone marrow aspirate again appeared hypoplastic. As shown in Table 3, the patient's marrow yielded no colonies, however when mixed with a normal control, there was modest augmentation of colony formation. A repeat bone marrow in November, 1977 was normocellular with abundant myeloid, erythroid and megakaryocytic cells. This marrow sample yielded 5 colonies per 2×10^5 cells and showed considerable augmentation when cocultured with a normal marrow.

Follow-up bone marrow obtained in February 1978 appeared completely normal. This marrow was co-cultured with three of the previous donors to determine whether there was disappearance of the inhibitory cell population. As shown in Table 4, the patient's marrow yielded 10 colonies per 2×10^5 cells and

		Colonies as % of normal control
10/19/77	Normal marrow	$100 \pm 22\%$
	Patient	0
	Mixing study	$163 \pm 13\%$
11/16/77	Normal marrow	$100 \pm 15\%$
	Patient	$19 \pm 4\%$
	Mixing study	$177 \pm 50\%$

Table 3. Post-treatment marrow cultures

Values are expressed as percent of control. Colonies ± 1 S.E. The normal marrows yielded 54 ± 12 and 26 ± 4 colonies/ 2×10^5 marrow cells respectively

showed inhibition with donors A and B but no decrease in colonies by co-culture with donor D. Serum obtained from the patient at this time showed no evidence of lymphocytotoxicity against the three marrow donors nor any reactivity in a 20 cell lymphocyte panel.

	Normal	Patient	Mixing study	
			Expected	Observed
A	42.4 ± 2.7	10 ± 2	52.4	29.0 ± 1.1
B	40.8 ± 2.6	10 ± 2	50.8	30.4 ± 4.0
D	31.8 ± 1.8	10 ± 2	41.8	41.6 ± 2.0

Table 4. Marrow cultures done after recovery

Values are colony counts ± 1 S.E. using 2×10^5 marrow cells/culture. Co-cultures contained 2×10^5 cells from both the patient and normal donor. Patient serum obtained ad the time of study showed no reactivity with peripheral lymphocytes from any of the three donors

Discussion

The great majority of patients with severe bone marrow aplasia succumb within six to 12 months of diagnosis [16, 17]. In general, those patients with less than 500 neutrophils or 20,000 platelets/μl show the worst prognosis with virtually no chance of spontaneous recovery. By all criteria, the present patient manifested a grave prognosis as judged by less than 100 neutrophils, less than 10,000 platelets and repeated marrow studies which showed virtually no hemopoietic cells. Initial therapy included corticosteroids and androgens, however most investigators agree that these measures are of limited value in severe aplastic anemia.

Several groups have shown that bone marrow transplantation is an effective means of therapy for selected patients with severe aplasia [7, 19, 27]. Treatment includes intensive conditioning therapy with irradiation or cytotoxic drugs and transplantation with histocompatible marrow cells. Between 75 to 80% of recipients show recovery of hemopoiesis with 45 to 50% long term survivors. It is of interest that despite matching at HLA loci and the use of mixed leukocyte cultures, as many as 25% of recipients failed to engraft. In part, these problems may relate to an altered marrow microenvironment. Such problems could result from an abnormal microvasculature or to immunoreactive cells directed against hemopoietic precursors.

Based on animal experiments, Crosby and Knospe have suggested that changes in the marrow architecture might effect the seeding and growth of hemopoietic stem cells [12]. Moreover, Ascensao and Kagan have recently shown that aplastic anemia may be associated with certain suppressor cells which inhibit in vitro granulocyte differentiation [2, 10]. Selective removal of these T lymphocytes caused enhanced colony formation. Similar findings have been noted in erythroid cultures wherein peripheral blood lymphocytes from patients with aplastic anemia inhibit normal erythroid differentiation [9]. Since many of the donors for such experiments have received multiple transfusions, the possibility still remains that many such results are secondary to anti-HLA antibodies.

Despite these reservations, certain clinical observations are in accord with an immune etiology for some cases of bone marrow aplasia. A number of patients have now shown recovery of their own hemopoietic cells following high dose conditioning therapy and transplantation with incompatible bone marrow cells [20, 22, 25, 28]. Although it remains conjectural whether some nondectectable stromal cell has survived the transplantation, this seems unlikely in view of the HLA incompatibility. Moreover, in one recent case, the administration of cyclophosphamide alone was associated with a dramatic recovery from severe aplasia [3]. Recently Speck, et al. [23], have reported their combined experience using antithymocyte globulin for the treatment of aplastic anemia. Nine of 14 patients treated with ALG and incompatible marrow showed significant improvement in neutrophil and platelet counts over one year of study. Similar partial recovery was noted in 10 of 15 patients treated with ALG alone.

In the present case severe bone marrow aplasia developed approximately one month following an upper respiratory infection. The presence of atypical lymphocytes, elevated heterophile titer and high titer IgM antibodies directed against EB virus are compatible with the diagnosis of infectious mononucleosis. This disorder is known to be associated with multiple signs of aberrant immunity; however aplastic anemia is rarely seen [13, 29]. In this patient, bone marrow colony studies showed absent colony formation by the patient's marrow. Although neither the patient serum nor peripheral blood lymphocytes were inhibitory to normal colony growth, co-culture of her marrow with normal marrow cells markedly reduced colony formation. These results were not secondary to HLA sensitization as the studies were instituted 3 and 11 days after initial transfusion. Moreover the patient's serum was nonreactive in lymphocytotoxicity assays.

Owing to a worsening clinical picture and the absence of a HLA-compatible donor, antithymocyte globulin was instituted before the results of the in vitro cultures were known. By the eighth day of treatment, the patient showed a slight increase in circulating granulocytes. Three weeks after treatment the neutrophils rose to normal levels with stabilization of the platelet count in the 25,000 to 30,000 range. Over the ensuing months there was a gradual increase in red cell and platelet production such that the counts had stabilized by November 1977.

Repeat bone marrow studies immediately after treatment showed continued aplasia and absent colony forming cells, however there was a disappearance of the inhibitory effect on co-culture. Two months after treatment a small number of colony forming cells were noted and again augmentation of normal colony growth was seen. When studied in February 1978, the marrow colony forming cells were still reduced as has been noted in other cases of aplasia [11]. Admixture of the patient's marrow with donor A and B showed inhibition however, donor D was unaffected. Lymphocytotoxicity tests with serum obtained after recovery were negative with all three donors and with a 20 member cell panel. This suggests the inhibition of colony growth seen after recovery was not due to sensitization to major HLA antigens. Since some degree of specificity was noted (2 of 3 donors), it is possible that the results reflects an immune response to minor histocompatibility [21] or neutrophil [15] antigens.

The events observed in this patient suggest that EB virus infection may be associated with severe bone marrow aplasia. Although the possibility of spontaneous remission cannot be excluded, the results of in vitro culture studies and the seeming response to antithymocyte globulin are in accord with an immune basis to this patient's aplasia. Further studies with such in vitro techniques prior to transfusion therapy and correlation with the response to antithymocyte globulin will be necessary to establish the frequency of immune mediated aplasia.

Acknowledgements

The studies were supported by NIH grants RO1CA 15237-03 and RO1AE 11739-03. The authors gratefully appreciate the technical assistance of Mrs. Florence Boegel and the supply of antithymocyte globulin by the Upjohn Company.

References

1. Amos, D. B., Pool, P.: HLA Typing. In: Manual of Clinical Immunology. Rose, N. R., Friedman, H. (ed.). Washington: American Society for Microbiology, 1976, pp 797–804
2. Ascensao, J., Pahwa, R., Kagan, W., Hansen, J., Moore, M., Good, R.: Aplastic anemia: evidence for an immunological mechanism. Lancet *I*, 669–671 (1976)
3. Baran, D. T., Griner, P. F., Klemperer, M. R.: Recovery from aplastic anemia after treatment with cyclophosphamide. New Engl. J. Med. *295*, 1522–1523 (1976)
4. Boyum, A.: Isolation of mononuclear cells and granulocytes from human blood. Scand. J. Lab. Clin. Invest. *21*, (Suppl 97), 77–89 (1968)
5. Burgess, A. W., Wilson, E. M. A., Metcalf, D.: Stimulation by human placental conditioned medium of hemopoietic colony formation by human marrow cells. Blood *49*, 573–583 (1977)
6. Dicke, K. A., Lowenberg, B.: In vitro Analysis of Pancytopenia: Its Possible Relevance to the Clinical Course and the Preleukemic State in the Aplastics. In: Proceedings of the Eighth Leukocyte Culture Conference. Lindahl-Kiessling, K., Osaba, D. (ed.). New York: Academic Press, 1974, pp. 417–424
7. Gale, R. P., Cline, M. J., Fahey, J. L., Feig, S., Opelz, G., Young, L., Territo, M., Golde, D., Sparkes, R., Naeim, N., Julliard, G., Haskell, C., Fawzi, F., Sarna, G. and Falk, P.: Bone marrow transplantation in severe aplastic anemia. Lancet *II*, 921–923 (1976)
8. Greenberg, P. L., Schrier, S. L.: Granulopoiesis in neutropenic disorders. Blood *41*, 753–769 (1973)

9. Hoffman, R., Zanjani, E. D., Lutton, J. D., Zalusky, R., Wasserman, L. R.: Suppression of erythroid-colony formation by lymphocytes from patients with aplastic anemia. New Engl. J. Med. 296, 10–13 (1977)

10. Kagan, W. A., Ascensao, J. A., Pahwa, R. N., Hansen, J. A., Goldstein, G., Valera, E. B., Incety, G. S., Moore, M. A. S., Good, R. A.: Aplastic anemia: presence in human bone marrow of cells that suppress myelopoiesis. Proc. Natl. Acad. Sci. U.S.A. 73, 2890–2894 (1976)

11. Kern, P., Heimpel, H., Heit, W., Kubanek, B.: Granulocytic progenitor cells in aplastic anemia. Br. J. Haemat. 35, 613–632 (1977)

12. Knospe, W. H. and Crosby, W. H.: Aplastic anemia: A disorder of the bone marrow sinusoidal microcirculation rather than stem cell failure. Lancet I, 20–22 (1971)

13. Koppes, G. M., Ratkin, G. A. jr.: Pancytopenia and "capillary leak syndrome" with infectious mononucleosis. South Med. J. 69, 145–148 (1976)

14. Kurnick, J. E., Robinson, W. A., Dickey, C. A.: In vitro granulocytic colony-forming potential of bone marrow from patients with granulocytopenia and aplastic anemia. Proc. Soc. Exp. Biol. Med. 137, 917–920 (1971)

15. Lalezari, P., Radel, E.: Neutrophil-specific antigens: immunology and clinical significance. Semin. Hematol. 11, 281–290 (1974)

16. Lewis, S. M.: Course and prognosis in aplastic anemia. Brit. M. J. 1, 1027–1031 (1965)

17. Lynch, R. E., Williams, D. M., Reading, J. C., Cartwright, G. E.: The prognosis in aplastic anemia. Blood 45, 517–528 (1975)

18. Pike, B. L., Robinson, W. A.: Human bone marrow colony growth in agar gel. J. Cell Physiol. 76, 77–84 (1970)

19. Santos, G. W., Sensenbrenner, L. L., Burke, P. J., Colvin, M., Owens, A. H. jr., Bias, W. B. and Slavin, R. E.: Marrow transplantation in man following cyclophosphamide. Transplant. Proc. 3, 400–404 (1977)

20. Sensenbrenner, L. L., Steele, A. A., Santos, G. W.: Recovery of hematologic competence without engraftment following attempted bone marrow transplantation for aplastic anemia: Report of a case with diffusion chamber studies. Exp. Hemat. 5, 51–58 (1977)

21. Singer, J. W., Brown, J. E., James, M. C., Doney, K., Warren, R. P., Storb, R., Thomas, E. D.: Effect of peripheral blood lymphocytes from HLA-matched and -mismatched marrows: Effect of transfusion sensitization. Blood 52, 37–46 (1978)

22. Speck, B., Cornu, P., Jeannet, M., Nissen, C., Burri, H. P., Groff, P., Nagel, G. A., Buckner, C. D.: Autologous marrow recovery following allogeneic marrow transplantation in a patient with severe aplastic anemia. Exp. Hemat. 4, 131–137 (1976)

23. Speck, B., Gluckman, E., Haak, H. L., van Rood, J. J.: Treatment of aplastic anemia by antilymphocyte globulin with and without allogeneic bone marrow infusions. Lancet 2, 1145–1148 (1977)

24. Stohlman, F. jr.: Aplastic anemia. Blood 40, 282–286 (1972)

25. Territo, M. C.: Autologous bone marrow repopulation following high dose cyclophosphamide and allogeneic marrow transplantation in aplastic anemia. Br. J. Haemat. 36, 305–312 (1977)

26. Thomas, E. D., Collins, J. A., Herman, E. D. jr. and Ferribee, J. W.: Marrow transplants in lethally irradiated dogs given methotrexate. Blood 19, 217–228 (1962)

27. Thomas, E. D., Storb, R., Clift, R. A., Fefer, A., Johnson, F. L., Neiman, P. E., Lerner, K. G., Glucksberg, H. and Buckner, C. D.: Bone marrow transplantation. New Engl. J. Med. 292, 832–843, 895–902 (1975)

28. Thomas, E. D., Storb, R., Giblett, E. R., Longpre, B., Weiden, P. L., Fefer, A., Witherspoon, R., Clift, R. A., Buckner, C. D.: Recovery from aplastic anemia following attempted marrow transplantation. Exp. Hemat. 4, 97–102 (1976)

29. Worlledge, S. M., Dacie, J. V.: Hemolytic and other Anemias in Infectious Mononucleosis. In: Infectious Mononucleosis. Carter, R. L., Penman, H. G. (eds.). Oxford: Blackwell Scientific Publications, 1969, pp. 82–98

5 Problems of an Immunological Pathogenesis in Aplastic Anemia

5.1 Some Possible Mechanisms Leading to Autoimmune Destruction of Bone Marrow Cells

E. Thorsby

Some possible immunological mechanisms which may be involved in causing injury or destruction of autologous bone marrow will be briefly reviewed in the following. No attempts will be made to cover available literature.

Evidence for Autoimmune Destruction of Bone Marrow Cells and Tissue

For some time autoimmune mechanisms have been postulated to be involved in some of the idiopathic aplastic anemias (AA). Mainly, three pieces of evidence can be listed for this concept:

1. Immunosuppressive treatment may lead to recovery.
2. Rejection of a bone marrow allograft may lead to recovery.
These two arguments are closely related. Mathé et al. [18] first reported complete hemopoietic restoration of AA with the use of anti-lymphocyte serum (ALS) and HLA mismatched bone marrow cells, without graft-versus-host (GVH) disease. This has been confirmed in later studies, which also showed that ALS or cyclophosphamide alone may lead to complete remission [6, 21, 22, 24]. It is not yet known whether the infusion of allogeneic bone marrow cells, which will be rejected, potentiate the effects of the immunosuppressive treatment in restoring autologous bone marrow functions.

3. Autologous lymphoid cells may inhibit hematopoietic stem cells in vitro.
Ascensao et al. [1] first showed that in a patient with AA pretreatment of the patient's bone marrow with ALS and complement prior to culture lead to an increase in colony forming ability. More important, by co-culturing cells from the bone marrow of the patient with bone marrow from normal donors, they demonstrated reduction of the normal colony forming ability. The inhibitory cells were probably lymphocytes [see also 15]. Later, Hoffman et al. [13] showed that peripheral blood lymphocytes of patients with AA would inhibit erythroid colony formation of normal bone marrow cells. No inhibition was found of lymphoid cells from normal individuals or multiply transfused patients not having AA. Supportive evidence of this concept has also been presented by Haak et al. [12].

Taken together, these observations suggest that autoimmune mechanisms may play a role in the pathogenesis of at least some cases of aplastic anemia. However, studies in more well-defined systems are needed before this can be firmly established. On the other hand, this concept is difficult to reconcile with the observations that in most cases a bone marrow transplant from a monozygotic twin to a patient with AA without immunosuppression, will lead to recovery [23]. One would expect that possible autoimmune reactions would also destroy the grafted genetically identical bone marrow cells. Possibly, the autoimmune mechanisms may have terminated at the time of transplantation, or the grafted normal bone marrow cells might include potential suppressor lymphocytes necessary for counter-balancing normal autoimmune phenomena (see later).

Immunological Mechanisms Which May Lead to Tissue Injury

The immunological effector mechanisms are broadly divided into two main categories; *humoral* responses mediated by antibodies and *cellular* responses mediated by intact cells. However, there are many important overlaps between these two categories, which are both the results of complicated interactive cellular events.

Some of the different types of immune responses which may be generated against *autologous* cells may be illustrated by the immune responses which today are known to play a role in the rejection of *allogeneic* cells, schematically pictured in Figure 1 [see for example 20 for further details and references].

Almost all cell-membrane molecules which are antigenic are thymus-dependent antigens. Thus, immune responses against foreign cells are dependent upon an initial activation of the T-helper (T_H) sub-set of T lymphocytes with receptors for the foreign antigenic determinants. The activation of T_H "helps" activation of B lymphocytes with receptors for membrane antigens on the same foreign cells, whereby many B cells are transformed into antibody-producing plasma cells. Antibody (IgM and certain sub-classes of IgG) may, after combination with cell-membrane antigens, activate the complement cascade, causing antibody-in-

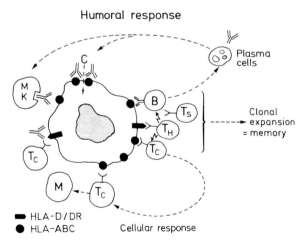

Fig. 1. The immune response against foreign cells (highly schematical). Activation of T-helper (T_H) cells is an important trigger mechanism for initiating the response. T_H then help the specific B lymphocyte (B) and cytotoxic T cell (T_C) clones to be activated. The effector mechanisms include antibody-induced complement-dependent (C) cytotoxicity, antibody-induced cellmediated cytotoxicity by Fc receptor bearing macrophages and monocytes (M) as well as other lymphoid cells (K), T cell cytotoxicity and cytotoxicity by T cell activated ("armed") macrophages. The T-suppressor cells (T_S) may regulate the response through feed-back inhibition

duced, complement-dependent cytotoxicity. Secondly, certain lympoid cells (K = killer cells etc.) and macrophages (M) have receptors for the Fc part of IgG antibodies, and may be activated through combination with target cell-bound antibody; antibody-induced, cell-mediated cytotoxicity. Thirdly, antibody and complement (C3) bound to the target cell will induce phagocytocis by macrophages. These are all examples of antibody-mediated immune responses.

Another sub-set of T lymphocytes which may be activated by foreign antigenic determinants are the cytotoxic T cells (T_C). Their activation appears also to be helped or amplified by T_H. After combination with the target cell, activated T_C will be cytotoxic; T cell-mediated cytotoxicity. Sensitized T cells may also produce different factors (lymphokines) which may activate macrophages etc. These are examples of cell-mediated immune responses.

Whether the T and B cells are activated by the same or different allo-antigenic determinants is a much discussed issue. In human allogeneic combinations, T_H seems mainly to be activated by the HLA-D antigenic molecules, while the B and T_C cells are mainly activated by the HLA-ABC antigens [see 3 for discussion]. This illustrates that T cell "help" may be provided by activation with other alloantigens than those which the B cell and T_C recognize [17], which is important for the understanding of some autoimmune phenomena (see later).

Recognition of the HLA antigens does not only play an instrumental role in human allogeneic immune responses. Recent studies have demonstrated that recognition of self-HLA molecules may be part of any T cell immune response against foreign antigens. Thus, the self-HLA-D molecules of macrophages appear to be involved in activating T_H cell responses to foreign soluble antigen, and the self-HLA-ABC molecules seem to play a similar role in activating T_C cells against virus infected target cells, chemically modified cells etc. [25 for references]. Two explanations have been put forward to explain the HLA restriction of T cell immune responses: (a) The foreign antigen in complex with self-HLA may produce new antigeneic determinants which are recognized as foreign (= "altered self"), or (b) T cells carry two receptors, one of low avidity for self-HLA, the other for foreign antigen (x), and both must combine to trigger a response. Which of these explanations is correct is unknown, but most recent evidence seems to favor the latter.

Another mechanism which may lead to tissue injury, and which is of relevance for autoimmune reactions, is antigen-antibody complex induced activation of complement and platelet aggregation [see 20 for details].

One or more of the above-mentioned immunological mechanisms have been assumed to be involved in the pathogenesis of certain autoimmune diseases. For example, there are good reasons to believe that autoantibody-induced complement and cell-mediated cytotoxicity are responsible for many cases of autoimmune haemolytic anemias and that autoimmune T cells are of major importance in causing experimental autoimmune encephalomyelitis. Further, complexes between autoantibodies and certain cell or tissue constituents (DNA etc.) are believed to be involved in the pathogenesis of systemic lupus erythemathosus [see 2 for further details].

Another sub-set of T cells which have recieved great interest lately is the T suppressor cell (T_s)[1] which may have an inhibitory or regulatory influence on both antibody production and cell-mediated immune responses [see 19]. Both antigen-specific and non-specific T_S have been described, with certain differences between them. How they function and which are the target cells for their suppressive effects are largely unknown. However, they may be considered an important part of a regulatory feed-back system of immune responses.

It would appear from this brief outline that the immune response to a given antigen is the result of complicated interactions between different sub-types of lymphoid cells, participating in a immunological network [14]. This is schematically illustrated in Figure 2, which also demonstrates that T_H, T_C and T_S express different Ly antigenic phenotypes.

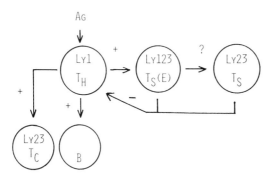

Fig. 2. The immunological network. Antigen activation (via macrophages) of the Lyl (T_H) cell will (a) help B cells to produce antibodies, (b) help Ly23 (T_C) cells to develop into cytotoxic effector cells and (c) induce the Ly123 ($T_{S[E]}$) to develop into active suppressor cells (through differentiation to Ly23 T_S?), which again will inhibit the activity of the T_H cell. In this way both "positive" ($+$) and "negative" ($-$) signals are delivered during the activating process, resulting in a homeostatic regulation of the immune response [modified from 11]

A cell whose place in this picture is not yet clear is the natural killer (NLK) cell, which is named from the fact that no prior sensitization appears to be necessary for its generation (whereby it differs from T_C etc.). Some of its other characteristics are listed in Table 1 [see 4, 16 for references]. Whether this cell may be involved in causing autoimmune phenomena is not known. However, it may be an important candidate for resistance to bone-marrow grafting [16].

Table 1. Natural killer cells

Lymphoid cells (from non-sensitized individuals) which are:

a) Cytotoxic particularly for hematopoetic/lymphoid target cell lines
b) Thymus independent (present in "nude" mice)
c) Relatively radio-resistant; influenced by injections BCG and CP (Corynebacterium parvuum)
d) Show age-dependent variations
e) Have unknown specificity, if any (?)
f) May be responsible for the hybrid resistance of parental-to-Fl hybrid bone marrow transplants

---

1 In order to avoid confusion, the name *suppressor cell* should be reserved for the particular sub-set of T lymphocytes which inhibit other lymphoid cells. Until it has been demonstrated that the same sub-set is also responsible for inhibition of haemopoietic cells, another non-committed name should be used for the inhibitory latter cell.

Mechanisms Causing Loss of Self Tolerance

It has been a dogma for many years that immunological reactions against self-constituents would normally not take place. Burnet [7] postulated that during intra-uterine development, all clones of lymphocytes reactive with *self* would be eliminated. Autoimmunity would be due to, through somatic mutations, the appearence of "forbidden clones" of lymphocytes with receptor for self.

More recent studies have shown that Burnet's hypothesis needs some modification. It has been demonstrated that lymphocytes exist, also among normal healthy individuals, which are able to combine with and be activated by self-constituents. For example, Bankhurst et al. [5] found lymphocytes able to combine with normal thyreoglobulin in healthy individuals. Further, several studies have shown that in vitro it is possible to generate an immune response against autologous target cells, both by T cell proliferation and cytotoxicity [see for example 9]. Also, as mentioned above, all T cells may have one receptor for self-HLA molecules, albeit the avidity would be too low for activation by self-HLA to occur under normal circumstances.

These and other observations have led to a modified self-tolerance concept. Thus, self-tolerance may be due to:

1. Genetically determined self-tolerance. Lymphocytes with receptors of sufficient avidity to be activated by (un-modified) self-constituents are not formed. This may be true for many cell and tissue constituents.

2. One of the sub-sets of lymphocytes participating in the necessary interactive events leading to anti-self immune responses are rendered tolerant; i.e. the lymphocytes with high avidity anti-self receptors are either functionally inactivated or eliminated after confrontation with self-consituents in foetal life. Many studies have schown that the T-helper cells, but not the B cells with specificity for a given antigen, may be rendered tolerant, leading to inability to produce antibodies against the thymus dependent antigen [26]. This may be another mechanism causing self-tolerance and explain why the presence of potentially self-reactive B lymphocytes usually do not result in autoimmune phenomena.

3. Potential anti-self reactivity may be counter-balanced by specific T suppressor cells. Of course the T_S cell is a good candidate for controlling self-tolerance and much experimental evidence exists demonstrating that they may be of importance for induction and maintenance of tolerance [10, 19, 26]. Possibly, they may be instrumental in counter-balancing potential T cell anti-self reactivity.

Following this, at least three mechanisms may be responsible for autoimmunity [see also 2]:

1. Bypass of need for self-reactive T_H (Fig. 3). Virus-infections, chemical agents (drugs) etc. may cause the development of foreign (viral) or modified cell-membrane components for which specific T_H cells exist. This activation of T_H may provide the necessary "help" for potential self-reactive B (and T_C) cells to be triggered, resulting in autoimmunity. As mentioned above, alloimmune reactions suggest that T_H cells need not be activated by the same antigenic determinants as those self-determinants for which B and T_C may have receptors, to "help" the latter cell types to be activated. However, it seems to be necessary that the

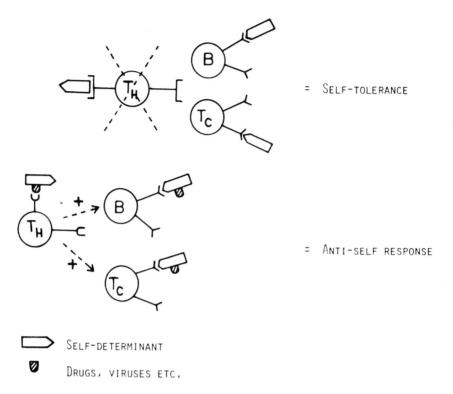

= SELF-TOLERANCE

= ANTI-SELF RESPONSE

⇨ SELF-DETERMINANT

▨ DRUGS, VIRUSES ETC.

Fig. 3. By-pass of need for self-reactive T_H cells. An important mechanism behind self-tolerance appears to be inactivation or elimination of self-reactive T_H cells, whereby potentially self-reactive B and T_C cells do not receive a necessary "helper" signal for activation (top part of figure). By introducing new determinants on self-molecules (chemical agents, viral antigens etc.) T_H with specificity for these foreign determinants may help self-reactive B and T_C to develop autoimmune responses (bottom part)

different determinants usually be present on the same cells. Also certain T and B cell mitogens (for example LPS) might bypass the need for specific T_H cells.
2. Loss of anti-self T_S control. The best evidence for the lack of T_S control being important in causing autoimmune reactions is the data suggesting that a reduced number of T_S cells in NZ B mice may be responsible for their autoimmune phenomena [2, 8]. Perhaps the known genetic control of T_S functions may partly explain the genetic predisposition to autoimmune phenomena in certain families. How and whether particular environmental factors may specifically cause inhibition of T_S functions is unknown.
3. Activation of T and B cells by self-antigens usually segregated from immune recognition. Certain self-constituents (for example of the lens) may usually be prevented from contact with immunocompetent lymphocytes, thus neither inducing tolerance nor immune responses. Under particular circumstances, segregation may be lost leading to autoimmune reactions. This mechanism has been suggested to be the cause of the T cell anti-self reponse generated in experimental autoimmune encephalomyelitis [27].

Which of these mechanisms (if any) that may lead to the development of aplastic anemia is unknown. To account for an organ-specific autoimmune destruction of bone marrow cells one would need an immune response of B and T cells directed against self-determinants only or mainly present on these cells. The induction of such specific autoimmunity might, for example, be due to certain drugs having particular affinity for bone marrow cells or their supporting tissue.

Conclusion

Recent research has provided new and better insight into the multitude of different immune mechanisms which may lead to tissue injury, both in allogeneic combinations and autologous situations. It appears that an immunological attack is the result of a highly complicated interplay of different sub-sets of lymphoid cells forming an immunological network. Self-tolerance may be the result of lack or non-responsiveness of one or more of the interacting sub-sets in the network with specificity for self-constituents, or an inhibitory effect of specific suppressor cells. Autoimmune mechanisms may thus be caused by activation of pre-existing potentially self-reactive cells through by-pass of the suppressed link in the network. The evidence of autoimmune phenomena being responsible for some cases of aplastic anemia is, however, mainly of an indirect nature, and nothing is yet known as to which of the possible mechanisms may be involved.

References

1. Ascensao, J., Kagan, W., Moore, M., Pahwa, R., Hansen, J., Good, R.: Aplastic anaemia: Evidence for an immunological mechanism. Lancet *I*, 669 (1976)
2. Allison, A. C.: Autoimmune Diseases: Concepts of Pathogenesis and Control. In: Autoimmunity. Talal, N. (ed.). New York: Academic Press, 1977, pp. 91–139
3. Bach, F. H., Bach, M. L., Sondel, P. M.: Differential function of major histocompatibility complex antigens in T-lymphocyte activation. Nature *259*, 273–281 (1976)
4. Baldwin, R. W.: Immune surveillance revisited. Nature *270*, 557 (1977)
5. Bankhurst, A. D.: Lymphocytes binding human thyroglobulin in healthy people and its relevance to tolerance for autoantigens. Lancet *I*, 226 (1973)
6. Baran, D. T., Griner, P. F., Klemperer, M. R.: Recovery from aplastic anemia after treatment with cyclophosphamide. New Engl. J. Med. *295*, 1522–1523 (1976)
7. Burnet, F. M.: The Clonal Theory of Acquired Immunity. London and New York: Cambridge Univ. Press, 1949
8. Cantor, H., McVay-Boudreau, L., Hugenberger, J., Naidorf, K., Shen, F. W., Gershon, R. K.: Immunoregulatory circuits among T-cell sets. II. Physiologic role of feedback inhibition in vivo: Absence in NZB. J. Exp. Med. *147*, 1116–1125 (1978)
9. Cohen, I. R., Wekerle, H.: Autoimmunity, Self-recognition, and Blocking Factors. In: Autoimmunity. Talal, N. (ed.). New York: Academic Press, 1977, pp. 231–265
10. Cunningham, A. J.: Self-tolerance Maintained by Active Suppressor Mechanisms. In: Transplantation Reviews. Gøran Møller (ed.). Copenhagen: Munksgaard, Vol. *31*, 1976, pp. 23–43
11. Eardley, D. D., Hugenberger, J., McVay-Boudreau, L., Shen, F. W., Gershon, R. K., Cantor, H.: Immunoregulatory circuits among T-cell sets. I. T-helper cells induce other T-cell sets to exert feedback inhibition. J. Exp. Med. *147*, 1106–1115 (1978)

12. Haak, H. L., Goselink, H. M., Veenhof, W., Pellinkhof-Stadelmann, S., Kleiverda, J. K., te Velde, J.: Acquired aplastic anaemia in adults. IV. Histological and CFU in transplanted and non-transplanted patients. Scand. J. Haematol *19*, 159–171 (1977)

13. Hoffman, R., Zanjani, E. D., Lutton, J. D., Zalusky, R., Wassermann, L. R.: Suppression of erythroid-colony formation by lymphocytes from patients with aplastic anemia. New Engl. J. of Med. 10–13 (1977)

14. Jerne, N. K.: Towards a network of the immune system. Ann. Immunol. (Paris) *125c*, 373–389 (1974)

15. Kagan, W. A., Ascensao, J. A., Pahwa, R. N., Hansen, J. A., Goldstein, D., Valera, E. B., Incefy, G. S., Moore, M. A. S., Good, R. A.: Aplastic anemia: Presence in human bone marrow of cells that suppress myelopoiesis. Proc. Natl. Acad. Sci. USA, *73*, 2890–2894 (1976)

16. Kiessling, R., Hochman, P. S., Haller, O., Shearer, G. M., Wigzell, H., Cudkowicz, G.: Evidence for a similar or common mechanism for natural killer cell activity and resistance to hemopoietic grafts. Eur. J. Immunol. *7*, 655–663 (1977)

17. Lake, P., Mitchison, N. A.: Regulatory mechanisms in the immune response to cell-surface antigens. Cold Spring Harbor Symp. Quant Biol. *41*, 589–594 (1977)

18. Mathé, G., Amiel, J. L., Schwarzenberg, L., Choay, J., Trolard, P., Schneider, M., Hayat, M., Schlumberger, J. R., Jasmin, C.: Bone marrow grafts in man after conditioning by antilymphocyte serum. Brit. med. J. *II*, 131–136 (1970)

19. Möller, G. (ed.): Transpl. Rev. *26*, (1975)

20. Roitt, I.: Essential Immunology. Oxford–London: Blackwell Scientific Publications 1977

21. Speck, B., Gluckman, E., Haak, H. L., van Rood, J. J.: Treatment of aplastic anaemia by antilymphocyte globulin with and without allogeneic bone-marrow infusions. Lancet *II*, 1145–1148 (1977)

22. Speck, B., Cornu, P., Sartorius, J., Nissen, C., Groff, P., Burri, H. P., Jeannet, M.: Immunologic aspects of aplasia. Transpl. Proc. *X*, 131–134 (1978)

23. Thomas, E. D., Storb, R., Clift, R. A., Fefer, A., Johnson, F. L., Neiman, P. E., Lerner, K. G., Glucksberg, G., Buchner, C. D.: Bone marrow transplantation. New. Engl. J. Med. *292*, 832–843 (1975)

24. Thomas, E. D., Storb, R., Giblett, E. R., Longpre, B., Weiden, P. L., Fefer, A., Witherspoon, R., Clift, R. A., Buchner, C. D.: Recovery from aplastic anemia following attempted marrow transplantation. Exp. Hemat. *4*, 97–102 (1976)

25. Thorsby, E.: The biological function of HLA. Tissue Antigens *11*, 321–328 (1978)

26. Weigle, W. O.: Cyclical production of antibody as a regulatory mechanism in the immune response. Adv. Immunol. *21*, 87–111 (1975)

27. Weigle, W. O.: Cellular Events in Experimental Autoimmune Thyroiditis, Allergic Encephalomyelitis and Tolerance to Self. In: Autoimmunity. Talal, N. (ed.). New York: Academic Press, 1977, pp. 141–170

5.2 Immunologic Function in Aplastic Anemia

R. P. Gale, R. Mitsuyasu, Coralee Yale

Introduction

Aplastic anemia is a hematologic disorder characterized by decreased or absent bone marrow function. While toxic agents, drugs, radiation, and viruses have been implicated in a small proportion of cases, the etiology is in most instances unknown. Potential mechanisms of abnormal marrow function include defective or absent hematopoietic stem cells, microenvironmental abnormalities, and immunologic factors.

There have been several studies of immune function in aplastic anemia, but most have been limited to small numbers of patients [1–4, 6–11, 13, 15–19, 21, 23, 24, 27, 28, 30–33, 36]. Larger series are summarized in Table 1. Most studies have reported normal immune function, but there are occasional reports of decreased T and B lymphocytes, monocytopenia, depressed skin test reactivity, and decreased immunoglobulins (Ig). Since aplastic anemia is a heterogenous disease, it is likely that much of the conflicting data in the literature results from the small number of patients studied. To answer this question we prospectively studied immune function in 73 patients with severe aplastic anemia.

Table 1. Selected studies of immune function in patients with aplastic anemia (nl = normal)

Author	Ref.	N°. patients	Lymphocytes Total	T	B	Mono-cytes	Mitogens PHA	Con-A	MLC Responder	Stimulator	Ig	Skin tests
Twomey	32	22	nl	–	–	↓↓	nl	nl	nl	↓↓	–	–
Morley	23,24	20	nl	nl	↓↓	↓↓	nl	–	–	–	nl	↓↓
Elfenbein	9	40	↓	↓	↓	↓↓	nl	nl	nl	nl	nl-↓	↓↓
Harada	15	18	nl	nl	–	–	nl	–	–	–	–	nl
Mickelson	21	53	–	–	–	–	–	–	nl-↑	↓↓	–	–

↓ = mild decrease in ≤ 50%, ↓↓ = marked decrease in > 50% of patients, blanks = not tested

Methods and Materials

Study Group

The study group consisted of 73 consecutive patients with aplastic anemia referred to UCLA between September, 1973, and June, 1978. All patients had severe aplasia defined as a marked decrease ($\leqq 40\%$) in bone marrow cellularity and $\geqq 2$ of the following criteria: (1) granulocytes $\leqq 0.5 \times 10^9/l$;

(2) platelets $\leqq 20 \times 10^9/l$; (3) reticulocytes (corrected) $\leqq 1$ per cent [5]. Thirty-seven patients were referred for bone marrow transplantation and 10 for a trial of antithymocyte globulin (ATG).

Hematologic Studies

Hematologic studies were performed by standard technics [35]. White blood cell (WBC) and platelet counts were determined with a Coulter Counter (Coulter Corp. Hialeah, FLA). WBC $\leqq 2.0 \times 10^9/l$ and platelets $\leqq 50.0 \times 10^9/l$ were checked by direct counting in a hemocytometer. Differential counts were performed on Wrights stain prepared slides, and a minimum of 200 WBC were analyzed. Reticulocytes were determined by supravital staining and corrected for the hematocrit. Bone marrow analysis was performed on $\geqq 2$ Jamshidi needle biopsies from separate sites.

Immunologic Studies

T-lymphocytes were assayed by rosette formation with sheep red blood cell (SRBC, E rosettes) [12]. B-lymphocytes were determined by surface membrane immunolobulin (SMIg) staining with a fluorescence conjugated goat anti-human antiserum and by rosette formation with rabbit antibody coated SRBC (EA-rosettes) [12]. Monocytes were analyzed by staining with naphtyl-ASD-esterase and by ingestion of latex particles.

Levels of circulating IgG, A, and M were determined by laser nephelometry (Hyland Labs, Costa Mesa, CA). Antibodies to blood group antigens and complement fixing antibodies to cytomegalovirus (CMV) were determined by standard technics.

Mitogen responsiveness was determined using a semi-micro technic [12]. Peripheral blood mononuclear cells were incubated with phytohemagglutinin (PHA-M; Difco Labs, Detroit, MI) or concanavalin-A (con-A; Calbiochem, Los Angeles, CA), pulsed with ^{3}H-TdR, and harvested at 96 h. Results were expressed as a reponse index of the maximum CPM of the patient/maximum CPM of a normal control (RI_{cpm}). In most instances, maximum proliferation of patient and normal was observed at the same mitogen concentration.

Mixed lymphocyte cultures (MLC) were performed as described [29]. Briefly, 2×10^5 responder cells were incubated with an equal number of irradiated stimulator cells for 5 days and ^{3}H-TdR incorporation determined. Lymphocytes from a minimum of 2 unrelated controls were used in each experiment. In 37 patients an HLA-identical sibling was tested at the same time as the patient. Reactivity in MLC was expressed both as the absolute CPM and stimulation index. Responder cell reactivity in MLC was evaluated by comparing the response of patient and their HLA-identical sibling to an unrelated stimulator cell. Stimulatory capacity in MLC was evaluated by using lymphocytes of the patient and the HLA-identical sibling to stimulate lymphocytes from one or more unrelated controls. Results were expressed either as the ratio of the stimulation indices (RI_{si}) or a ratio of the CPM (RI_{cpm}).

Lymphocytotoxins were assayed against a panel of lymphocytes from 90–120 normal unrelated donors [22]. Results were expressed both as the per cent of the panel killed and as the index strength.

Skin testing was performed by the intradermal injection of purified protein derivative (PPD), mumps, trichophyton, monilia, and streptokinase/streptodornase (SK/SD). Inflammatory responsiveness was assayed by topical application of croton oil. Selected patients received a sensitizing dose of dinitrochlorobenzene (DNCB-2000 µg), followed in 14 days and 6 months by a challenge with 50, 100, and 200 µg. Skin tests were examined at 24×48 h and those with $\geqq 10$ mm induration were scored as positive.

Statistical Analysis

Statistical analysis was performed using the BMDP programs of the UCLA Health Sciences Computing Facility (BMDP2D, BMDP6D, and BMDP7D). The analysis of differences in MLC reactivity between patients and their siblings was performed using paired t-tests.

Results

Seventy-three patients with severe aplastic anemia were studied. Clinical features are indicated in Table 2. The median age was 23 years, and there were twice as many males as females. Eleven patients (15%) had hepatitis-associated aplasia and 9 (12%) had drug-associated aplasia. No etiology was identified in 45 patients (62%). Approximately two-thirds of patients were treated with androgens and/or corticosteroids. Most patients were studied early in the course of their disease with a median time from diagnosis to study of 38 days. Ninety per cent of patients were studied within the first 3 months.

Table 2. Patient characteristics

Age[a] (yr)	23 (2–61)
Sex	47 Male, 26 Female
Etiology	
Idiopathic	45 (62%)
Hepatitis	11 (15%)
Drug	9 (12%)
NE[b]	8 (11%)
Androgens[c]	41/63 (65%)
Corticosteroids	37/61 (61%)
Diagnosis to Study[a]	38d (4-3120 d)

[a] Median (range)
[b] Not evaluable
[c] Number treated/number evaluable (%)

Hematologic features are indicated in Table 3. All patients had severe aplasia with hypocellular or absent myeloid activity on bone marrow biosy, and decreased WBCs, PMNs, and platelets. Two-thirds had decreased reticulocytes and approximately one-half had decreased lymphocytes and monocytes.

Table 3. Hematologic parameters in patients with aplastic anemia

Parameter ($\times 10^9$/l)	Patients (4)		Normals (291)[a]		Patients
	Mean[b]	Range	Mean	95% CI[c]	<95% CI
WBC	2.1±0.1	0.7–4.5	7.0	4.3–10.0	63 (98%)
PMN	0.48±0.06	0–2.73	3.65	1.83–7.25	63 (98%)
Lymphocyte	1.5±0.1	0.2–3.6	2.5	1.5–4.0	23 (36%)
Monocyte	0.19±0.02	0.02–0.14	0.43	0.20–0.95	34 (53%)
Platelet	14.0±1.7	2.0–65.0	241	140–340	64 (100%)
Hemoglobin[d]	8.4±0.3	3.0–13.0	15.0	12.0–18.0	57 (89%)
Reticulocyte[e]	0.5±0.1	0–3.0	1.6	0.8–2.5	39 (61%)

[a] Reference
[b] Mean±SEM
[c] 95% confidence interval
[d] g/dl
[e] %corrected

Table 4. Lymphocyte subpopulations in patients with aplastic anemia

		Patients			Normals		Patients	
		No.	Mean[a]	Range	Mean	95% CI[b]	<95% CI	>95% CI
Per cent	Lymphocytes	64	70.3±2.3	8–99	36.1	19.6–52.7	0	55 (86%)
	T	30	67.1±3.0	27–88	65.7	56.0–76.0	5 (17%)	0
	B[c]	27	15.9±2.2	2–46	6.9	4.0–10.0	0	17 (63%)
Total	Lymphocytes	64	1.5±0.1	0.2–3.6	2.5	1.5–4.0	31 (48%)	0
	T	30	1.0±0.1	0.1–2.7	1.6	0.8–3.0	12 (40%)	0
	B	27	0.2±0.1	0–0.9	0.2	0.0–0.4	3 (11%)	3 (11%)

[a] Mean±SEM
[b] 95% confidence interval
[c] SMIg (+)

Immunologic features are indicated in Tables 4–6. The per cent of lymphocytes and B-cells was increased in approximately two-thirds of patients. The per cent of T-cells was normal. Total lymphocytes were decreased in approximately 60 per cent, and T-lymphocytes were decreased in one-third. B-lymphocytes were normal in 80 per cent. One-third of patients had an increased proportion of lymphocytes and plasma cells in the bone marrow.

Immunoglobulins were normal in most patients (Table 5). Four patients had decreased IgG, 5 decreased IgM, and 5 decreased IgA. Six patients had elevated levels of IgG.

Table 5. Immunoglobulins in patients with aplastic anemia

	Patients (42)		Normals		Patients	
	Mean[a]	Range	Mean	95% CI[b]	<95% CI	>95% CI
IgG	1117±76	365–2649	999	602–1397	4 (10%)	6 (14%)
IgM	126±64	29–350	152	47–256	5 (12%)	0
IgA	252±25	35–790	342	121–563	5 (12%)	0

[a] Mean±SEM
[b] 95% confidence interval

Reactivity to PHA and con-A are indicated in Table 6. While there was no difference in mean PHA reactivity between patients and normals ($RI_{cpm} = 0.93$), 6 patients had RI_{cpm} values of less than 0.50 and 5 of these had RI_{cpm} values of less than 0.25. Eight patients showed increased PHA reactivity with RI_{cpm} values greater than 1.5, and 6 of these had values exceeding 2.0. Similar proportions of patients exhibited altered con-A reactivity.

Results of MLC studies are indicated in Table 6. The mean SI for patients as responder cells was 51.0 while that of normal siblings was 33.8. The RI_{si} was 1.51. While this difference is statistically significant in the paired t-test (t2.88, $P<0.005$), it is explained by lower background CPM amongst patients since the absolute CPM of both groups were comparable and the RI_{cpm} was 1.06.

Table 6. Mitogen and mixed lymphocyte culture reactivity (MLC) in patients with aplastic anemia compared to HLA-identical siblings

	Response index[a]	
	RI_{si}	RI_{cpm}
PHA	–	0.93
Con-A	–	0.92
MLC-responder	1.51	1.06
MLC-stimulator	0.92	0.62

[a] See text for details

RI_{si} = patient SI/normal SI, RI_{cpm} = patient CPM/normal CPM

The stimulatory capacity of patients and their siblings are indicated in Table 6. The RI_{si} comparing patients and siblings was 0.92 (12.5 vs 13.6) and the RI_{cpm} was 0.62. These differences are not statistically significant. Furthermore, a highly significant correlation was observed comparing patients and siblings as stimulator cells ($r = 0.764$, $p < 0.0005$) (Fig. 1).

Thirty-eight patients were skin tested with 2 or more antigens. Twenty-two (58%) failed to react to any antigen and 9 (25%) reacted to 1 antigen. A similar degree of decreased skin test reactivity was observed in less than 10 per cent of normals. Only one of 7 patients could be sensitized to DNCB.

Antibody titers to blood group antigens (isohemagglutinins), CMV, and to HLA (lymphocytotoxins) were determined in 60 patients. Isohemagglutinin titers were normal. Several patients had antibodies to non-ABH antigens including Rh (D, E), Lewis (Le[a]), Kidd (Jk[a]), BU[a], I, P_1, and Bg[a]. Thirty-five of 48 patients had a low to absent (<1:8) antibody titers to CMV and 13 had

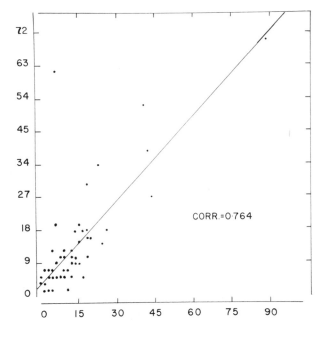

Fig. 1

titers $\geq 1:16$. Twenty of 45 patients (44%) had positive lymphocytotoxins. Eighteen patients had a lymphocytotoxin index strength greater than 20 per cent and 12 had an index strength greater than 50 per cent.

Discussion

We studied immune function in 73 patients with aplastic anemia. While some patients had normal immune function despite severe marrow failure, we were able to identify a substantial proportion of patients who exhibited abnormal immune function in one or more assays. These abnormalities included: (1) decreased lymphocytes (50%), monocytes (50%), and T-lymphocytes (30%); (2) decreased or increased IgG, A, or M (10%); (3) increased lymphocytes and plasma cells in the marrow (33%); (4) decreased (20%) or increased (10%) mitogen responsiveness; (5) lymphocytotoxins (20%); and (6) depressed skin test reactivity and sensitization to DNCB (80%).

Some of our results differ from those reported by others. For example, we found significant lymphocytopenia with a secondary decrease in T-lymphocytes in 50 per cent of patients. Lymphocytopenia is not a common feature in several series [19, 33], and two investigators have reported normal levels of T-lymphocytes in small numbers of patients [15, 24]. Morley and coworkers reported decreased B-lymphocytes in patients with "chronic hypoplastic marrow failure", a finding not confirmed in the present study [24]. Mickelson and coworkers reported that lymphocytes from patients with aplastic anemia were less effective in stimulating allogeneic cells in MLC than lymphocytes from their HLA-identical siblings [21]. We were unable to confirm this observation but rather found a highly significant correlation in the stimulatory capacity between patient and sibling lymphocytes. Similar data have been reported by Elfenbein and coworkers [9].

Our data suggest that while some patients with aplastic anemia have normal immune function, selective defects can be found in a substantial proportion of patients. Whether these defects antedate the development of aplasia or was a consequence of it remains to be determined.

Supported by Grants CA-23175, CA-12800, and CA-15688, from the National Cancer Institute, and Grant HB-62971 from the National Heart, Lung, and Blood Institute, and Grant RR-3 from the National Institutes of Health, and Grant RR-0865 from the U.S. Public Health Service. Robert Peter Gale is a Scholar of the Leukemia Society of America.

References

1. Abels, D., Reed, W. B.: Fanconi-like syndrome: Immunologic deficiency, pancytopenia and cutaneous malignancies. Arch. Dermatol. *107*, 419 (1973)
2. Albert, E., Thomas, E. D., Nisperos, B., Storb, R., Camitta, B. M., Parkman, R.: HLA-antigens and haplotypes in 200 patients with aplastic anemia. Transplantation *22*, 528 (1976)
3. Arimori, S., Kobashi, H., Tada, S., Ichikawa, Y., Koriyama, K.: The lymphocytes of the patients with hypoplastic anemia. Jap. J. Clin. Hematol. *16*, 763 (1975)

4. Brookfield, E. G., Singh, P.: Congenital hypoplastic anemia associated with hypogammaglobulinemia. J. Peds. 85, 529 (1974)
5. Camitta, B. M., Thomas, E. D., Nathan, D. G., Santos, G., Gordon-Smith, E. C., Gale, R. P.,
   Rappeport, J. M., Storb, R.: Severe aplastic anemia: A prospective study of the effect of early
   marrow transplantation on acute mortality. Blood 48, 63 (1976)
6. Corneo, G., Cortellaro, M., Miaolo, A. T.: Aplastic anemia: some clinical and laboratory aspects
   and an evaluation of androgen glucocorticoid treatment. Acta Haemat. 46, 50 (1971)
7. Dreyfus, B., Varet, B., Sultan, C.: Aplasie medullaire avec hypogammaglobulinemie. Nouv. Rev.
   Franc. Hemat. 9, 33 (1969)
8. Eldor, A., Gale, R. P., Brautbar, H., Lourie, A., Slavin, S., Eliakim, M.: Aplastic anemia
   following acute viral hepatitis associated with hypogamma-globulinemia attempted treatment by
   bone marrow transplantation. Harefuah 89, 105 (1975)
9. Elfenbein, G. J., Kallman, C. H., Tutschka, P. J., Bias, W. B., Braine, H. B., Saral, R., Santos, G.
   W.: Immune function in patients with aplastic anemia (AA). Exp. Hematol. 6, Suppl. 3, 28
   (1978) (Abstr.)
10. Fayos, J. S., Outerino, J., Calabuig, T., Valdes, D., Paniagua, G.: Significacion citofuncional de la
    publacion celular linfoide en las pancitopenias aplasticas. Sangre (Barc) 19, 200 (1974)
11. Flad, H. B., Hochapfel, G., Fliedner, T. M., Heimpel, H.: Blastentransformation und
    DNS-Synthese in Lymphozytenkulturen von Patienten mit Aplastischer Anaemie (Panmyelopa-
    thie). Acta Haemat. 44, 21 (1970)
12. Gale, R. P., Opelz, G., Kiuchi, M., Golde, D. W.: Thymus-dependent lymphocytes in human
    bone marrow. J. Clin. Invest. 56, 1491 (1975)
13. Haas, R. J., Kretschmer, V., Fliedner, T. M.: Phytohaemagglutinin-Stimulation von Lymphozy-
    ten bei Patienten mit aplastischer Anaemie. Med. Klin. 65, 724 (1970)
14. Han, T., Dadey, B.: T and B lymphocytes: Exclusive role as responders and stimulators in human
    "one-way" mixed lymphocyte reaction. Immunology 31, 643 (1976)
15. Harada, M., Omura, T., Tanaka, K., Hattori, K.: A study on cell-mediated immunity in patients
    with aplastic anemia. Jap. J. Clin. Hematol. 16, 868 (1975)
16. Hathaway, W. E., Brangle, R. W., Nelson, T. L., Roeckel, I. E.: Aplastic anemia and
    alymphocytosis in an infant with hypogrammoglobulinemia: Graft-versus-host reaction? J. Peds.
    68, 713 (1966)
17. Huhn, D., Fateh-Mughadam, A., Demmler, K., Kranseder, A., Ehrhart, H.: Haematologische
    und immunologische Befunde bei Knochenmarkaplasie. Klin. Wschr. 53, 7 (1975)
18. Korn, D., Gelderman, A., Cage, G., Nathanson, D., Strauss, A. J. L.: Immune deficiencies,
    aplastic anemia, and abnormalities of lymphoid tissue in thymoma. New Eng. J. Med. 276, 1333
    (1967)
19. Lewis, S. M.: Aplastic anemia: Problems of diagnosis and treatment. J. Roc. Soc. Physician Lond.
    3, 253 (1969)
20. Lohrmann, H. P., Nivakovs, L., Graw, R. G. jr.: Stimulatory capacity of human T and
    B lymphocytes in mixed leukocyte culture. Nature 250, 144 (1974)
21. Mickelson, E. M., Fefer, A., Thomas, E. D.: Aplastic anemia: Failure of patient leukocytes to
    stimulate allogeneic cells in mixed leukocyte culture. Blood 47, 793 (1976)
22. Mittal, K. K., Mickey, M. R., Singal, D. P., Terasaki, P. I.: Serotyping for homotransplantation.
    XVIII Refinement of microdroplet lymphocyte cytotoxicity test. Transplantation 5, 913 (1968)
23. Morley, A., Forbes, L.: Impairment of immunological function in aplastic anemia. Aust. N. Z. J.
    Med. 4, 53 (1974)
24. Morley, A., Holmes, K., Forbes, K.: Depletion of B-lymphocytes in chronic hypoplastic marrow
    failure (aplastic anemia). Aust. N. Z. J. Med. 4, 538 (1974)
25. Opelz, G., Kiuchi, M., Takasugi, M.: Reactivity of lymphocyte subpopulations in human mixed
    lymphocyte culture. J. Immunogenet. 2, 1 (1975)
26. Plate, J. M. G., McKenzie, I. F. C.: "B"-cell stimulation of allogeneic T-cell proliferation in
    mixed lymphocyte cultures. Nature (New Biol.) 245, 247 (1973)
27. Sampson, J. P., DeGast, C. C., Nieweg, H. O.: Immunological responsiveness in idiopathic and
    drug-induced panmyelopathy: Discrepancy between sensitization with DNCB and haemocyanin.
    Br. J. Haematol. 26, 227 (1974)
28. Sasportes, M., Bernard, A., Dausset, J.: Abnormal mixed leukocyte culture reaction in bone
    marrow aplasia. Br. Med. J. 2, 28 (1973)

29. Sengar, D. P. S., Terasaki, P. I.: A semi-micro mixed-lymphocyte culture test. Transplantation *11*, 260 (1971)
30. Strauss, R. G., Bove, K. E., Lake, A., Kisker, C. T.: Acquired immunodeficiency in hepatitis-associated aplastic anemia. J. Peds. *86*, 910 (1975)
31. Szmigiel, A., Turowski, G., Rzepeckiw, B., Aleksandrowicz, J., Skotnicki, A. B.: The effect of myasthenia thymus fragments transplantation on the immune state of patients with proliferative and aplastic diseases of the haematopoietic system. Acta Med. Pol. *16*, 61 (1975)
32. Twomey, J. J., Waddell, C. C., Krantz, S., O'Reilly, R., L'Esperance, P., Good, R. A.: Chronic mucocutaneous candidiasis with macrophage dysfunction, a plasma inhibitor, and co-existent aplastic anemia. J. Lab. Clin. Med. *85*, 968 (1975)
33. Twomey, J. J., Douglass, C. C., Sharkey, O.: The monocytopenia of aplastic anemia. Blood *41*, 187 (1973)
34. Uchiyama, T., Nagai, K., Yamagishi, M., Takatsuki, K., Uchino, H.: Pokeweed-mitogen induced B cell differentiation in idiopathic aplastic anemia associated with hypogrammaglobulinemia. Blood *52*, 77 (1978)
35. Wintrobe, M. M.: Clinical Hematology, 3rd Ed. Philadelphia: Lea and Febiger, 1977
36. Yoda, Y., Abe, T., Komiya, M.: Immunological studies in aplastic anemia. Jap. J. Clin. Hematol. *17*, 628 (1976)

5.3 Serum Inhibitors of Granulocyte CFU-C in Aplastic Anaemia

A. J. Barrett, A. Faille

Introduction

Aplastic anaemia is characterised by pancytopenia with profound bone marrow hypoplasia. This is associated with a reduced number of granulocyte precursor cells (CFU-C) in the bone marrow [6, 2]. While the end-result of a variety of processes may produce this apparent stem cell failure in aplastic anaemia, an immunological mechanism may be responsible in at least some patients [3, 1, 5].

The purpose of this study was to search for possible immunological abnormalities in a large series of patients with aplastic anaemia by examining the lymphocytes and the serum for inhibitory effects on normal CFU-C growth. Some of the characteristics of serum inhibitory to CFU-C growth found in aplastic anaemia patients are discussed.

Patients and Methods

A. Patients and Treatment

This study concerns patients with aplastic anaemia referred to the Hopital Saint Louis, Paris between 1975 and 1977.

Of 41 patients with aplastic anaemia, 33 fitted the diagnostic criteria for severe aplastic anaemia (neutrophils less than 0.5×10^9/l, platelets less than 20×10^9/l, reticulocytes less than 1%, with hypocellular bone marrow aspirates and biopsy). All patients had been transfused with blood products before referral and most patients had received androgens.

Twenty patients had been referred for consideration for bone marrow transplantation, and subsequently received a bone marrow graft from a sibling who was HL-A identical and compatible by mixed lymphocyte culture. Marrow was transplanted in accordance with the Seattle protocol [9]. Twenty-one patients who had no compatible bone marrow donor received ALG (Merieux) in a dose of 15 mg/kg/day for five days. Response to ALG was defined as a rise in median neutrophil count to over 1×10^9/l within 30 days. Responding patients also showed a rise in platelet and reticulocyte counts and a decreased requirement for transfusions, but these features varied from case to case.

B. Granulocyte Culture Studies

Blood and bone marrow were taken for colony culture studies before any specific treatment was started and at various times after ALG treatment or marrow transplantation. As controls, sera were collected from normal subjects not immunised by pregnancy or transfusion, and from patients with various haematological disorders who had been given multiple transfusions. Other control sera were taken from multiparous women found to have high titres of polyspecific HL-A antibodies. All sera were stored at $-70°$ C until they were tested.

I. Co-Culture Assay

Washed mononuclear cell fractions of blood or bone marrow cells from patients with aplastic anaemia were used to co-culture with normal bone marrow mononuclear cell fractions. Three incubations were set up for each patient tested:

1. Patients cells (2.0×10^6)
2. Target marrow cells (2.0×10^6)
3. 1.0×10^6 patient cells + 1.0×10^6 Target marrow cells.

The cultures were incubated in McCoy 5A culture medium containing 20% foetal calf serum for 24 hours before being washed and put into agar culture in triplicate at a concentration of 2×10^5 cells/culture plate. CFU-C inhibition by patients cells was calculated using the formula:

$$\text{Inhibition index} = \frac{\text{Observed count}}{\text{Expected count}} \text{ of target bone marrow}$$

$$= \frac{\text{mean CFU-C tube 3} - \left(\dfrac{\text{mean CFU-C}}{2} \text{ tube 1} \right)}{\dfrac{(\text{mean CFU-C tube 2})}{2}}$$

II. Serum Inhibition Assay

Serum was assayed for its inhibitory effect on normal colony growth by incubating 6×10^5 bone marrow cells in the presence of 10% patient's serum with 10% rabbit complement for 1½ hours at 37° C. The marrow cells were washed and resuspended in agar and medium and plated in triplicate at a final concentration of 2.10^5 cells per plate. As control sera from normal adult subjects were tested in the same manner, the ratio of colony growth after exposure to test serum was used to calculate an inhibition index.

III. Absorption of Sera

Nine sera were absorbed on platelets or myeloblasts and re-tested for their inhibitory activity:

1. *Platelet absorption.* A pool of platelets from 40 blood donors was used. Washed platelets were preserved at 4° C in buffered saline until used. For absorption 2×10^9 platelets/ml of serum were incubated for 18 hours at 37° C with frequent agitation.

2. *Myeloblast absorption.* Myeloblast removed from three patients with untreated acute myeloblastic leukaemia were used. 2×10^8 washed cells per ml of serum were incubated over 18 hours at 37° C with frequent agitation.

The sera were centrifuged and filtered through 20µ'millipore' filters. Control (unabsorbed) sera were incubated and filtered in the same manner before testing in parallel with absorbed sera.

IV. CFU-C Culture

Bone marrow and blood CFU-C assays were carried out using the method of Pike and Robinson [7]. In particular triplicate cultures were prepared for each marrow, and feeder layers of 1.10^6 buffy coat leucocytes were used as a source of colony stimulating factor (CSF). Marrow samples were separated on a ficolltriosil density gradient according to Boyum [4] 1968, and the washed mononuclear cells were cultured at a concentration of 2.10^5 cells/plate. Cultures were incubated in a humidified atmosphere containing 5% CO_2 for 10 days, and colonies (aggregates of more than 50 cells) and clusters (aggregates of 5 to 49 cells) were counted using an inverted objective microscope.

Results

A. Co-Culture

Figure 1 shows the results of co-culture of blood lymphocytes from normal and aplastic subjects with normal bone marrow mononuclear cells.

Normal subjects showed a wide range of observed/expected values between 0.65 and 1.32 for colonies, and between 0.76 and 1.45 for total aggregates. Lymphocytes from three subjects were tested after incubation in liquid culture for

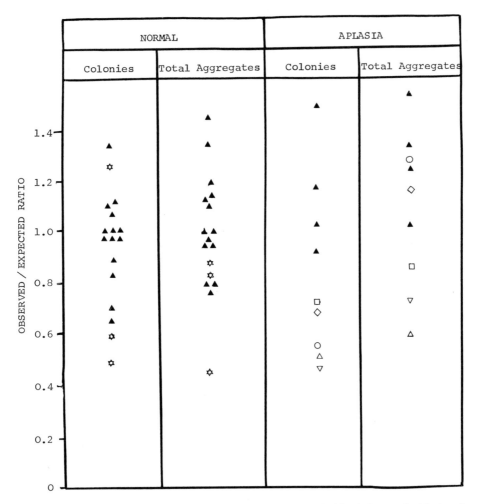

Fig. 1. Co-culture of normal bone marrow with lymphocytes from the blood of normal subjects and patients with aplastic anaemia, represented as observed/expected CFU-C growth. ☆ = Co-culture with normal lymphocytes incubated in mixed lymphocyte culture for six days. □ △ ▽ ○ □ 5 patients with aplasia who shared observed/expected values for colony growth of less than 75%, showing lack of correlation with total aggregate observed/expected values

six days in a proportion of 2:1 with random irradiated lymphocytes from other donors. Cells from this mixed lymphocyte culture tested against normal bone marrow mononuclear cells in the same way, showed an increase ability to inhibit colony and total aggregate growth. Lymphocytes from nine patients with aplastic anaemia were tested. Three showed inhibitions of less than 0.65 of control values but a wide range of results was found both for colonies and total aggregates.

While inhibition was observed with aplastic anaemia lymphocytes it did not exceed that seen with sensitised normal lymphocytes.

B. Serum CFU-C Inhibition

Incubation of normal bone marrow with sera from untransfused subjects produced a range of values between 76% and 128% of control (mean $103 \pm 29:2$ SD). This range was obtained by pooling results from two series of experiments where the 100% control value was taken in the first series as the unincubated marrow, and in the second series as the growth after incubation with autologous serum. The 100% value used thereafter was that obtained with autologous serum. By choosing sera that contained ABO agglutinins directed against the ABO group of the target marrow the effect of blood group antibodies on CFU-C was investigated. Figure 2 shows that no difference occurred in colony growth when the serum was mis-matched with the marrow.

Serum from patients who had been transfused or from subjects sensitised by multiple pregnancy showed marked inhibition of CFU-C in nine out of ten tested.

Similarly 14 patients with aplastic anaemia were found to have serum inhibitory to CFU-C, some patients showing almost complete inhibition of colony growth.

C. Properties of Inhibitory Sera (Figure 2)

I. Reproducibility

Eight aplasia sera tested on three different bone marrows showed reproducible observed/expected values. Only one patient showed a borderline result with a range of 0.66 ± 0.12.

II. HL-A Specificity

Six sera were tested on marrow from their HL-A identical MLC compatible donor. In every case inhibition was still found despite full HL-A and D locus compatibility.

III. Effect on Autologous Bone Marrow

Nine sera were tested on the patients own bone marrow. No CFU-C inhibition was seen in any case.

IV. Effect of Complement (Table 1)

Four sera were tested with and without the presence of complement by heating the test sera + rabbit sera to 56° C for 15 minutes. Less inhibition occurred in the absence of complement.

V. Absorption Studies

Figure 3 shows the effect of platelet and myeloblast absorption on the presence of polyspecific HL-A antibodies and colony inhibition in the sera tested. Three patters of absorption were found:
1. Myeloblasts and platelets both removed CFU-C inhibition.
2. Myeloblasts alone removed CFU-C inhibition.
3. Inhibition was not removed by either myeloblasts or platelets.

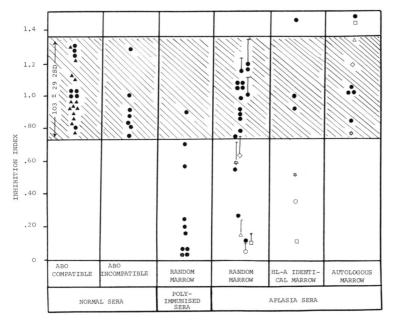

Fig. 2. Serum inhibition of CFU-C in normal, polytransfused, and aplastic anaemia subjects. ▲ = Unincubated bone marrow taken as 100% control growth. ● = Marrow incubated in autologous serum or a single batch of AB serum taken as 100% control growth. The normal range of serum inhibition (103 ± 29,2SD) was calculated from the pooled data of and ☆ □ △ ○ = Serum from aplasia patients who showed inhibition against random marrow, who were also tested against HL-A compatible or autologous bone marrow

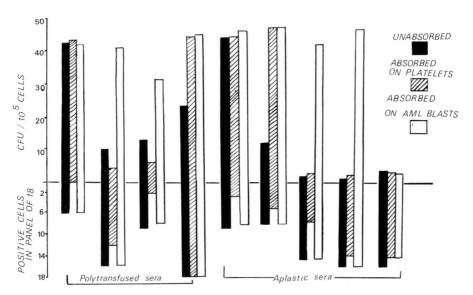

Fig. 3. CFU-C inhibition and anti HL-A activity of sera from four polytransfused individuals and five aplastic anaemia patients before and after absorption on platelets or myeloblasts (see text)

	Inhibition index	
	With complement	Without complement
Inhibitory sera 1	0.24	0.36
2	0.02	0.18
Non-inhibitory sera 3	0.83	0.85
4	1.1	1.05

Table 1. Effect of complement on inhibitory activity of two aplastic anaemia sera

Control non-inhibitory sera did not show altered characteristics with CFU-C after absorption despite the removal of HL-A antibodies. Similarly inhibition of CFU-C was removed by myeloblasts without alteration of the HL-A antibodies present in the serum.

VI. Relationship of Inhibition to Haematological Features (Table 2)

The haematological characteristics of patients with sera inhibitory to CFU-C were compared with the group without inhibition. No significant difference occurred between two groups. In particular no difference in lymphocyte count or bone marrow lymphocyte count was observed.

VII. Relationship of Serum Inhibitors to Result of ALG Treatment

Six out of seven patients tested who showed a clinical response to ALG were found to have serum inhibitors to CFU-C, while only two non-responding patients were found to have a serum inhibitor (Table 3). In both these cases the degree of inhibition was moderate.

Table 2. Relationship of CFU-C inhibition to haematological findings in 34 patients with aplastic anaemia. Inhibiting sera were defined as those showing an index of less than 0.7 against random bone marrow samples

Patients	Lymphocyte count $\times 10^9$/l	Lymphocyte percent in BM	Neutrophil count $\times 10^9$/l	Colonies per ml marrow	Total aggregates /10^5 cells plated	Colony: cluster ratio
(10) Inhibitors mean ± SD	1.84 ± 0.75	64 ± 31	0.7 ± 0.5	66 ± 58	15.7 ± 15	0.42 ± 0.31
(24) NO Inhibitors mean ± SD	1.75 ± 0.79	53 ± 29	0.4 ± 0.3	65 ± 60	22 ± 24	0.51 ± 0.32
Difference[a]	NS	NS	NS	NS	NS	NS

[a] Unpaired t-tests
NS = not significant (p = >0.1)

Case	Response (rise in neutrophils to over 1×10^9/l within 30 days)	Inhibition index		
		Pre ALG	10 days after ALG	30–60 days after ALG
1	+	0.01	0.01	0.01
2	+	0.16	0.75	–
3	+	0.23	0.58	1.0
4	+	0.25	–	–
5	+	0.53	0.85	0.98
6	+	0.6	–	–
7	+	0.75	–	–
8	–	0.58	–	–
9	–	0.70	0.55	–
10	–	0.76	0.9	0.76
11	–	0.8	–	–
12	–	0.84	–	–
13	–	0.86	0.84	0.78
14	–	0.90	1.18	0.98
15	–	0.94	–	–
16	–	0.97	–	–

Table 3. Relationship of CFU-C inhibitors to neutrophil response to ALG treatment, and effect of treatment on CFU-C inhibitors, in 16 patients in whom serum was tested for inhibitors before treatment

Significant inhibition (index<0.74) italic

VIII. Relationship of Serum Inhibition to Outcome of Bone Marrow Graft

Table 4 shows the relationship of serum CFU-C inhibitors to the outcome after marrow transplantation. Of five patients who showed failure of any haemopoietic reconstitution all had serum inhibitors. Eight patients rejected their graft between 24 to 90 days after transplantation, three of these had inhibitors. Only two out of nine patients who had sustained marrow engraftment had serum inhibitors.

There was some association of the presence of inhibitors with the degree of HL-A sensitisation of the patient as measured by the lymphocytotoxicity against a panel of 24 cells. However, the correlation was not absolute. In particular three patients with CFU-C inhibitors did not have any detectable HL-A antibodies and six patients with HL-A antibodies did not have detectable CFU-C inhibition.

Discussion

These studies enlarge previous observations of lymphocytes in aplastic anaemia inhibitory to bone marrow stem cells [1, 5]. However, we were unable to demonstrate clear cut inhibition by co-culturing normal bone marrow mononuclear cell fractions with aplasia lymphocytes because of the wide normal variations that occurred. Colonies appeared to be more susceptible to inhibition than clusters as measured in the counts of total aggregates, but no difference was

Table 4. Relationship of colony inhibition in serum taken ten days before transplantation with the presence of HL-A antibodies and marrow graft rejection

Patient sex (age)	Conditioning	Marrow cells × 10⁸/kg	Outcome	CFU-C inhibition	HL-A antibodies
F (12)	Cy	6.0	No take	+	−
F (7)	PAPA Cy	6.4		+	+ + +
M (16)	Cy	1.8		+	+ + +
M (18)	Cy	1.2		+	+ + +
F (30)	Cy	4.0		+	−
M (8)	PAPA Cy	5.0	Rejection day 24–90	−	+
F (24)	Cy	2.5		−	+
M (29)	Cy	1.6		−	−
F (4)	Cy	5.6		−	−
M (23)ᵃ	Cy	3.3		+	+ + +
F (26)ᵇ	Cy	3.0		+	+
F (15)	PAPA Cy	5.0		−	−
M (4)	Cy	8.4		+	+ +
M (23)ᵃ	Cy TBI	3.5	Take	−	+ +
M (24)	Cy	3.0		−	−
F (26)ᵇ	Cy	2.0		−	+
M (27)	Cy TBI	3.3		+	+ + +
M (10)	PAPA Cy	4.4		−	+ +
M (27)	Cy	1.9		+	−
M (20)	PAPA Cy	2.4		−	+
M (18)	Cy	2.4		−	−
M (20)	PAPA Cy	2.2		−	−

ᵃ ᵇ } Same patients regrafted

CFU-C inhibition equals less than 74% of control colony growth, HL-A antibodies measured against a panel of 24 normal lymphocyte donors, + + + = more than 10 panel cells positive, + + = 5–10 panel cells positive, + = 1–5 panel cells positive, − = non-positive

seen between aplasia and normal lymphocytes in this respect. It is important to note that normal lymphocytes from mixed lymphocyte cultures inhibited colony growth to the same degree as aplasia lymphocytes. The wide variation seen in the normal range, and the ability of in vitro sensitised normal lymphocytes to inhibit CFU-C makes it impossible to demonstrate a specific immunological abnormality in aplastic anaemia by this coculture technique.

We also demonstrated serum inhibitory to CFU-C in aplasia patients and in non-aplastic HL-A sensitised individuals. The inhibitor found in these sera may have been an antibody since we demonstrated complement-dependence and removal of the inhibitor by absorption on myeloblasts. We showed that the inhibitors did not have HL-A A,B or D locus specificity since CFU inhibition was demonstrated on the marrow of HL-A and mixed lymphocyte culture identical sibling donors, and since absorption on myeloblasts failed to remove HL-A antibodies while removing CFU-C inhibition. In addition some sera showed CFU-C inhibition without the presence of HL-A antibodies and vice versa. The ability of myeloblasts to remove the inhibitor suggests that it may be an antibody directed against myeloid precursor cells. The data suggests that blood transfusion or sensitisation by pregnancy may induce antibodies to CFU-C myeloid antigens simultaneously with the development of anti-HL-A antibodies. This process appears to occur as frequently in non-aplasia subjects as in aplastic anaemia patients. Because of this the relationship of serum inhibitors to the pathophysiology of aplastic anaemia is not established. The failure of inhibitors to prevent the growth of autologous (but aplastic) bone marrow is further evidence against an auto-immune antibody being involved. It could be argued however, that the aplastic bone marrow is already maximally inhibited by the patients antibodies in vivo making it impossible to demonstrate further inhibition in vitro.

We are left with the possibility that in addition to "physiological" serum CFU inhibitors there exists in some patients with aplastic anaemia inhibitors more specifically related to their disease which need not necessarily be antibodies. There is some evidence of this:

1. In two inhibitory aplasia sera myeloblast absorption failed to remove CFU-C inhibition.

2. Serum inhibitors were associated with a clinical response to ALG treatment and tended to disappear after treatment, suggesting a relationship between CFU-C inhibition and immunological abnormalities in aplastic anaemia.

3. Gordon has reported that while aplasia sera inhibits CFU-C in the absence of compliment inhibitory sera from non-aplasia patients do not (see Chapter 5.4 in this volume).

4. In a case of post-hepatitis aplastic anaemia recently described serum CFU inhibitors were associated with marrow graft rejection from an identical twin. A successful regraft with cyclophosphamide preparation was associated with the disappearance of the inhibitor [8].

In conclusion we have been unable to confirm any specific immunological effect of lymphocyte or sera from aplastic anaemia on normal bone marrow CFU-C. While evidence from a variety of in vivo and in vitro data points strongly towards an immunological cause for bone marrow failure in at least a subgroup of patients with aplastic anaemia, our studies do not elucidate an immunological process. Fractionation and further absorption of aplastic sera is required to try and identify possible inhibitors specific to the disease, together with studies of patients with aplastic anaemia who have never been transfused or become pregnant. Only the demonstration of inhibitors in these patients will provide convincing proof of an abnormal immunological process in aplastic anaemia.

Acknowledgements

This work was carried out during tenure of an Anglo-French Leukaemia Research Fund Fellowship by A. J. Barrett.

References

1. Ascensão, J., Pahwa, R., Kagan, W., Hansen, J., Moore, M. A. S. and Good, R.: Aplastic anaemia: Evidence for an immunological mechanism. Lancet *I*, 669 (1976)
2. Barrett, A. J., Faille, A., Balitrand, N., Ketels, F. and Gluckman, E.: Granulocyte colony culture in aplastic anaemia. J. Clin. Path. *32*, 660–665 (1979)
3. Benestadt, H. B.: Aplastic anaemia: Consideration on the pathogenesis. Acta Medica Scand. *196*, 255–262 (1974)
4. Boyum, A.: Separation of leucocytes from blood and bone marrow. Scand. J. Clin. Lab. Invest. *21*, Suppl 97, 77–89 (1968)
5. Kagan, W. A., Ascensão, J. A., Pahwa, N., Hansen, J. A., Goldstein, G., Valera, E. B., Incefy, G. S., Moore, M. A. S. and Good, R. A.: Aplastic anaemia: Presence in human bone marrow of cells that suppress myelopoiesis. Proceeding of the National Academy of Sciences *73*, 2890–2894 (1976)
6. Kern, P., Heimpel, H., Heit, W. and Kubanek, B.: Granulocytic progenitor cells in aplastic anaemia. Brit. J. Haem. (1977)
7. Pike, B. L. and Robinson, W. A.: Human bone marrow colony growth in agar gel. J. of Cellular Physiology 76–77 (1970)
8. Royal Marsden Hospital Bone Marrow Transplantation Team: Failure of syngeneic bone marrow graft without pre-conditioning in post-hepatitis marrow aplasia. Lancet *II*, 742 (1977)
9. Storb, R., Thomas, E. D., Weiden, P. L., Dean Buckner, C., Clift, R. A., Fefer, A., Fernando, L. P., Giblett, E. R., Goodell, B. W., Johnson, F. L., Lerner, K. G., Neiman, P. E. and Sander, J. E.: Aplastic anaemia treated by allogeneic bone marrow transplantation: A report on 49 new cases from Seattle. Blood *48*, 817–841 (1976)

Discussion

Porzsolt: You reported about HLA-antibodies in your inhibitory sera. It would be interesting if there was some correlation between those antibodies and the HLA-type of the blood transfusions your patients got.

Barrett: I think we showed that HLA-antibodies have nothing to do with the inhibition itself. When we were detecting antibodies they were outside the HLA-system.

5.4 Serum Inhibitors in Aplastic Anaemia

M. Y. Gordon

Introduction

Aplastic anaemia is the result of reduced numbers or malfunction of haemopoietic stem cells. In many cases the condition can be related to an exogenous toxin, but in some 50% of cases, no such insult can be identified. To account for these cases, other pathogenetic mechanisms have been invoked.

Several lines of clinical and experimental evidence point to an environmental defect which may be mediated immunologically. Such a mechanism could account for the graft failure or rejection noted in approximately 25% of patients [13] particularly since this may occur even when the donor and recipient are identical twins [14]. Similarly, autologous reconstitution following immunosuppression and failed engraftment [8, 15, 11, 13] could result from elimination of suppressor cells, although the provision of a critical cell type by the grafted marrow must be considered.

Lymphocytes from approximately one third of aplastic patients block early myeloid differentiation [9, 5] and aplastic marrow shows improved colony-forming ability in vitro following treatment with anti lymphocyte globulin (ALG) [1, 6].

Serum from aplastic patients also inhibits colony-formation by normal bone marrow in vitro [6, 2, 3]. However, Barrett et al. [2, 3] found that sera from transfused and multiparous individuals also had this effect while we [6] found inhibition by aplastic sera alone. Since the two test systems differed in detail, we have made a comparison of the methods used in order to confirm these results.

To determine whether the inhibiting effect of aplastic sera could be reversed by a stimulus to granulopoietic proliferation, experiments were set up using cells preincubated in the presence of Colony Stimulating Activity (CSA).

The relevance of the in vitro findings to bone marrow function in vivo has been assessed by investigating sera before and after immunosuppression and successful bone marrow transplantation.

Materials and Methods

Sera

Serum samples were collected from aplastic patients, transfused non-aplastic individuals, those with isoantibodies induced by multiple pregnancies and from normal volunteers. Sequential samples were also taken from grafted aplastic patients at graft rejection and during haemopoietic revocery following a further transplant. All sera were stored frozen until tested and serial samples from a particular individual were all tested in the same experiment.

Bone Marrow Culture

The colony assay system used is essentially that described by Robinson and Pike [10] but with the addition of lysed rat erythrocytes to the upper agar layer [6].

Serum Testing

Normal bone marrow suspensions of $2-4 \times 10^6$ cells/ml were incubated in culture medium, containing 20% of the test serum, for 3 hours at 37° C. The suspensions were then diluted 10 – fold and transferred to petri dishes for measurement of the colony-forming cells. In experiments assessing the effect of complement on these cultures, 20% (marrow) autologous serum was added to the exposure phase. Controls for this experiment used 20% heat inactivated (56° C for 30 mins) autologous serum.

Degrees of inhibition by the test sera were estimated by comparison of the colony yield with that from marrow which had been preincubated without the addition of test serum.

Interaction Between Sera and Colony Stimulating Activity (CSA)

Leukocyte feeder layers containing different numbers of cells (10^5-10^6) were set up and each plate was used to condition 1 ml medium. Five days later, 5×10^5 normal marrow cells were added to each plate and incubated for 18 hours before the addition of 20% test serum. After a further 3 hours incubation, the cells were harvested and plated in the usual way.

Results

The effects of preincubation with aplastic or control sera on colony-forming cells are shown in Figure 1. In the presence of complement, inhibition was found in transfused and multiparous sera as well as in the aplastic sera. Only the normal sera failed to inhibit whether complement was present or not; only the aplastic sera inhibited irrespective of the addition of complement.

Five of the aplastic sera were heat inactivated and tested with unheated sera in the same experiment. The results (Table 1) show that heating reduced the inhibitory effect by up to 50%. This experiment was carried out without the addition of (marrow) autologous complement.

Preincubation of the cells with low levels of CSA (conditioned medium) reduced the susceptibility of the cells to inhibiting serum (Fig. 2). Higher CSA levels abolished the effect completely but the effect of normal serum was unchanged.

Table 1. Effect of heat inactivation on colony inhibition by aplastic sera

	% Inhibition of colony formation	
	Untreated serum	Heat inactivated serum
1	59	38
2	61	39
3	64	35
4	73	65
5	72	66

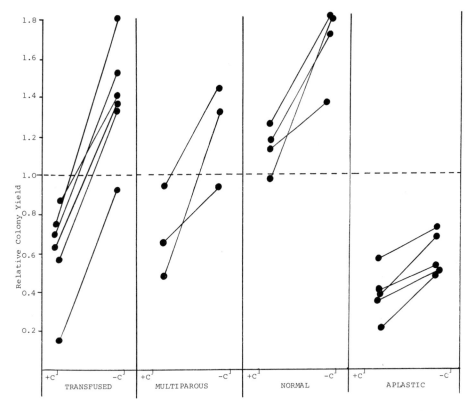

Fig. 1. Effects of preincubation with sera from aplastic, transfused, multiparous or normal individuals on the colony forming ability of normal human bone marrow. $+C^1$ – complement present. $-C^1$ – complement absent

Figure 3 shows measurements of serum inhibition and colonyforming ability in a patient grafted on two previous occasions for post-hepatitis aplasia. On each occasion, the donor was her identical twin and the first graft was given without any conditioning treatment. The second graft was given after a course of azathioprine which abolished the inhibitor in her serum. Both the first and second attempts at transplantation produced only transient improvements in haemopoiesis and at the second rejection there were high levels of inhibitory activity in her serum. Azathioprine and cyclophosphamide were used in preparation for the third graft. The level of inhibition fell and remained low as the graft, measured here by its colony-forming ability, became established.

In the second patient (Fig. 4) a recipient of allogencic marrow, the second graft was successful. Antilymphocyte globulin (ALG) was included in the conditioning regime on the second occasion and again the colony-forming cells rose as the inhibitor declined. The third case (Fig. 5) was prepared for her first graft with ATG and cyclophosphamide and the graft was successful. The results from this patient were similar to the other two cases described.

250

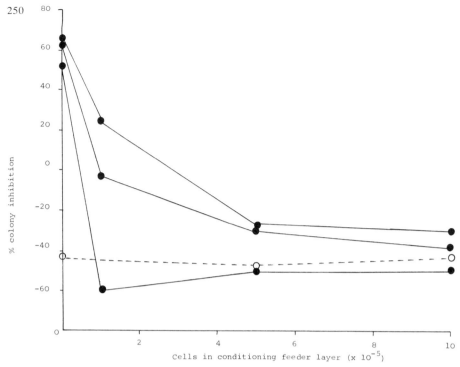

Fig. 2. Effects of sera from ● aplastic and ○ normal individuals on colony forming cells previously exposed to colony stimulating activity

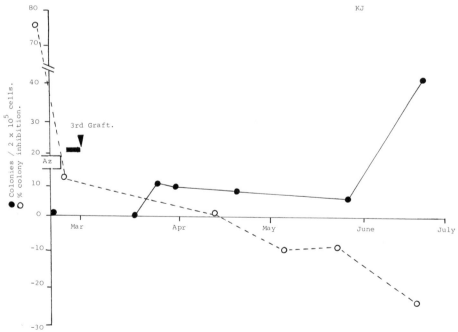

Fig. 3. ○ Serum inhibitors and ● bone marrow colony forming ability is a recipient of her 3rd graft of syngeneic marrow. A$_z$ – azathioprine; ▬▬ cyclophosphamide; ▼ bone marrow infusion

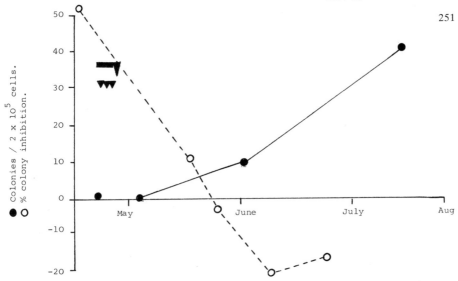

Fig. 4. ○Serum inhibitors and ● bone marrow colony forming ability is a recipient of her 2nd graft of allogeneic marrow. ▼Antilymphocyte globulin; ▬▬cyclophosphamide; ▼ bone marrow infusion

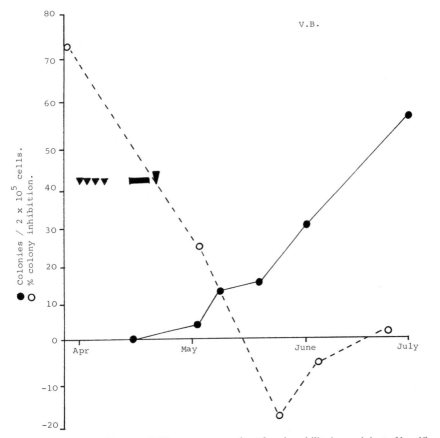

Fig. 5. ○ Serum inhibitors and ● bone marrow colony forming ability is a recipient of her 1st graft of allogeneic marrow. ▼Antilymphocyte globulin; ▬▬ cyclophosphamide; ▼ bone marrow infusion

Discussion

The results confirm findings reported in two independent studies [2, 3, 6] particularly since tests on sera obtained from both Paris and London are included. The data show that complement dependent inhibition can be induced by transfusion or pregnancy since no inhibition was found in sera of normal individuals. Although it may be argued that the inhibiting effect of aplastic sera may, at least in part, result from a history of transfusion, this study establishes the presence of a complement independent component in aplastic sera which was not found in any other sera tested. Furthermore, inhibition has been demonstrated by sera from aplastic patients who had not been transfused [6]. Complement independent inhibition by aplastic sera is also supported by the results using heat inactivated sera when the inhibition was not completely abolished by destruction of complement in the test sample.

The inhibiting effects of aplastic sera can be explained by competition with Colony Stimulating Activity since preincubation with CSA abolished inhibition (Fig. 2). This experiment also revealed the presence of stimulatory activity in the aplastic sera which attained the level of stimulation in the control normal serum.

The measurements on sera from aplastic patients who had previously rejected their grafts suggests that the inhibitor played a role in their failure. The results also suggest that a prolonged reduction in inhibitor level is necessary for sustained engraftment. Graft take was evident in both patients on the previous occasions. In one of these, inhibitor levels were shown to be low at the time of engraftment [14] and in both inhibitor levels were high at the time of rejection.

It is tempting to believe that the serum inhibitors reflect a pathological mechanism which can account for both the development of aplasia and graft rejection. Such a mechanism is not incompatible with the immunological suppression of stem cells indicated by the studies of Kagan et al. [9], Ascensao et al. [1] and Hoffman et al. [7]. Kagan et al. have shown that aplastic marrow contains lymphocytes which can suppress myeloid differentiation and removal of these cells on treatment of the marrow with ALG [1, 6] releases the cells from this block.

An immunopathogenesis of aplastic anaemia is but one of the mechanisms which have been suggested. The strongest argument against any environmental defect being the cause of aplasia is the undoubted success of many cases of bone marrow tranplantation [12]. It is not inconceivable however, that immunosuppressive therapy destroys the cells responsible for producing the inhibitor thus allowing repopulation of the marrow, in some cases even by autologous cells [8, 15, 11, 13].

Boggs and Boggs [4] have suggested that the stem cell compartment is depleted by inappropriate differentiation at the expense of stem cell replication. It is difficult to reconcile this proposition with cases of autologous reconstitution [8, 15, 11, 13] which indicate that the stem cell compartment is qualitatively as well as quantitatively defective. Barret et al. [2, 3] have reported in this volume that reduction of the inhibitor by ALG allows recovery which again indicates a block in myeloid differentiation.

The experiments investigating the interaction between CSA and inhibitory aplastic serum shows that each can counteract the effect of the other. This result indicates that CSA and the inhibitor have opposite effects. Since CSA is a stimulus to differentiation, it is not unreasonable to assume that the aplastic inhibitor acts as a block to differentiation possibly by competing with CSA. The unmasking of a stimulating activity in aplastic serum when it was tested against cells pretreated with CSA supports the concept that aplastic anaemia involves an abnormal balance of stimulatory and inhibitory factors, which may result from the presence of suppressor cells in the marrow.

Acknowledgements

The work was supported by a grant from the Medical Research Council. The author wishes to thank the medical and nursing staff of Bud Flanagan Ward, The Royal Marsden Hospital; Drs. N. M. Blackett and A. J. Barrett for their continued encouragement and Miss. V.B. Shepherd for her excellent technical assistance.

V.B.S. is supported by the Bud Flanagan Fund of The Royal Marsden Hospital.

References

1. Ascensao, J., Pahwa, R., Kagan, W., Hansen, J., Moore, M., Good, R.: Aplastic anaemia: Evidence for an immunological mechanism. Lancet *1*, 669 (1976)
2. Barrett, A. J., Faille, A., Saal, F., Balitiane, N., Gluckman, E.: Marrow graft rejection and CFU-C inhibition by serum in aplastic anaemia. J. Clin. Path. In press
3. Barrett, A. J.: Serum inhibitors as an indicator for the pathomechanisms of aplastic anaemia. This volume (1978)
4. Boggs, D. R., Boggs, S. S.: The pathogenesis of aplastic anaemia: A defective pluripotent hematopoietic stem cell with inappropriate balance of differentiation and self replication. Blood *48*, 71 (1976)
5. Good, R. A.: Aplastic anaemia – suppressor lymphocytes and haemopoiesis. New. Eng. J. Med. *296*, 41 (1977)
6. Gordon, M. Y.: Circulating inhibitors of granulopoiesis in patients with aplastic anaemia. Brit. J. Haematol. *39*, 491 (1978)
7. Hoffman, R., Zanjani, E. D., Lutton, J. D., Zalusky, R., Wasserman, L. R.: Suppression of erythroid – colony formation by lymphocytes from patients with aplastic anaemia. New. Eng. J. Med. *296*, 10 (1977)
8. Jeannet, M., Speck, B., Rubinstein, A., Pelet, B., Wyss, M., Kummer, H.: Autologous marrow reconstitutions in severe aplastic anaemia after ALG pretreatment and H-LA semiincompatible bone marrow cell transfusion. Acta Haemat. *55*, 129 (1976)
9. Kagan, W. A., Ascensao, J. A., Pahwa, R. N., Hansen, J. A., Goldstein, G., Valeza, E. B., Incefy, G. S., Moore, M. A. S., Good, R. A.: Aplastic anaemia: Presence in human bone marrow of cells that suppress myelopoiesis. Proc. Natl. Acad. Sci. U.S.A. *73*, 2890
10. Robinson, W. A., Pike, B. L.: Colony Growth of Human Bone Marrow Cells in vitro. In: Haemopoietic Cellular Proliferation. Stohlman, F. (ed.). New York: Grune and Stratton, 1970, p. 249
11. Speck, B., Cornu, P., Jeannet, M., Nissen, C., Burri, H. P., Groff, P., Nagel, G. A., Buckner, C. D.: Autologous marrow recovery following allogeneic marrow transplantation in a patient with severe aplastic anaemia. Exp. Haemat. *4*, 131 (1976)
12. Storb, R., Thomas, E. D., Weiden, P. L., Buckner, C. D., Clift, R. A., Fefer, A., Fernando, L. P., Giblett, E. R., Goodell, B. W., Johnson, F. L., Lerner, K. G., Nieman, P. E., Sanders, J. E.: Aplastic anaemia treated by allogeneic bone marrow transplantation: A report on 49 new cases from Seattle. Blood *48*, 817 (1976)

13. Territo, M. C., for the U.C.L.A. Bone Marrow Transplantation Team: Autologous bone marrow repopulation following high dose cyclophosphamide and allogeneic marrow transplantation in aplastic anaemia. Brit. J. Haemat. *36*, 305 (1977)

14. The Royal Marsden Hospital Bone Marrow Transplantation Team: Failure of Syngeneic bone marrow graft without preconditioning in past-hepatitis marrow aplasia. Lancet *2*, 742 (1977)

15. Thomas, E. D., Storb, R., Giblett, E. R., Longpre, B., Weiden, P. L., Fefer, A., Wetherspoon, R., Clift, R. A., Buckner, C. D.: Recovery from aplastic anaemia following attempted marrow transplantation. Exp. Haemat. *4*, 97 (1976)

5.5 Aplastic Anemia After Bone Marrow Transplantation for Severe Combined Immunodeficiency

D. Niethammer

I was asked to present the case of a boy suffering from severe combined immunodeficiency. After bone marrow transplantation he developed an aplastic anemia which could be a true example of an aplasia caused by an immunological process.

Case Report

The history of the child, the results of typing for histocompatibility, and the immunological parameters before and six months after the bone marrow transplantation have already been reported elsewhere [4] as well as the details of the aplastic anemia [5].

In short: The boy was born by caesarian section and isolated immediately afterwards in a plastic isolation system. The test results proved that he suffered from severe combined immunodeficiency. Therefore, at the age of nine months, he was transplanted with the marrow of the mother who was identical in the HLA-B and HLA-D-antigens and only mismatched in one HLA-A-antigen. No sibling was available and the mother was selected as a donor because of the nonreactivity in the mixed lymphocyte culture [4].

As a consequence of the transplantation severe graft-versus-host reaction occurred which the child survived probably due to the fact that he was germfree at that point. A split chimerism developed since only the engraftment of the maternal lymphocytes could be proved. Six months after the transplantation full immunocompetence had developed. At about the same time an autoimmune

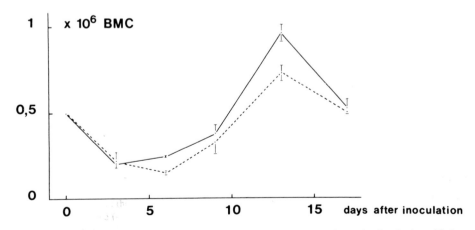

Fig. 1. Proliferation of the bone marrow of the mother in diffusion chambers after incubation with the ATG in the presence (—) and absence (– –) of the recipient's serum. 0.5×10^6 cells were inoculated into each chamber and triplets were used for every point (Results by Dr. Körbling, Department of Clinical Physiology, University of Ulm)

hemolytic anemia as well as a neutropenia were observed. During the following time the cellularity of the bone marrow declined and nine months after the transplantation the child was completely aplastic.

A second transplantation was performed and we used a proceeding which had been successfully used by another group in a comparable case [2]. Maternal marrow was incubated with the antithymocyte globulin described by Dr. Netzel this morning [3, 7]. As shown in Figure 1 neither the antilymphocyte globulin nor the patient's serum inhibited the growth of the bone marrow in the diffusion chamber. The marrow was given intraperitoneally. No engraftment was achieved. A third transplantation was performed, where the untreated marrow was administered intravenously in a high dose (5.3×10^8 nucleated cells/kg body weight) but it failed again. In both cases no preconditioning regimen was used since the lymphocytes were already donor-derived.

Discussion

This report demonstrates the development of aplastic anemia nine months after the first bone marrow transplantation for severe combined immunodeficiency and seven months after appearance of the split chimerism. The most likely cause of the boy's aplasia is an immunological mechanism. The existance of a humoral factor cannot be excluded but it seems not very likely judging from the in-vitro experiments. A better explanation could be that suppressor cells had developed which were directed originally against the child's hematopoesis. These cells then might have also prevented the engraftment of the maternal hematopoesis after the following transplantations. Because of the small amount of lymphocytes present in the bone marrow as well as in the circulating blood no experiments could be performed to prove this hypothesis. Another possible cause of aplastic anemia could be some kind of genetic resistance as it has been demonstrated in rodents [1]. Such a mechanisms would have prevented also the engraftment of the maternal hematopoesis.

Suppressor cells or genetic resistance – both mechanisms might have been overcome by pretreatment with cyclophosphamide. This was finally done in a case of severe combined immunodeficiency where aplastic anemia developed after an unrelated donor had been used [6]. Again a split chimerism had been induced by previous transplantations. Cyclophosphamide and subsequent administration of marrow from the same donor led to a final engraftment of the hematopoesis. On the other hand cyclophosphamide was not used in the other case of split chimerism and aplastic anemia mentioned before [2].

In summary it can be stated that the aplasia encountered in this child as well as in the other two mentioned cases seem to be true examples of an aplastic anemia caused by an immunological process. But the split chimerism makes it impossible to deduct that similar mechanisms are encountered in other patients with aplastic anemia.

Supported by the Deutsche Forschungsgemeinschaft (SFB 112)

References

1. Cudkowicz, G., Bennet, M.: Peculiar immunobiology of bone marrow allografts. I. Graft rejection by irradiated responder mice. J. Exp. Med. *134*, 83 (1971)
2. Meuwissen, H. J., Gatti, R. A., Teresaki, P. I., Hong, R., Good, R. A.: Treatment of lymphopenic hypogammaglobulinemia and bone marrow aplasia by transplantation of allogeneic marrow. Crucial role for histocompatibility matching. N. Engl. J. Med. *281*, 691 (1969)
3. Netzel, B., Rodt, H., Hoffmann-Fezer, G., Thiel, E., Thierfelder, S.: The effect of crude and differently absorbed anti-human T-cell globulin on granulocytic and erythropoietic colony formation. Exp. Hemat. *6*, 410 (1978)
4. Niethammer, D., Goldmann, S. F., Haas, R. J., Dietrich, M., Flad, H.-D., Fliedner, T. M., Kleihauer, E.: Bone marrow transplantation for severe combined immunodeficiency with the HLA-A-incompatible but MLC-identical mother as a donor. Transpl. Proc. *8*, 623 (1976)
5. Niethammer, D., Bienzle, U., Rodt, H., Goldmann, S. F., Körbling, M., Flad, H. D., Netzel, B., Haas, R. J., Fliedner, T. M., Thierfelder, S., Kleihauer, E.: Aplastic anemia as the consequence of split chimerism after bone-marrow transplantation for severe combined immunodeficiency. Submitted for publication
6. O'Reilly, R. J., Dupont, B., Pahwa, S., Grimes, E., Smithwick, E. M., Pahwa, R., Schwartz, S., Hansen, J. A., Siegal, F. P., Sorell, M., Svejgaard, A., Jersild, C., Thomsen, M., Platz, P., L'Esperance, P., Good, R. A.: Reconstitution in severe combined immunodeficiency by transplantation of marrow from an unrelated donor. N. Engl. J. Med. *297*, 311 (1977)
7. Rodt, H., Netzel, B., Niethammer, D., Körbling, M., Kolb, H. J., Thiel, E., Haas, R. J., Fliedner, T. M., Thierfelder, S.: Specific absorbed anti-thymocyte globulin for incubation treatment in human bone marrow transplantation, Transpl. Proc. *9*, 187 (1977)

Discussion

Thierfelder: If your hemolytic anemia was antibody-derived, do you know the specifity of the antibody?

Niethammer: There were several different kinds of antibodies. First of all, there were antibodies against anti-D because the child was rhesus-positive and the mother negative, a situation which led to an intrinsic rhesus-incompatibility. In addition there were polyspecific antibodies which were also directed against rhesus-negative cells, and especially against the mother's cells.

Thierfelder: So this is an autoimmune phenomenon.

5.6 The Contribution of in Vitro Cultures to the Elucidation of the Pathogenesis of Aplastic Anaemia

H. L. Haak, H. M. Goselink, L. Sabbe, W. F. J. Veenhof, A. J. J. C. Bogers, J. L. M. Waayer

In this paper we shall consider aplastic anaemia (AA) as the syndrome, characterized by pancytopenia and a hypocellular bone marrow, without evidence of malignancy, metabolic disease or cytostatic treatment.

After diagnosis, the patients' course is variable; in half of the cases it is characterized by a slow and often incomplete recovery, in others it leads to progressive pancytopenia and death under conventional care. On clinical grounds alone it is already clear that this syndrome is probably heterogenous in nature.

Experimentally, a first indication in this direction was found in 2 patients, who were treated for AA in 1975 (Fig. 1). After cocultivation of normal donor and recipient marrow we found suggestive evidence of inhibition of the normal marrow in the first case before a bone marrow graft was given. Experiments repeated on the day of grafting (after 4 successive days of cyclophosphamide 50 mg/kg i.v.) showed a similar inhibition, albeit slightly less. The graft was rejected at day +60.

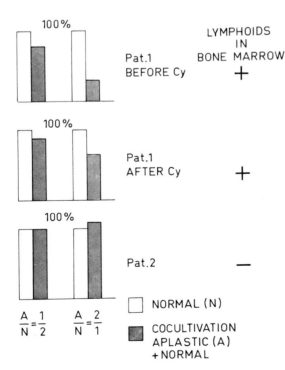

Fig. 1. Cocultivation of bone marrow from 2 patients with severe aplastic anaemia with normal histocompatible marrow. The experiment in patient 1 was repeated after four successive days of cyclophosphamide 50 mg/kg (Cy) on the day of BM-grafting. Both before and after this treatment a dose dependent inhibition of the normal marrow was observed. In patient 2 no such inhibition was found. The presence of a conspicuous amount of lymphocytes in the marrow biopsy is indicated

In the other patient no inhibition was found in several experiments. He finally died of persistent pancytopenia after two courses of ALG-treatment.

These results could be interpreted as evidence in favour of the existence of a CFU-inhibiting cell in the first case, while no such mechanism appeared to be present in the second.

At the same time, Jan te Velde and I found the first indication that the tendency for haematological recovery, reflected in the survival of patients, was closely associated with the amount and distribution of an inflammatory infiltrate present in the marrow biopsies [24].

In several conventionally treated patients, low levels of CFU-c were found over long periods of time correlating with a conspicuous amount of lymphoid cells in the infiltrate. This prompted us to remove ERFC (Spontaneous sheep red blood cell rosette forming lymphocytes) from the marrow suspensions. In several instances, we obtained a significant increase in colony counts, which could be inhibited by adding the ERFC back to the autologous suspension [6].

Using different techniques Ascencao et al. [1], Kagan et al. [9] reported essentially similar results. This led to the conclusion that in some cases of AA a cell population, probably belonging to the T-cells inhibits the outgrowth of CFU-c.

The clinical relevance of these findings in illustrated by the fact that several cases of AA have recovered after intensive immune suppressive treatment [23].

This concept has been challenged by Singer et al. [21] who found that CFU-c inhibition was related to transfusion history and histocompatibility. In this particular system we used autologous marrow and T-cell combinations, obviating possible influences of disparity in histocompatibility-antigens.

Studies by L. Sabbe et al. (1978, unpublished) showed that in vitro lymphocyte stimulation tests in many AA-patients were abnormal. These results could be explained by assuming a disbalance in lymphocyte subpopulations. Remarkably, this disbalance was swiftly corrected in those patients who respondend clinically to ALG-treatment.

These findings focussed our attention on the interdependence of the immune apparatus and haematopoietiesis. Goodman a.o. [5] suggested on the basis of a large body of evidence in rodents [17, 11, 4, 16] that a thymus derived "amplifier" cell might exist, that controls the proliferation of haematopoietic tissue, possibly by interfering in the function of the micro environmental "niches" where the stemcell lodge, proliferate and differentiate.

This cell appears to belong to mature lymphocyte T-subpopulation and is found also in thymus and lymphnodes. Its influence can be detected in situations when the stem cell-proliferation and -differentiation are impaired either by irradiation [11] or by a genetical defect, e.g. in W/W^v mice [25].

Human and rodent T-cells stimulate CFU-c and BFU-e, both without and after stimulation by mitogens [2, 19, 10, 13]. In humans a possible regulating role of an intact immune apparatus on haematopoiesis is indicated by the haematological defects found in certain thymomas.

The fact that the addition of peripheral blood cells increases the take-ability of human BM-grafts [22] cannot be explained by the addition of stem cells alone ($<10\%$ of the bone marrow stem cells). It may be due to the addition of some "regulating" mature lymphocytes. The increased take-ability of MHC-compatible BM after addition of syngeneic thoracic duct cells in dogs [3] points into the same direction.

This line of thought led us to experiments using normal marrow and lymphocyte subpopulations.

Lymphocytes were fractionated in γFCR enriched and -depleted fractions according to the technique of Van Oers, et al [14, 15], using a monolayer of lysed O-pos. red cells coated with anti-D serum. We used anti-D serum, that had been tested in EA-rosette formation with myeloid cells, monocytes and lymphocytes. Bone marrow was used either not fractionated or after nylon adherence in these experiments. CFU-c cultures were performed according to Iscove et al. [8], the only modification was that 10^4 cells in 0.1 ml were plated in each of 6 wells of a flat bottomed plastic Sterilin microtitre plate. Figure 2 shows, that inhibition was found in those cultures where non-adherent BM-cells were cocultivated with γ-FCR positive cells, in contrast to cocultures with γ-FCR negative cells.

This effect was more pronounced after incubating peripheral lymphocytes with concanavalin A for 2 days. It is to be noted that the γ-FCR (EA-pos.) positive fraction also inhibited MLR and PHA cultures of autologous and allogeneic lymphocytes in accordance with the findings of Shou et al. [20], Moretta et al. [12], and Sakane et al. [18].

Results in normal individuals showed that "purified" peripheral γ-FCR positive cells inhibited the CFU-c outgrowth of non-adherent and lymphocyte-depleted marrow to some extent, in contrast to the γ-FCR negative cells. Control smears showed.

Table 1 shows that in a small number of pancytopenic patients no evidence of inhibition by autologous lymphocytes is observed, except in one case of pre-leukaemia. However, the pattern emerging from the interactions of T-cell subpopulation with colony forming cells in the pancytopenic patients differs from the pattern observed in the patients in remission and in the controls These preliminary findings suggest some regulating activity of T-cell subpopulations on haematopoietic progenitor cells in the normal state, which might be disturbed in manifest AA.

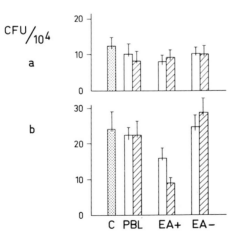

Fig. 2. Cocultivation of normal bone marrow with peripheral blood lymphocytes (PBL), and PBL separated into FCR enriched (EA+) and depleted subpopulation (Ea−). Hatched columns: the same using PBL etc. after 48 hours of incubation with Con-A and subsequent removal of Con-A with α-methyl-mannoside. **a** Cocultivation of unseparated marrow with PBL and subpopulation, **b** the same, using non-adherent bone marrow cells. C: control mean and s.d. of 6 cultures, containing 10^4 BM cells$\pm 10^4$ PBL each

Table 1. Pattern of in vitro interactions of T-cell subpopulations with colony forming cells in patients in various stages of aplastic anaemia and controls

Pat.	Control[a] BM susp.	1:1 cocultivation with autologous (PBL-)cells[b]			
		"T"	"T-non-γ"	"Tγ"	
1.	20±4	26±3	27±5	20±2	pancytopenia
2.	13±3	8±1	10±2	12±2	pancytopenia
3.	12±2	18±4	20±2	13±3	pancytopenia
4.	25±3	16±2	13±2	13±2	pancytopenia + preleukaemia
5.	29±3	33±4	27±4	33±3	p.n.h.[c]
6.	19±3	21±2	23±3	14±3	remission
7.	26±5	23±5	24±5	10±2	remission
Controls	85±8	81±6	77±7	59±5	
	66±4	58±3	58±4	40±4	
	36±6	48±3	47±6	17±3	

[a] BM cells after removal of mononuclear cells by Fe-carbonyl ingestion and combined E-Ea-rosette sedimentation. CFU: aggregates >20 cells per 10^4 cells; mean of 6 cultures [15]
[b] Peripheral blood cells obtained after FE-carbonyl ingestion and sheep E-rosette sedimentation. T-cells were obtained from the pellet. The separation of "T-non-γ" and "Tγ" cells was performed according to Van Oers et al. [14].
[c] Paroxysmal nocturnal hemoglobinuria

We have no evidence that similar influences occur in vivo. The hypotheses considering "auto-immunity" in AA have postulated the existence of aberrant immune competent cells, that block or kill maturating haematopoietic cells, perpetuating the marrow failure after its induction.

In the light of our results, it is tempting to speculate about the existence of a cellular regulation of the stem cell-micro-environmental interactions by at least two lymphocyte subpopulations (Fig. 3).

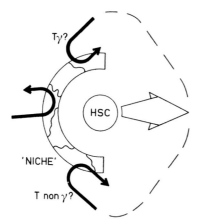

Fig. 3. Hypothetic "niche", consisting of sedentary stromal elements and recirculating cells. Among the latter, T-cell-subpopulations may have an important influence on the lodging of haematopoetic stem cells (HSC) (pluripotent and committed) and the regulation of their proliferation and differentiation

In some cases of aplastic anaemia induced by toxic exposure or viral infection both haematopoiesis and immune apparatus may be disturbed. It might well be that the recovery is determined not only by the number of residual stem cells but also by a disbalance between cells with "helper" and "suppressor" activity present in the immune apparatus and circulating through the bone marrow taking part in the micro-environmental mosaique, that forms the "stem cellniche".

It should be stressed that most of our studies are done in patients several weeks or months after diagnosis. This implies that the results are more relevant to the persistence of the bone marrow failure than to its cause. Further studies are required to establish the interaction of different subpopulation of immune competent cells and haematopoietic precursor cells.

References

1. Ascensao, J., Kagan, W., Moore, M., Pahwa, R., Hansen, J., Good, R.: Aplastic anaemia: Evidence for an immunological mechanism. The Lancet 27, 669–671 (1976)
2. Barr, R. D., Whang-Peng, J., Perry, S.: Regulation of human hemopoietic stem cell proliferation by syngeneic thymus-derived lymphocytes. Acta Haemat. 58, 74–78 (1977)
3. Deeg, H. J., Storb, R., Weiden, P. L., Torok-Storb, B., Graham, T., Thomas, E. D.: Resistance to marrow grafts in dogs mediated by antigens close to but not identical with DLA-A, B and D and overcome by infusion of thoracic duct lymphocytes. Exp. Hemat. 6, 23 (1978)
4. Frindel, E., Croizat, H.: The relationship between CFU kinetics and the thymus. Annals of New York Academy of Sciences 249, 468–476 (1975)
5. Goodman, J. W., Basford, N. L., Shinpock, S. G., Chambers, Z. E.: An amplifier cell in hemopoiesis. Exp. Hemat. 6, 151–160 (1978)
6. Haak, H. L., Goselink, H. M., Veenhof, W., Pellinkhof-Stadelmann, S., Kleiverda, J. K., te Velde, J.: Acquired aplastic anaemia in adults. IV. Histological and CFU studies in transplanted and non-transplanted patients. Scand. J. of Haemat. 19, 159–171 (1977)
7. Hoffman, R., Esmail, D. Z., Lutton, J. D., Zalusky, R., Wasserman, L. R.: Suppression of erythroid-colony formation by lymphocytes from patients with aplastic anaemia. New Engl. J. of Med. 6, 10–13 (1977)
8. Iscove, N. N., Senn, J. S., Till, J. E., McCulloch, E. A.: Colony formation by normal and leukemic human marrow cells in culture: Effect of conditioned medium from human leukocytes. Blood 37, 1–5 (1971)
9. Kagan, W. A., Ascensao, J. A., Pahwa, R. N., Hansen, J. A., Goldstein, G., Valera, E. B., Incefy, G. S., Moore, M. A. S., Good, R. A.: Aplastic anemia: Presence in human bone marrow of cells that suppress myeolopoiesis. Proceedings of National Academy Science 73, 2890–2894 (1976)
10. Lau, L., McCulloch, E. A., Till, J. E., Price, G. B.: The production of hemopoietic growth factors by PHA-stimulated leukocytes. Exp. Hemat. 6, 114–121 (1978)
11. Lord, B. I., Schofield, R.: The influence of thymus cells in hemopoiesis: Stimulation of hemopoietic stem cells in a syngeneic in vivo situation. Blood 42, 395–404 (1973)
12. Moretta, L., Webb, S. R., Grossi, C. E., Lydyard, P. M., Cooper, M. D.: Functional analysis of two human T-cells subpopulations: Help and suppression of B-cell responses by T-cells bearing receptors for IgM or IgG. J. of Exp. Med. 146, 184–200 (1977)
13. Nathan, D. G., Chess, L., Hillman, D. G., Clarke, B., Breard, J., Merler, E., Housman, D. E.; Human erythroid burst-forming unit: T-cell requirement for proliferation in vitro. J. of Exp. Med. 147, 324–339 (1978)
15. Van Oers, M. H. J., Zeijlemaker, W. P.: The mixed lymphocyte reaction (MLR) stimulatory EA-rosette forming lymphocytes in humans. Europ. J. of Immunol. 7, 143–150 (1977)
15. Van Oers, M. H. J., Zeijlemaker, W. P.: The mixad lymphocyte reaction (MLR) stimulatory capacity of human lymphocyte subpopulations. Cellular Immunology 31, 205–215 (1977)
16. Pritchard, L. L., Goodman, J. W.: Dose-dependence of the augmentation of hemopoiesis by thymocytes. Exp. Hemat. 6, 161–171 (1978)

17. Resnitzky, P., Zipori, D., Trainin, N.: Effect of neonatal thymectomy of hemopoietic tissue in mice. Blood *37*, 634–646 (1971)
18. Sakane, T., Green, I.: Human suppressor T-cells induced by concanavalin a: Suppressor T-cells belong to distinctive T-cell subclasses. J. of Immunol. *119*, 1169–1178 (1977)
19. Shah, R. G., Caporale, L. H., Moore, A. S.: Characterization of colony-stimulating activity produced by human monocytes and phytohemagglutinin-stimulated lymphocytes. Blood *50*, 811–821 (1977)
20. Shou, L., Schwartz, S. A., Good, R. A.: Suppressor cell activity after concanavalin a treatment of lymphocytes from normal donors. J. of Exp. Med. *143*, 1100–1110 (1976)
21. Singer, J. W., Brown, J. E., James, M. C., Doney, K., Warren, R. P., Storb, R., Thomas, E. D.: The effect of peripheral blood lymphocytes from patients with aplastic anemia on granulocytic colony growth from H. L. A. matched and mismatched marrows: the effect of transfusion sensitization. Blood *52*, 37–46 (1978)
22. Storb, R.: Personal communication (1978)
23. Speck, B., Gluckman, E., Haak, H. L., van Rood, J. J.: Treatment of aplastic anaemia by antilymphocyte globulin with or without marrow infusion. Clin. Hematol. *7*, 611–621 (1978)
24. te Velde, J., Haak, H. L.: Acquired aplastic anaemia in adults: Histological investigation of methacrylate embedded bone marrow biopsy specimens. Correlation with survival after conventional treatment in 15 adults patients. Brit. J. of Hematol. *35*, 61–69 (1977)
25. Wiktor-Jedrzejczak, W., Ahmed, A., Sharkis, S. J. and Sell, K. W.: Surface phenotype and other properties of a theta-sensitive cell which regulates hematopoiesis. Exp. Hematol. *6*, (Suppl. 3) 40 (1978)

5.7 Interpretations of Culture Data in Aplastic Anemia: Evidence for a Suppressor Cell Involvement

M. A. S. Moore

Aplastic anemia is a syndrome characterized by marrow hypoplasia and severe pancytopenia. While most cases of aplastic anemia have been attributed to congenital or acquired stem cell defects, it seems likely that several diseases of different etiology and pathogenesis exist. The availability of in vitro techniques for detection of myeloid and erythroid progenitor cells has provided an experimental approach for determining the mechanisms leading to aplastic anemia. Using these systems, evidence has been obtained for a reversible immunologically-mediated suppression of hematopoiesis in certain cases of aplastic anemia [1, 8]. In these studies, the severe defect in myeloid colony formation in vitro was corrected by removal of small lymphocytes from the patients' marrow by velocity sedimentation or treatment with anti-thymocyte globulin and complement. These candidate suppressor cells were also identified by their ability to inhibit normal CFU-c when marrow from the patient was co-cultured with marrow from a normal donor in the CFU-c assay. These preliminary studies on immunologic mechanisms leading to aplasia have been confirmed by others [5, 6] and extended to a demonstration of suppression of erythroid colony formation by aplastic peripheral blood mononuclear cells. Further clinical support for the existence of lymphocyte-mediated suppression of hematopoiesis is provided by the small proportion of aplastic patients who have received either HL-A identical or haploidentical marrow transplants and after rejecting the allograft, spontaneously recovered their autologous marrow function [14]. Considerable criticism has been directed at the in vitro systems used to demonstrate suppressor cell mechanisms in aplastic anemia. In particular, co-culture inhibition has been attributed to transfusion sensitization to major or minor histocompatibility differences in experimental animals [15] and clinical studies [13]. In the latter studies, co-culture studies on transfused aplastic patients using HLA-matched, MLC compatible sibling bone marrow target cells produced variable inhibition which correlated with a direct chromium release assay, a known test for sensitization to minor transplantation antigens.

The present communication will attempt to review critically the methodology currently available in vitro to classify the nature of the hematopoietic defect(s) in aplastic anemia and will concentrate particularly on the nature of the "suppressor" phenomenon observed in certain cases.

Material and Methods

39 patients with pancytopenia and hypoplastic marrow (18 females, 21 males, ranging in age from 8 to 81 years) were studied at the Walter and Eliza Hall Institute for Medical Research, Melbourne,

Australia, between 1971–1975, at the Memorial Sloan-Kettering Cancer Center, and The New York Hospital-Cornell Medical Center between 1976–1978. The mean white cell count of the patients at the time of study was $1,700/mm^3$, platelets $23,600/mm^3$, and hemoglobin 8.3 gm %.

One ml aspirations were obtained from the posterior iliac crest of patients and normal donors. Informed consent was obtained from all subjects.

Assay procedure. Colony (>40 cells/clone) and cluster (3–5 cells/clone) formation of bone marrow CFU-c were stimulated by exogenously supplied colony stimulating activity (CSA). $0.5–2 \times 10^5$ cells/ml in 0.3% agar culture medium from Difco Laboratories, Detroit, Michigan (with enriched McCoy's 5A medium [Grand Island Biological Co.] containing 10% heat-inactivated fetal calf serum [Microbiological Associates, Inc., Bethesda, MD]) were stimulated by CSA derived from human blood leukocytes in a 0.5% agar feeder layer (prepared 3–5 days before addition of target cells and test material) or human placental cell conditioned medium. In vitro cultures were incubated at 37° C in a humidified atmosphere of 5% CO_2 in air. Colonies and clusters were scored after 7 days of incubation.

Erythroid culture. Bone marrow cells were mixed with alpha MEM media and layered onto 9% Ficoll-Metrizoate, and spunt at 2000 rpm for 30 minutes. The interlayer was harvested and washed three times with alpha medium. Cells were plated at 2×10^5 cells/ml in alpha medium containing 0.8% methylcellulose (Dow), 1% bovine serum albumin (Calbiochem), 30% fetal calf serum (Flow), and 1 unit/ml of human urinary erythropoietin or 2U/ml of Connaught Step III anemic sheep plasma erythropoietin [7]. The cultures were incubated at 37° C in 7.5% CO_2 in a humidified incubator. Using an inverted microscope colonies were scored at 7 days for CFU-e and 14 days for BFU-e. Colonies were placed onto glass slides and stained with Wright's Giemsa and Benzidine.

Velocity sedimentation separation. The procedure of Miller and Phillips [10] was used to separate subpopulations of cells in normal and aplastic marrow. Low density cells (density <1.077 gm/cm^3) obtained by Ficoll-Hypaque separation were allowed to sediment at 1 xg at 4° C through a gradient of 0.4–2.0% bovine serum albumin in isotonic phosphate-buffered saline. After 4 hours, fractions were collected containing cells having different sedimentation rates.

Preparation of cell free inhibitors activity. Populations of separated cells at a concentration of 1×10^6 cells/ml were incubated at 37° C in petri dishes in 1 ml enriched McCoy's 5A medium with or without 10% heat inactivated (56° C for 30 min) FCS. The conditioned medium (CM) was harvested from cultures at various intervals for up to 8 days and spun at 1800 rpm for 10 minutes. Supernatant was passed through a Millipore filter (pore size 0.45 μ) and stored at $-20°$ C until assayed. We prepared extracts by suspending cells at concentrations of $1.0–10.0 \times 10^6$ cells/ml in serum free enriched McCoy's 5A medium and lysing the cells by rapid freezing and thawing (one to three times to * 0° C). Extracts were passed through a Millipore filter (pore size 0.45 μ) and assayed directly or frozen until use.

Suppressor cell co-culture assay. Marrow cells from a patient and a normal donor were suspended at 1×10^6 cells/ml in McCoy's modified medium and mixed in equal proportions. The original unmixed cell suspensions and the patient-normal cell mixtures were then plated immediately as described above in the CFU-c assay. Other aliquots of unmixed suspensions and the patient-normal mixture were incubated in McCoy's for 12 hours at 37° C in 10% CO_2, 90% humidified air and then plated in the CFU-c assay.

ATG treatment. Five million marrow cells from patients and normal controls respectively were incubated for 30 minutes at 37° C in 5% CO_2, 95% humidified air with 0.5 ml of complement (C) (pooled human serum diluted 1/10 with McCoy's), and with either 0.5 ml of McCoy's or 0.5 ml of anti-thymocyte globulin (ATG) (Upjohn lot 17,900 diluted 1/200 with McCoy's). The cells were washed twice and viability assessed with trypan blue. They were then resuspended to 2×10^5 viable cells/ml and cultured in the CFU-c assay as described.

Results

1. Myeloid and erythroid progenitor cells in aplastic anemia. The incidence of CFU-c in aplastic anemic bone marrow (Table 1) was consistently between 1–10% of normal, and in view of the marked hypocellularity of the marrow

Table 1. CFU-c, CFU-e and BFU-e in aplastic anemic bone marrow

	Number	Marrow/10^5	
		CFU-c	CFU-clusters
Aplastic anemia	38	1.8±0.2	18± 2
Aplastic anemia	1	0	3000
Normal	46	31 ±1.5	270±18
		BFU-e	CFU-e
Aplastic anemia	8	2.8±1.7	7.4± 2.9
Normal	32	23.0±5.0	71 ± 8.0

aspirates, the absolute reduction in CFU-c was significantly greater. In almost all cases the relative proportion of colony forming cells to cluster forming cells was within the normal range and colony maturation was normal. In one of the 39 cases studied marrow culture revealed a growth pattern of excessive numbers of clusters exhibiting defective maturation, and absence of colony formation (Table 1). This growth pattern, typical of the majority of cases of acute myeloblastic leukemia, and seen also in a proportion of cases of preleukemia [11, 12], was not associated with hematological or clinical signs of leukemia. Further observation of the progression of this patient's aplasia was not possible since an HLA matched bone marrow transplant was performed. Erythroid progenitor cell assays were performed on 7 patients and revealed an incidence of BFU-e and CFU-e approximately 10% of normal values (Table 1).

2. Suppressor cell involvement in aplastic anemia. Co-culture of normal allogeneic bone marrow in semi-solid agar culture with or without a 12 hour preincubation in suspension culture gave a variable recovery of CFU-c with two standard deviations of the mean value of CFU-c recovered ranging from −40% to +50%. Co-culture of bone marrow from 14 aplastic patients with normal allogeneic bone marrow revealed significant suppression of CFU-c in 20% of cases with no preincubation and in 54% of cases after a 12 hour preincubation of the cells in suspension culture. In two of these cases complete inhibition of colony formation was observed. The effect on colony formation of anti-thymocyte globulin (ATG) plus complement treatment of the aplastic marrow prior to agar culture was investigated in those cases where suppression was evident in co-culture. In only one case was a significant augmentation of colony formation observed.

An alternative approach then was used to investigate further the influence of aplastic anemic bone marrow on normal CFU-c proliferation. In the co-culture experiments, the possibility existed that the suppression observed was due to transfusion-induced sensitization to major or minor histocompatibility antigens. This classic immune reactivity is mediated by intact cells and cannot be duplicated by cell extracts or conditioned media. The possible existence of suppressive mediators contained in, or secreted by, marrow cell populations in patients with aplastic anemia was tested in the following manner. Extracts or conditioned media were obtained from 2×10^6 control or aplastic bone marrow cells/ml and were either added directly to agar cultures of normal bone marrow or were used in preincubation studies prior to agar culture. The results obtained in four patients are shown in Table 2. In three of these patients, extracts obtained from 2×10^6 marrow cells did not inhibit colony or cluster formation when added at 10% v/v to agar cultures of 2×10^5 normal unrelated or HLA identical bone marrow cells. In the case of a fourth patient, significant inhibition of normal CFU-c proliferation was found and this inhibition was reproducible when the aplastic marrow extract was retested. This patient received an HLA matched bone marrow graft and no inhibitory activity was detected in the patient's bone marrow following pretransplant immunosuppression or in the post-engraftment stage (Table 2).

Table 2. Influence of aplastic anemic bone marrow extract on normal CFU-c proliferation

Patient	Stage	Target bone marrow		
		Colonies + clusters/2×10^5		
		Control	+ A.A. extract	Origin
A	Pretransplant	199 ± 8	202 ± 2	unrelated
B	Pretransplant	142 ± 5	136 ± 10	HLA identical
		142 ± 5	135 ± 6	HLA identical
		178 ± 8	154 ± 12	unrelated
C	Pretransplant	256 ± 12	255 ± 9	HLA identical
		142 ± 5	138 ± 5	unrelated
D	Pretransplant	142 ± 5	$82 \pm 8(-42\%)$[a]	unrelated
	Immunosuppression	316 ± 6	312 ± 4	unrelated
	Post-transplant	283 ± 11	264 ± 5	unrelated

[a] % significant inhibition

In order to determine the cell population containing the inhibitory activity, marrow cells from one patient (patient D-Table 2) were separated by a density cut procedure at 1.070 gm/cm^3 and adherence to plastic. Extracts of the separated cells were assayed for colony inhibition against normal non-adherent light density bone marrow cells stimulated by blood leukocyte feeder layers. Only extract from cells present in the non-adherent light density fractions demonstrated significant inhibitory activity. Cell separation procedures using velocity sedimentation and spontaneous sheep red blood cell rosetting were performed to further characterize the inhibitory activity producing cell. In one patient studied, inhibitory activity was detected in a population of slowly sedimenting cells (3.0–4.5 mm/hr), which could be separated from the majority of marrow CFU-c. Dilution of the extract or conditioned media prior to assay demonstrated that the cells sedimenting at 3.0–4.0 mm/hr contained more inhibitory activity than more rapidly sedimenting cells. Inhibitory activity was confined to the non-rosetting fraction following separation of marrow with neuraminadase treated sheep red blood cells.

Suggestive evidence of a suppressor cell mechanism in aplastic anemia was obtained from a serial study of marrow CFU-c in a patient who received an HLA matched bone marrow transplant (Table 3). At the time of initial study no colony formation was evident in marrow cultures with low numbers of apparently normal

Table 3. Patient P. – Aplastic anemia

Days	Marrow/10^5		WBC$\times 10^3$	PMN$\times 10^3$
	Colonies	Clusters		
0	0	10	3.0	0.8
30	Prednisone			
37	13	44	3.7	0.9
	Cyclo., Ara-C., 6-TG			
44	11	37	3.0	1.9
	HLA matched BMT			
51	5	50	3.3	0.2
58	12	59	6.8	0.5
65	5	76	7.6	1.7
72	3	74	15.0	1.6
79	16	48	18.0	0.5
87	3	30	18.0	1.1
91	0	2	18.0	7.5

cluster. Following the onset of pretransplant immunosuppression, colony and cluster incidence returned to the low-normal range one week after prednisone therapy and remained elevated two weeks later despite Cytoxan, ara-C and 6-thioguanine treatment. Engraftment of HLA matched bone marrow was obtained with the CFU-c incidence persisting in the low range of normal for 4–5 weeks.

Discussion

The marked reduction in erythroid and myeloid progenitor cells seen in almost all cases of aplastic anemia is compatible with a profound defect presumably operative at the pluripotential stem cell level. The existence of an occasional case of true aplastic anemia progressing to acute leukemia is supported by the one example out of 39 cases in this study which displayed unequivocal evidence of myeloid leukemic cluster formation in marrow culture.

Evidence for a subgroup of aplastic anemia attributable to an active suppressor mechanism must be reviewed critically in the light of the methodologies employed. It is clear that marrow co-culture studies, even when undertaken with HLA-matched target marrow, may demonstrate CFU-c inhibition due to transfusion sensitization of the patient to minor histocompatibility differences. It should be noted that all patients in this study had a prior history of extensive transfusions. The duration of co-culture, the cell concentration used and the aplastic to target marrow cell ratio are frequently uncontrolled variables particularly since the cellular composition of aplastic marrow or blood differs so markedly from normal. It is also the case that the human CFU-c assay is understimulated and is particularly sensitive to enhancement or inhibition of a specific and non-specific nature. It is clear now that adherent cell populations in marrow and blood can modulate granulopoiesis and CFU-c proliferation by elaboration of two opposing activities, colony stimulating factor (CSF) and prostaglandin E. [9] Increasing numbers of monocytes or macrophages particularly in an activated state may markedly inhibit CFU-c proliferation due to their Prostaglandin biosynthesis. Further inhibitory interactions have been reported in which lactoferrin released by mature granulocytes directly inhibits monocyte-macrophage CSF production and indirectly influences prostaglandin biosynthesis [3, 9]. Thus, co-culture studies of marrow populations containing variable properties of CFU-c, adherent cells and granulocytes may demonstrate colony inhibition or stimulation unrelated to the pathophysiology of the aplasia. In an attempt to circumvent these difficulties, we explored techniques for separation of marrow cells on the basis of adherence, density, size and E-rosetting capacity, followed by extraction or conditioning procedures to isolate soluble inhibitory activity. A precedent for this was established by Broxmeyer et al. [2] who demonstrated a subpopulation of cells in leukemic bone marrow which release a diffusible leukemia inhibitory activity (LIA) which suppress normal CFU-c proliferation. The suppressor cells observed in the one case of aplastic anemia studied by cell separation appeared to be similar to the LIA producing cells in leukemic marrow in so far as they were non-adherent, slowly sedimenting, light density, E-rosette negative and Fc positive. The difference between the inhibitory activity released by aplastic marrow cells and the leukemic inhibitory cell was that the former inhibited CFU-c at all stages of the cell cycle whereas LIA was S-phase specific. Furthermore, leukemic CFU-c were not inhibited by LIA whereas the aplastic inhibitor suppressed normal, leukemic and the patients own CFU-c. A similar suppressor cell population has been reported by Broxmeyer et al. in the marrow of patients with neutropenia of a variety of etiologies [4]. The existance of such an inhibitory cell population in the marrow of some patients with aplastic anemia cannot be attributed to prior transfusion sensitization since an extracable inhibitory activity with similar properties could not be identified in control patients without aplastic anemia but with a history of multiple transfusions [2].

The available evidence suggests that any suppressor cell activity in aplastic anemia is unlikely to involve classic immunological mechanisms. Abrogation of suppression by treatment of aplastic marrow with anti-thymocyte or anti-lymphocyte globulin plus complement is difficult to interpret since the specificity of such antisera is poorly understood relative to minor subsets of non-T hematopoietic cells with regulatory potential. Similarly, the recovery of normal CFU-c in the marrow of one patient following the onset of pretransplant immunosuppression does not prove that an immunological suppressor cell mechanism need exist. The existence of an autoaggressive process in certain patients with aplastic anemia may involve an inappropriately activated hemopoietic regulatory cell or the development of a suppressor cell population not of a classic B or T cell type but possibly related to the natural killer cell population. Finally, it remains unclear as to whether such suppressor cell populations are acquired as a consequence of the aplastic state or are instrumental in inducing it.

This work was supported by NIH grants CA 17353 and CA 20194.

References

1. Ascensao, J., Pahwa, R., Kagan, W. A., Hansen, J. A., Moore, M. A. S. and Good, R. A.: Aplastic anemia: evidence for an immunologic mechanism. Lancet *I*, 669–671 (1976)
2. Broxmeyer, H. E., Jacobsen, N., Kurland, J., Mendelsohn, N. and Moore, M. A. S.: In vitro suppression of normal granulocyte stem cells by inhibitory activity derived from leukemic cells. J. Natl. Cancer Inst. *60*, 497–511 (1978)
3. Broxmeyer, H. E., Smithyman, A., Eger, R. R., Meyers, P. A. and de Sousa, M.: Identification of lactoferrin as the granulocyte-derived inhibitor of colony stimulating activity (CSA)-production. J. Exp. Med. *148*, 1052–1067 (1978)
4. Broxmeyer, H. E., Pahwa, R., Jacobsen, N., Grossbard, E., Pahwa, S., Meyers, P. A., Kapoor, N., Miller, D., Good, R. A. Ralph, P. and Moore, M. A. S.: In vitro inhibitory activity obtained from cells of patients with neutropenias of varying etiology. (submitted)
5. Haak, H. L. and Goselink, H. M.: Mechanisms in aplastic anemia. Lancet *I*, 194 (1977)
6. Hoffman, R., Zanjani, E. D., Lutton, J. D., Zalusky, R. and Wasserman, L. R.: Suppression of erythroid-colony formation by lymphocytes from patients with aplastic anemia. New. Eng. J. Med. *296*, 10–13 (1977)
7. Iscove, N., Sieber, F. and Winterhalter, K.: Erythroid colony formation in cultures of mouse and human bone marrow: Analysis of the requirement for erythropoietin by gel filtration and affinity chromatography on agarose-Con A. J. Cell. Physiol. *83*, 309–320 (1974)
8. Kagan, W. A., Ascensao, J. A., Pahwa, R. J., Hansen, J. A., Goldstein, G., Valera, E. B., Incefy, G. S., Moore, M. A. S. and Good, R. A.: Aplastic anemia: Presence in human bone marrow of cells that suppress myelopoiesis. Proc. Natl. Acad. Sci. (USA) *73*, 2890–2894 (1976)
9. Kurland, J. I., Broxmeyer, H. E., Pelus, L. M., Bockman, R. S. and Moore, M. A. S.: Role for monocyte-macrophage-derived colony stimulating factor and prostaglandin E in the positive and negative feedback control of myeloid stem cell proliferation. Blood *52*, 388–407 (1978)
10. Miller, R. and Phillips, R.: Separation of cells by velocity sedimentation. J. Cell. Physiol. *73*, 191–201 (1969)
11. Moore, M. A. S.: In vitro Studies in the Myeloid Leukaemias. In: Advances in Acute Leukaemia. Cleton, F. J., Crowther, D. and Malpas, J. S. (eds.). Amsterdam: ASP-Biological and Medical Press, 1974, p. 161–183
12. Moore, M. A. S.: Marrow culture – A new approach to classification of leukemias, Blood Cells *I*, 149–156 (1975)
13. Singer, J. W., Brown, J. E., James, M. C., Doney, K., Warren, R. P., Storb, R. and Thomas, E. D.: The effect of peripheral blood lymphocytes from patients with aplastic anemia on granulocytic colony growth from HLA matched and mismatched marrows: The effect of transfusion sensitization. Blood (in press)
14. Thomas, E. D., Storb, R. and Giblett, E. R.: Recovery from aplastic anemia following attempted marrow transplantation. Exp. Hemat. *4*, 97–102 (1976)
15. Torok-Storb, B. J., Storb, R., Graham, T. C., Prentice, R. L., Weiden, P. L. and Adamson, J. W.: In vitro erythropoiesis: The effect of normal versus "transfusion-sensitized" mononuclear cells. Blood *52*, 607–611 (1978)

5.8 Co-culture Studies in Transfused and Untransfused Patients with Aplastic Anemia

J. W. Singer

Studies on Untransfused Patients with Aplastic Anemia (AA)

Several recent reports have presented the results of co-cultivation of marrow or peripheral blood mononuclear cells from AA patients with normal marrows and have suggested that the inhibition of growth of either granulocytic or erythroid colonies from normal marrows by lymphocytes from AA patients indicated that cell-mediated suppression of marrow growth caused the disease [2]. We have previously reported that only by inhibition on co-culture studies, using mononuclear cells from untransfused AA patients could immunologic mediation of the disease be presumed [2]. Mononuclear cells from AA patients who received blood products inhibited growth of granulocytic colonies from HLA mismatched marrows with regularity but only occasionally (approximately 15%) significantly inhibited the growth of CFU-C from an HLA matched family member. We interpreted our initial studies to indicate that sensitization to HLA antigens could account for some of the suppression seen with mismatched marrows and that sensitization to minor histocompatibility antigens could be a presumed cause of the inhibition noted with HLA matched marrows. A strong correlation was found between the relative stimulation or inhibition of marrow growth by co-cultured lymphocytes from the transfused AA patients with HLA identical marrows and direct chromium release activity; a test previously shown to indicate sensitization in aplastic anemia patients undergoing marrow transplantation [3].

We have now performed co-culture studies using HLA matched marrows on thirteen untransfused patients with severe AA. Nine of thirteen had idiopathic AA; two were probably post hepatitic, one was possibly drug related, and one was a Fanconi type. The duration of the disease varied between one week and five years; however, eleven were studied within one month of the initial diagnosis. Ten of the thirteen had severe depression of a least two of three hematopoietic cell lines and met the criteria of the International Cooperative Aplastic Anemia Study Group for severe aplastic anemia [1].

All studies were performed by previously reported methods [2]. The results of the co-culture studies are shown in summary form in Table 1. Lymphocytes from eleven out of thirteen patients either had no effect or significantly enhanced the growth of co-cultured normal bone marrows. Lymphocytes from two patients significantly inhibited the growth of marrows from HLA matched family marrows ($p < 0.01$; Student T test). We interpreted the co-culture studies on these two patients to indicate possible immunologic mediation of the disease. Both of these two patients had idiopathic AA. One was treated with low doses of Prednisone

Table 1. Co-culture studies with peripheral blood lymphocytes from thirteen untransfused patients with aplastic anemia

Mean % increase in colony numbers with lymphocytes from 11 "noninhibitory" patients with AA[b]	Mean % increase in colony numbers with control lymphocytes	Mean % inhibition in colony growth from the two patients who inhibited in coculture against an HLA identical marrow
+36.1±7.0 (SEM) n=17	+40.3±9.2 n=19	Patient 737[b]−40±6(±SD) 773[b]−30±2

[a] p<0.01 by student T test when compared to control lymphocytes co-cultured in the same experiments.

[b] The % increase or decrease was calculated from the ratio of the mean number of colonies grown from normal marrow with AA or control lymphocytes divided by the number of colonies obtained from marrows without lymphocytes. The number of colonies formed by the lymphocytes plated without marrow was subtracted from the total obtained in co-culture before the ratios were calculated

from the time of diagnosis, four weeks before in vitro studies. He subsequently accepted a marrow graft without difficulty after preparation with cyclophosphomide. However, twelve months post-marrow grafting from his HLA identical sister, his hematopoietic cells were found to be male on cytogenetic analysis. Thus, he appears to have rejected his marrow graft and regenerated his own marrow. He is the only untransfused AA patient in the Seattle experience to do so. The second patient was diagnosed two weeks prior to study and received no therapy. A chromium release assay performed on this patient against his nineteen year old HLA matched sister was strongly positive at 18.2%. Unlike the other patient, he continues to show only donor cells one year after bone marrow grafting following cyclophosphomide preparation.

Our current data on thirteen untransfused aplastic anemia patients indicates an incidence rate for inhibition on co-culture of 2/13. From these studies, it is possible to project an incidence of possibly immunologically mediated aplastic anemia of not more than about 20%. Thus, our data suggest the majority of cases of severe aplastic anemia are not associated with an immune mechanism and therefore we would not expect immunosuppressive therapy to be of value.

Co-culture Studies on Transfused Patients with Aplastic Anemia

We have performed co-culture studies on 24 patients with severe aplastic anemia who have received one or more blood transfusions more than 24 hours prior to study. Most have received multiple transfusions. Mononuclear cells from the patients were collected by previously described methods, incubated with marrow cells from an HLA matched family member for two hours in suspension culture at lymphocyte marrow cell ratios of 2:1 and 1:1. The mixture was then plated for CFU-C growth using either a feeder layer of peripheral blood leukocytes or a PHA stimulated leukocyte conditioned medium. Lymphocytes from the marrow donor were incubated with the autologous marrow as a control. Colonies were scored under an inverted microscope at 40X as clusters of greater than 40 cells after fourteen days of culture. The mononuclear cell preparations were also plated without marrow and any colonies formed were subtracted from all results. The data was tabulated as "stimulation or inhibition ratios" and calculated by dividing the mean number of colonies produced by the marrow in the presence of lymphocytes from the AA patient by the mean number of colonies with normal lymphocytes. This method of calculation reflects lack of enhancement with the AA lymphocytes compared to autologous lymphocytes as well as true inhibition of CFU-C growth.

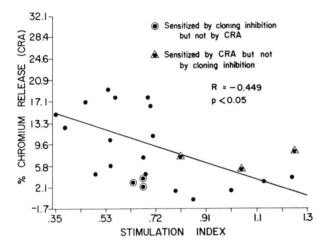

Fig. 1. Percentage of radioactive chromium released from HLA-matched lymphocytes by lymphocytes from AA patients versus the stimulation (inhibition) ratios of AA lymphocytes against HLA-matched marrows at lymphocyte to target cell ratios of 2:1

We interpret a stimulation index of less than 0.8 as being significantly different from 1.0 and indicating possible sensitization. Figure 1 shows the results of such studies on 24 transfused patients with aplastic anemia. The stimulation index is plotted against the percent chromium release activity. There was a significant correlation coefficient between the two tests ($R = -0.449$; $p < 0.05$). Two of three patients who were "sensitized" by the cloning inhibition assay but negative in the chromium release assay rejected a marrow graft after cyclophosphomide preparation. They were the only patients in this group to do so. The three patients who were positive in the chromium release assay but negative in the cloning inhibition assay did not reject grafts.

Summary

Co-culture studies have two potential values: (1) In untransfused patients with AA they may be able to detect the small population of patients who have immunologically mediated disease and who therefore may respond to immuno-suppresive therapy; (2) In transfused AA patients under consideration for allogenic marrow transplantation, inhibition in the coculture assay may indicate sensitization to minor histocompatibility antigens. Preliminary data indicates that this test may be more sensitive than the chromium release assay and should be added to a battery of prospective in vitro tests designed to detect those patients at risk for marrow graft rejection.

Supported in part by research funds of the Veterans Administration and by National Cancer Institute Grant No. CA 18029.

References

1. Camitta, B. M., Thomas E. D., Nathan, D. G., Santos, G., Gordon-Smith, E. C., Gale, R. P., Rappeport, J. M., Storb, R.: Severe aplastic anemia: A prospective study of the effect of early marrow transplantation on acute mortality. Blood 48, 63–67 (1976)
2. Singer, J. W., Brown, J. E., James, M. C., Dorey, K., Warren, R. P., Storb, R., Thomas, E. D.: Effect of peripheral blood lymphocytes from patients with aplastic anemia in granulocytic colony growth from HLA-matched and -mismatched marrows: Effect of transfusion sensitization. Blood 52, 37–46 (1978)
3. Warren, R. P., Storb, R., Weiden, P. L., Mickelson, E. M., Thomas, E. D.: Direct and antibody dependent cell-mediated cytotoxicity against HLA identical sibling lymphocytes. Correlation with marrow graft rejection. Transplantation 22, 631–635 (1976)

Discussion

Moore: All the patients we tested by the approach of extraction of inhibitory activity were transfused. What I would be intrigued about is whether we can show, using this method, that we would not be able to get inhibition simply as a result of sensitization by blood transfusion, which would require cell interaction. However, we looked at leukemic patients multiply transfused as controls and we can't show evidence of extractable inhibitory activity.

Heimpel: Dr. Singer, from your 24 patients, 3 inhibited the growth of granulocytic colonies from normal target marrow. Could you reproduce these results, when you repeated the test at another time or with other target marrows? In our own preliminary coculture assays, we felt that sometimes the results are erratic for reasons we just don't know. Could it be that you just have a distribution of results and some "positive" tests represent the extremes of this distribution? Did these patients have anything special as compared to the patients not showing inhibition?

Singer: Two of these 3 patients were tested against more than one marrow. Both inhibited all marrows strongly. There was no difference in those patients and other Seattle aplastics. What you must realize with Seattle aplastics is we do not have them around as aplastic very long; because they get transplanted, we are not able to serially study them.

Thorsby: Have you ever tested lymphocytes taken from the marrow?

Singer: No. I understand from your comments and from others a possible criticism and I think one of the difficulties here is gaining enough lymphocytes from marrow to be able to test multiple times. That was then reason we decided to go for peripheral blood.

Nissen: Have you looked at BFU-E inhibition? After doing this, I have given up CFU-c inhibition and I found that BFU-E inhibition was a repeated finding.

Singer: On the first 10 patients we looked at CFU-E inhibition and we found it related to HLA-antigens, i.e. there was less inhibition of matched than mismatched marrows. However, it was not nearly as consistent as with CFU-c. We have not specifically looked at BFU-E.

5.9 General Discussion

Moderators: S. Thierfelder, E. Thorsby

Thorsby: Perhaps I might start the discussion by mentioning a few of the requirements which should preferably be fulfilled before implying autoimmune mechanisms participating in the pathogenesis of AA:

1. If serum factors are found to inhibit certain hematopoietic cells in vitro, it should be established whether the effect is confined to the Ig fraction. If complement is not found to be necessary, the possibility exists that antibody-induced, cell-mediated cytotoxicity is involved and should be looked for.

2. If certain cells are found to be inhibitory, it should be clearly established that the cells involved are lymphoid, but also which sub-set: T lymphocytes, K cells, macrophages etc. It should, however, be noted that no effect of lymphoid cells prepared from peripheral blood of the patient does not exclude immune mechanisms, since the specific clones may mainly be present in the bone marrow.

3. Specificity for bone marrow cells should be looked for, by using other target cells as well (fibroblasts etc.)

4. If the above-mentioned criteria are fulfilled, it must be established that the same phenomena are not observed in vitro by using sera or cells of non-AA patients or healthy controls.

5. Allo-immune reactions (mixed lymphocyte interactions, HLA antibodies etc.) must be excluded in experiments mixing cells and sera of AA patients with those of healthy controls. The use of HLA identical (preferably sibling) combinations may partly circumvent this (but monozygotic twin combinations are of course particularly informative).

6. Another possibility is to show that antibodies and/or lymphoid cells taken during on-going disease (and stored for example in liquid nitrogen) would inhibit autologous haematopoietic cells obtained in remission, when serum and lymphoid cells have lost their inhibitory activity.

These and other requirements may be self-evident for most of you, but may be difficult to perform or are easily forgotten in clinical studies.

A further question here is whether the immunological phenomenon that one observes is a cause or is the result, a question which we always discuss when we discuss immunology. One should keep in mind that there are quite a few mechanisms leading to tisse injury, e.g. infectious, toxic or mechanical, activating macrophages in this particular area. If you find a lymphocyte infiltrate, then there is strong indication that immunological reactions are going on there now. If these lymphocytes are activated by antigens, they release a lot of factors which are called lymphokines, including interferon. I would be very careful to conclude from the fact that you find such factors produced by lymphocytes in aplastic anemia, that this has anything to do with pathogenesis.

Gordon-Smith: What, if anything, does ALG treatment tell us about the pathogenesis of aplastic anemia, of the type that is being successfully treated? Ad Dr. Thomas has said, this is a specially selected group of patients with aplastic anemia. I would like to ask the immunologists amoung us, what would the treatment with ALG do to the immunological system, if one refers to ALG as an immunosuppressor. Is it immunosuppressive?

Thorsby: There is no doubt that ALG will act as an immunosuppressor. I am wondering whether what you are observing with ALG and what you are observing by allogeneic bone marrow transplants could be the same thing. I am very intrigued by Dr. Netzel's observations that absorbed ALG seems to have a stimulating capacity on the colony forming unit. We also know that allogeneic cells are stimulatory, for example if you mix blood or bone marrow lymphocytes. There is also evidence that bone marrow transplantation will even be stimulatory in genetic combinations, when the host is not able to respond to the graft. If you use parental donor F_1 hybrids as recipients, you see that major proliferation still takes place on the host cells. It therefore could be that ALG and bone marrow transplantation both act as stimulatory agents for bone marrow or stem cell proliferation. This frequency of recovery by ALG which is approximately 20%, is comparable with spontaneous recovery of aplastic anemia which might also be about 20%, so you may just potentiate a spontaneous recovery. Has anybody tried to see whether allogeneic lymphocytes are stimulatory for colony formation by bone marrow cells?

Thierfelder: I would like to make a comment on the estimate of 20% of immunology pathogenesis of aplastic anemia which I think is quite a high estimate. One should keep in mind that the possible autoimmune reactions of antibody type in aplastic anemia do not fullfil the criteria one would expect from the term "autoantibody". Autoantibody is not only directed against the patient's own cells, but against similar cells of almost all normal individuals. The second fact is, that in 18 patients of Dr. Thomas which were not presensitized, there were 17 engraftments. Bone marrow transplantation is a very sensitive indicator system of whether antibodies are there. Therefore, the great majority of untransfused patients must have no antibodies. If one feels, that aplastic anemia is an autoimmune phenomenon, one would expect this autoantibody to be directed also towards the donor cells. In summary, a sensitive indicator system which shows that there are no antibodies and the postulate of autoimmune reaction against all cells, even from the donor, dismisses at least the humoral side of the autoimmune pathogenesis in aplastic anemia.

Thorsby: First I would like to qualify my statement. As far as I could understand from the materials which have been presented, I have a feeling that in only approximately 20% of the cases, I could "smell immunology".

Secondly, I agree with you that autoantibodies do not seem to have individual specificity. But I think that there are good experiments showing that even though you have preformed humoral antibodies, you will succeed with a graft. Bone marrow transplants have been successfully performed in the presence of ABO-incompatibility, after partial removal of these antibodies by plasmapheresis. If you do transplantation using strong immunosuppression, I could accept that even though there may be an autoimmune mechanism involved, you may establish a successful graft.

My third comment is to ATG. It is very hard to know what sort of effect that ALS or ATG might have. The data presented by Dr. Netzel are really interesting, because his paper seems to indicate, that some of the preparations in use have two separate effects: an effect on lymphocytes and a stimulatory effect on colony forming units. Until one can separate these two effects and look at them more carefully, it is very hard to conclude what might be the mechanism behind the clinical effects of ATG.

Moore: I would make a plea for standarization of ALG and ATG preparations. At the moment there is a collaboratory study in the United States comparing 5 of the Upjohn preparations plus the horse ATG from Speck in Basel, with all the immunological in vitro systems we have going. I only think that this question, of what your ATG preparations are recognizing, is causing some of the apparent contradictions in the data.

Shadduck: I have presented in this symposium the data on one patient who recovered after ATG. We have had a subsequent experience with the Upjohn ATG in a patient with T-cell lymphoma. This individual had diffuse seeding in the bone marrow of huge lymphoma nodules and severe pancytopenia that prohibited appropriate chemotherapy. This patient was splenectomised and treated with ATG for 2 weeks. Complete resolution of the lymphatic involvement in the marrow occured, but unfortunately, the patient developed severe aplastic anemia and died 24 days after treatment with near total aplasia of the marrow. This observation may contribute to the discussion on the various effects by Dr. Netzel and Dr. Thorsby. It underlines that some ATG preparations may be toxic to hemopoietic stem cells.

Moore: I would like just to make a comment. I have no idea if anything we are detecting in either co-cultures, things which inhibit CFUc, has anything to do with aplastic anemia, more to the effect that I think that they are phenomena that we have to recognize, if we use in vitro immunology. But to the second point, the question of course is a point well taken. I wonder if a transitory phase of aplasia which could be caused by any noxious factor could lead to secondary inflammatory reactions and non-specific release of lymphokines which serves to compromise then a recovery of the stem cell function.

Thorsby: I agree on that of course.

Heimpel: I think that hemopoiesis is different from many other tissues because physiologically it is a system with a large amount of cell renewal and also cell destruction. Relevant controls may not be normal individuals but those with accelerated cell destruction, e.g. by cytostatic agents in non-hemopoietic malignant diseases.

Thorsby: Yes, that is possible.

Heimpel: I have one question to the participants who worked with cocultures of bone marrow or with bone marrow extracts. The inhibition which is found by bone marrow material of aplastic anemia is due to abnormal cellular constituents of these marrows, because, as I see, you always compare this with normal marrows. However, in your assay you have to use similar cell numbers or similar tissue weights from aplastics and from controls. Could it be that with aplastic marrow, where there are very few hemopoietic cells, a certain cell type, for example normal lymphocytes, plasma cells or other normal mononuclear cells are just more representative and your abnormal result is a quantitative rather than a qualitative phenomenon?

Moore: I think you can exclude this in two ways. Firstly, we do identical cell separation to enrich for that particular subset of cells in normal bone marrow, and we also use a large number of controls of patients with leukemia of patients with solid tumors receiving chemotherapy. And the virtue of being able to work with an activity rather than a population of cells is that you can titrate over a range of cells in order to define the inhibitory activity. This would also overcome the question of whether we are just dealing with a quantitative difference based on whether there are low or high levels of those particular cells. But as a general point, it is a very good one that you made for cell mixing experiments. One should try to titrate the cells, cell to target cell ratio, over quite a wide range. Otherwise you could be falsely misled into thinking that there are qualitative differences.

Heimpel: If you do the separation procedures with an aplastic marrow and normal marrow, you still have different compositions of cells, because you may have different cell types with identical physical features.

Moore: You are basically isolating a population of cells of a particular density and cell volume.

Thorsby: I think it is important to state though that we have very little evidence that lymphoid cells as T-lymphocytes, B-lymphocytes, null-cells or whatever, participate in the normal physiological control of other cells than lymphocytes. Your comment is very good, but I doubt that the cells you are looking for might be lymphocytes.

Heimpel: Did you ever try to use normal non-hemopoietic human marrow as a control for marrow from aplastic anemia? Wouldn't this be a good control?

Moore: Fatty marrow? We tried very hard to get those fatty cells in the continuous marrow cultures, because they seem to be so nice for sustaining hematopoiesis. So hard in fact, we were deliberately isolating at least fat containing spicules to sustain hematopoiesis. Indirectly we have done that and they seem to be supportive.

Thorsby: I would like to state that there is not very good evidence for autoimmune mechanisms playing a major role in most cases of aplastic anemia. If you look at some of the data that have been presented it looks that approximately 20% of the patients would fulfill the criteria of an immunological disease. This would be in accordance with data that Dr. Thomas presented this morning that out of 10 patients who had received marrow grafts from identical twins, there were 8 takes and 2 failures. In summary, I am left with the material that has been presented, that approximately 20% of the cases is strong evidence of immunological phenomena.

6 Final Discussion

Final Discussion

Moderators: H. Heimpel, M. A. S. Moore

Heimpel: In the next hour we want to discuss possible mechanisms of aplastic anemia as they were brought up in the conference, and we want to analyse what the different methods can contribute to elucidate the pathogenesis of this disease.

Moore: In Figure 1, I tried to put down the cell populations involved in the hemopoietic and immune system, and regulating factors. We shall rub some of them out as we go along, when we find them not relevant to aplastic anemia. We might discuss two or three things that remain and which are relevant. We will start off by talking about the stem cell niche to find an anatomical concept. We have some information about the type of cells supporting stem cell replication from Mike Dexter's data. We know we have macrophages, endothelial cells and giant fat cells. We have a pluripotent stem cell, we have to consider signals that influence the cycling of the pluripotential stem cells. It is possible that by blocking cycle we may generate the disease of aplastic anemia. We have to consider what sort of receptors are on the pluripotential stem cell. We know there are receptors for antigens, there are β_1 receptors and these can all be utilized to trigger CFU_s into cycle.

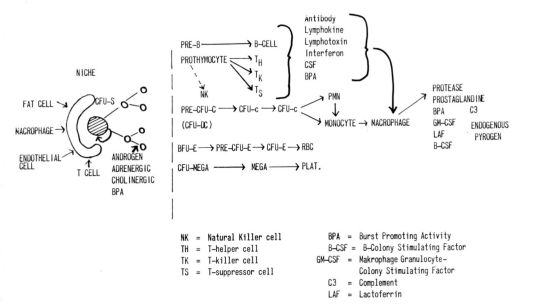

NK = Natural Killer cell	BPA = Burst Promoting Activity
TH = T-helper cell	B-CSF = B-Colony Stimulating Factor
TK = T-killer cell	GM-CSF = Makrophage Granulocyte-
TS = T-suppressor cell	Colony Stimulating Factor
	C3 = Complement
	LAF = Lactoferrin

Fig. 1. A schematic model of haematopoiesis. Glossary: CFU_s=colony forming unit spleen; CFU_{DC}=colony forming unit diffusion chamber, early granulocytic precursor; CFU_c=colony forming unit culture, granulocytic precursor; BFU_E=brust forming units early erythropoietic precursor; CFU_e=colony forming units erythropoietic, late erythropoietic precursor; BPA=burst promoting activity; CSA=colony stimulating activity; EPO=erythropoietin

We should also consider the concept of N. Iscove what he called burst-forming activity, mainly perhaps as a general factor required to sustain cells at this level rather than to support a differentiation. Then we have the various compartments which we can measure by various assays, BFU-E, CFU-E, the 7-day, 14-day diffusion chamber assays for the granulocyte-macrophage progenitor cells. We have not talked very much about the megakaryocyte pathway. It seems important to consider the T- and B-series, T-helper, T-killer and suppressor activity. Here is a series of molecules which are produced by macrophage nutrient proteases with a wide variety of effects, particularly plasminogen-activator, prostaglandine-E, a relevant suppressor of hematological proliferation, growth promoting factors, colony stimulating factors, activity that the macrophage provides for maintaining T- and B-cell proliferation, complement components, endogenous pyrogen; and we must also consider lymphokines, anti-bodies, lymphotoxines, and recognize that activated lymphocytes, particularly the T-cell series, produce colony stimulating factor and burst promoting activity. Getting to the question: is there any evidence that classic T- and B-lymphocytes have anything to do with hemopoiesis? One could argue that there is a late response to a T- oder B-cell proliferation which impinges certainly on the granulocyte-macrophage series and possibly on the erythroid series. Important humoral regulators are colony stimulating factors from the macrophage, the erythropoietin and thrombopoietin. Something we have not mentioned is the role of iron binding proteins in hematopoiesis, which could assume some considerable significane in the future, particularly lactoferrin, produced by the mature granulocytes. That's the picture. I will take the sponge and rub out, as we go along, anything that no longer appears to have relevance to aplastic anemia and we can amplify anything onto the scene that is remotely connected.

Heimpel: We would like to start by asking Dr. te Velde whether he can contribute from the morphology something to the table from his experience on bone marrow histology. Could you comment on this table? Do you believe that one mechanism would be more likely than the others?

te Velde: I think we have to combine the morphology with the clinical picture. I tried to define different populations of aplastic anemia patients on the basis of different distribution of the inflammatory infiltrate. In the group of patients in which we see the highest amount of leukemia, the stem cell defect should be high on the list. On the other hand we have patients in which we see an enormous amount of infiltrate in the bone marrow, which responds quickly to immune suppression. These patients die if you do not transplant them or give them immune suppression. I think we have a problem in defining the amount of infiltrate. Everybody has studied some patients in centers where they have been referred to, and there is an enormous selection of patients in the series we have seen from the Leiden University. Patients which have been put on the immune side have died before you could do anything, but these patients are so severely aplastic after you treat them with immune suppression. From the side of morphology I would say, it is useful first to define which patient will have aplastic anemia and plastic. If you look at the bone marrow, you get the impression that you have a population of at least two sorts of histological phenomena, the one is the stem cell defect, and the other is abnormal immune reaction.

Heimpel: You have discussed the two possibilities but you do not state which one is the minority and which one the majority. As I understand, you believe that there be at least two types of aplastic anemia

te Velde: I would suggest that the stem cell defect is the majority, but I think it's a case of 90 and 10%.

Heimpel: Does anyone want to comment on the evidence on histology?

Fliedner: I want to ask the Dutch colleagues whether they consider the inflammatory changes a consequence or a cause. You may look at it both ways. If you had a bone marrow oedema, it would cause a non-friendly environment for stem cells to replicate.

te Velde: I believe that the oedema is always secondary. There must have been something, either complement activation or other types of bone marrow damage. Most of the morphological changes we are looking whether the aplastic state is probably secondary.

Heimpel: Dr. Böttiger, is it true that the drugs you mentioned are also responsible for agranulocytosis and thrombocytopenia, and that they act on the more mature cells of these series through humoral antibodies? Is this evidence for an immune-mediated action directed to stem cells in case of aplastic anemia induced by these drugs?

Böttiger: Of course, it is a fact that several drugs act on the various cell lines of the marrow, all together or separately. It was mentioned here today that phenylbutazone caused more agranulocytoses than aplastic anemia. I was particularly interested in Dr. Wickramsinghes remarks that the drugs could alter the ultrastructure just a little. I think this might indicate that the effect is dose-dependent.

I would like to insist on the fact that what we call "idiopathic aplastic anemia" is something in which we did not find any evidence of drugs. However, more than half were possibly due to contact with toxics. It is quite possible, that if we looked more on "natural" toxics – not drugs – and on general toxics used in normal life, we would probably find more evidence for possible toxic bone marrow damage. People who work with insecticides, products used for washing and things like that. This may induce neutropenia or slight pancytopenia and it is quite possible, that most exposed individuals will show no or only a slight abnormality, but some people may develop aplastic anemia.

Heimpel: I think we would agree with you but this evidence wouldn't help too much to decide through which pathway – directly on stem cells or mediated by damage through the microenvironment – these xenobiotics act. In this context I would like to ask Dr. Camitta about viral etiology.

Camitta: I don't think that the evidence we have enables us to say how a virus causes aplastic anemia. I think Dr. Moore's interesting data in vitro suggest that the virus can damage the supporting structure. However, there is also the possibility of a direct low potent stem cell damage. I think the observations in post-hepatic aplastic anemia point to the latter. By the time the patient is diagnosed, the defect appears to be a stem cell defect and not an microenvironmental defect, because those patients getting a graft do just as well as other patients with other etiologies of aplastic anemia.

Heimpel: It can also be caused by immune mechanisms because all patients transplanted receive immune suppressive treatment.

Camitta: The virus is known to alter cell membranes and render them immunogenic, but no one has demonstrated this as yet. I think it is just a hypothesis.

Moore: I am not suggesting any irreversible damage by the virus to the microenvironment, that there could be a phase of continuous replication of the virus within some cellular constituent of the supporting matrix. This could be reversible, but the stem cells may have died out before the environment has recovered. The second point is, that there is some intriguing data particularly on C-type viruses. From some of the virus determined cell surface proteins one can raise anti-serum against them that cross-reacts with pluripotential stem cells in the mouse system. Thus, virus could be evoking an immune response against the stem cell.

Camitta: If you look at the marrow of the patients in post-hepatic aplasia, you usually will have some, although very low amounts of CFU-C or erythroid myeloid activity, detectable visually. There must be some stem cells that are still there. If the niche was damaged, you would not expect bone marrow transplant to repair that, unless bone marrow transplantation provided helper cells. At least by the time the patient gets to you, the damage exists no longer in the microenvironment. It is probably repaired at that point, and there is probably a deficiency in some cases absence of stem cells.

Moore: You don't think viruses may act through induction of interferon?

Camitta: I think viruses precede interferon. I am not sure that the amount of interferon released in the microenvironment is enough to cause bone marrow suppression.

Fliedner: You provoked the question as to whether there is another idea for experimental therapy in aplastic anemia. If cell mediated immune mechanisms, as T-cell suppression is relevant, it would provoke the idea to treat these patients with extracorporal irradiation of the blood. You should then get a good response with the removal of all those lymphocytes you can kill with extracorporeal irradiation. Would that be correct?

Niethammer: Is there evidence that there might be T-cell populations in the bone marrow which are different from the ones in peripheral blood? If true, you may not find any suppressor cells in the peripheral blood but you might find them in the bone marrow.

Fliedner: In chronic lymphocytic leukemia, one cannot remove lymphocytes from the bone marrow, even though one can remove them from the peripheral blood, the lymph nodes and the spleen. It may be important to know whether this type of lymphocytes suspected to be active in the bone marrow is in equilibrium with the blood. If there is an equilibrium, then extracorporeal irradiation of blood could have therapeutic effects. I wonder if anyone has tried such an approach.

Camitta: We have looked into the radio-sensitivity of cells which cause graft rejection. We found that, once these cells are formed and we can detect them by assays, they are resistant to radiation in vitro. Therefore I would suspect, as Dr. Thorsby has commented, that these other cells – even if they are in equilibrium – are also resistant to irradiation.

Fliedner: If you use extracorporeal irradiation of the blood, you can deliver accumulated doses of several thousands rad to the cells passing through the system. The question is, how would you define radiosensitivity. I would be satisfied if these cells did not undergo a second division. 1000 to 2000 rad do not prevent these lymphocytes to go into RNA-synthesis and even DNA-synthesis, but they will probably not go through a second mitosis.

Camitta: I would agree with that. I also should have added that some on these studies these radio-resistant cells are found only in the bone marrow and not in the peripheral blood.

Heimpel: Dr. Moore, would you believe that all of these possible immunological mechanisms by cells and serum inhibitors are really affecting the pluripotent stem cells? Or do they have a real suppressive activity which is working in the more mature compartments and prohibits differentiation of cells, analogous to B-cells differentiating into plasma cells?

Moore: It seems unlikely that one should get suppression at the level of committed cells, when after all, the primary level is at the pluripotent stem cell stage. But there is a presentiment that occured to me. In certain hemopoietic clonal stem cell disorders, like polycythemia vera, you can use the G6PDH heterocygote analysis to show that the peripheral blood elements are of one clone. One has actually shown that at the level of the CFU-C and BFU-E there are mixed populations. So, there is some block at the committed progenitor level, allowing only one clone to fully differentiate. This is a biologically fascinating phenomenon for which there is no obvious interpretation. Dr. Thomas, have you ever had an opportunity at looking at a patient with aplastic anemia, who was a G6PDH heterocygote to answer this question of clonality?

Singer: We actually had one which we found out after the graft.

Heimpel: If we consider Failkow's data on clonality, in CML and related disorders, B-cells are also of the transformed clone. Should we not find B-cell defects in some aplastic anemias, if really a rather early stem cell is affected?

Moore: The uniclonal B-cell population that he describes in CML is, as far as we know, functionally perfectly normal.

Heimpel: I am afraid we have to finish the discussion on immunological problems, because there are some other interesting problems to be discussed. I would therefore like the comment from Dr. Fliedner.

Fliedner: Dr. Thomas, we are all excited by your finding of 17 takes out of 18 patients transplanted without previous transfusions. Where do you see the effects of the pretreatment with cyclophosphamide? What do you think that drug does to the patients? What is the pathophysiological basis for this treatment? Even if there is no GvH, do you observe any other changes, e.g. in liver function tests? Were these patients in a gnotobiotic state? We have evidence, as you know, that the microbial flora would influence GvH-disease.

Thomas: Cyclophosphamide is given as an immunosuppressive agent. The basis of this action is not completely clear. Some patients were randomized in the study in a protective environment. I don't remember the number offhand. There is a very peculiar thing about these patients aside from the consistant engraftment. We have been reluctant to talk about it, because there were only 17 patients, but in fact we have seen much less GvH disease than we expected, based on our previous experience. The numbers may be so small that this is not yet significant. We would like some more numbers. This raises a whole second set of problems about sensitisation by transfusion, or foreign antigen administration, that might augment a graft versus host reaction. I think that's a little different from the question you were raising.

Fliedner: One of the reasons, why I asked about the possible action of cyclophosphamide relates to the question, how the stem cell goes into the sinusoidal system. What are the premises for a stem cell to seed in that microenvironment? Some of the pretreatments used may have not only an immune suppressive effect, but may also have effects on this capability.

Thomas: This touches on the conversation yesterday. We did not mention 20 multitransfused aplastic anemia patients, that we prepared with total body irradiation (800 rad). There are only 2 long term survivors in that group, one of them being observed for almost 6 years. But the relevant fact is that almost all of them were successfully engrafted. We stopped doing that because of other problems, but it is clear, that the preparative regimen, either with high doses of cyclophosphamide or with high dose total body irradiation does not interfere with engraftment. So the "niche" is intact after this preparative regimen, unless we are actually transplanting niches.

Heimpel: It has been mentioned yesterday, that aplastic anemia is most severe in children. The prognosis is worse without bone marrow transplantation in children, and a functioning graft is easier to obtain in children than in adults.

Camitta: I would like to clarify this: in our control studies there is no difference at all. Probability is 0.95 between survival of transplanted children and adults, but we have not enough patients to evaluate over the age of 25. The trend is that they do better without transplant but our numbers are not sufficient. Under the age of 25 transplantation is obviously better treatment.

Thomas: I would like to turn to Dr. Barrett and Dr. Gordon. It is not yet clear, what role is played by serum inhibitors. Is this something causal causing aplastic anemia, is this a product of non-hemopoietic cells? Would you like to tell us where your serum inhibitors fit in this picture, in this mosaic of different types of aplastic anemia mechanisms?

Barrett: The only way to resolve the dilemma is that there must be two different inhibitors: firstly, antibodies that are produced by multiple transfusions, which are directed against bone marrow CFU-c's. Secondly, "primary" inhibitors in a small number of patients, as evidenced by Dr. Gordon's twin patients and one of my patients who showed no absorption of the inhibition with myeloblasts. It does not have to be an antibody, it may well be the interferon that Dr. Moore has suggested.

Heimpel: Did you ever try to deplete your patients of this inhibitor by plasmaphoresis?

Barrett: No. However, we looked at the disappearance of bone marrow cells injected after transplantation, to see whether the patients that had inhibitors have a different disappearance of the CFU-c, and we found no difference.

Moore: Another possible component are the serum prostaglandine levels, because Dr. Gordon made a point that the inhibition could be counteracted by increasing the concentration of colony stimulating factor. This is exactly the sort fo balance interaction that appears to be involved with the monocytemacrophage system. It can be induced to produce prostaglandine E in response to increasing concentrations of colony stimulating factor and you have heard that aplastic patients have raised colony stimulati_ong factor.

Gordon: I think it is a very interesting suggestion, which is worth looking into. The other inferences such as lymphocytes from Dr. Barrett's experience don't correlate with the presence of inhibitors in serum.

Heimpel: We have a long lasting residual hemopoietic defect in the majority of non-transplanted patients with aplastic anemia, and we have documented well a lag period between drug exposure and manifestation of hemopoietic failure. May the BPA-data of Dr. Nissen provide an explanation for these observations? How could they be linked with the assumption of primary damage to pluripotent stem cells?

Moore: It is possible that burst promoting activity is of relevance not only in the replication of the earlier erythroid progenitor cells, but also plays a role somewhere here in the expansion of the pluripotent stem cell compartment. Some of the stem cells could have an impaired responsivness by possible loss of receptors which are required for BPA. If that is the case, then the patients who are in close remission should still have elevated levels of PBA; if not, then the levels should perhaps be in the normal range.

Nissen: In patients with almost complete remission, even having only mild thrombocytopenia, BPA was almost as high or just as high as when they had severe aplasia in contrast to cases with full autologous recovery. If there is a relapse after full recovery, BPA raises again.

Moore: That is very important. Have you been able to test the responsiveness to BPA of the marrow in aplastic anemia patients in which you get bursts?

Nissen: We did that. You can add as much BPA as you want to an aplastic marrow, without any increase of the burst incidence.

Iscove: That could even mean that they are hypersensitive. In other words, they are growing because they are more sensitive, so that one should have culture conditions where there is much less BPA than we have in plasma.

Moore: I think it is important to have a culture system suitable for dose response analysis of BPA, in order to decide whether there may be a target cell regulatory factor defect at the stem cell level, because I think that differences obviously present at the CSF and EPO level are not relevant to the disorder.

Iscove: I think when a brust appears in culture, it had to have been BPA.

Moore: Right. I am not suggesting a qualitative defect, but a different degree of responsiveness.

Iscove: We didn't discuss the connection of BPA to immune reactions yet. I have mentioned that we are seeing BPA in connection with immune responses in culture with mouse cells. We therefore discussed the possibility that the elevated serum BPA in human aplastic anemia might be the manifestation of an immune hyperreactivity. For that reason, Dr. Nissen looked at the serum in other autoimmune disorders.

Nissen: Patients we looked at did not have elevated BPA levels, but all were treated with corticosteroids, likely to suppress BPA production.

Kubanek: About 70% of your aplastic anemias had an elevated BPA, but you had no BPA in patients with pure red cell anemia, for instance. On the other hand, you stated first that patients with an immunological defect may represent a subpopulation of about 20%. I think it is hard to explain this discrepancy, if you link BPA with immune reactions.

Singer: I would like to ask Dr. Iscove if he thinks that the BPA might be similar to what Dr. Metcalf measured as CSF activity in mice undergoing GVHD since you cannot separate it from CSF activity totally.

Iscove: Yes, it looks as if all those activities are released. We have been looking at so-called primed T-cell populations. Thymocytes are injected into irradiated mice, then the spleen is brought into culture 7 days later with or without the priming antigene. In the presence of the priming antigen in vitro there is a tremendous release of PBA as well as granulocyte and macrophage colony stimulating activity. This agrees with Metcalf's suggestion.

Singer: A couple of years ago we recorded during GVHD levels of CSF which are apparently distinct from the monocyte type CSF.

Mangalik: How is BPA in patients after full hemopoietic recovery by successful allogeneic bone marrow transplantation? Cytotoxic drugs wipe the BPA out immediately. Patients who get a graft after cyclophosphamide all lose their BPA immediately.

Shadduck: I want to comment on certain physiological features of stem cell renewal systems. Stem cells have of course the potential for self-renewal as well as the potential for differentiation. One of the prime problems may be, that we are trying to force a markedly depleted stem cell population to differentiate and this may be exactly the opposite to what we should be doing. If there is some critical mass of 1% or 10% of stem cells that are necessary before differentiation should occur, then we should perhaps be attempting to put the stem cell compartment to rest, rather than trying to stimulate it with these various substances.

Schofield: I would like to comment on the transplantation studies. We are talking about the niche, which is really a conceptual thing. If a niche exists and we are going to transplant, then first of all the niche must be empty. I think that the data from Dr. Thomas, that takes occur after irradiation or after cyclophosphamide which would kill all cells, are quite relevant. One other report came out this morning: what appeared to be completely recovered aplastic anemia case showed itself under stress to be quite deficient. So the whole thing has to be driven very hard to give a normal histological appearance. I wonder in fact, if what we are seeing is incomplete filling of niches and it may be that the extra stimulators of burst formers is simply an extra factor being added to the system to drive it much harder.

Fliedner: From the whole symposion I learned that there is still some sort of a spectrum to what people in these particular series call aplastic anemia and how diagnosis relates to what one should do therapeutically and to what the outcome will be. My suggestion was that it would be worthwhile to have a sort of aplastic anemia characterisation work shop, where people actually bring typical or atypical cases.

Heimpel: Thank you for this idea. It is always a good idea if some symposium creates another.

7 Summing-up

Summing-up

E. C. Gordon-Smith

It is very difficult to sum up a meeting like this. We all recognise that the pathogenesis of aplastic anaemia has not been discovered by sitting around a table. I want to put up very briefly yet another flow diagram because it helps me in making a few "summary" remarks. This is a clinician's view of aplastic anaemia. What happens first is some event when a patient is exposed to a drug, a virus or something that is unknown. For a while at this meeting, I didn't think that you thought drugs caused aplastic anaemia. There seems to have been doubt that chloramphenical by itself was an etiological agent and much discussion about the role of viruses but I think that everybody finally agreed more or less that some event does take place. This event produces a response which, because disease follows, is either an abnormal response or a failure of a normal response. In other words, the response may be positive or negative and it might be immune or it might be chemical. Now we know that it is almost impossible to study this part of the disorder in man because there is a lag period between the event and the manifestation of the damage which takes place. The damage may be coincident with the event or it may be delayed and it may affect the stem cell or the environment. The weight of evidence and particularly that brought to light by the transplantation work of Don Thomas, suggests that it is the stem cell that is damaged but the evidence is not conclusive. This damage may be direct or indirect. By that I mean that the stem cell may be damaged directly as a result of a chemical or immune assault which took place in response to the event, or the event may have triggered a secondary response which causes the damage. We have discussed here various factors such as abnormal burst promoting activity (BPA) and the production of inhibitors which may be related to this kind of aplastic anaemia. Wherever the damage, either directly on the stem cell or on the environment, there is ultimately a failure of pluripotent stem cell function and that, of course, is what we tend to study in the kinetics of aplastic anaemia because that is what we have available in the patient.

In turn the failure of pluripotential stem cells produces quantitative and qualitative changes in the marrow. We can see the quantitative changes down the microscope, but one wonders sometimes listening to the work of people who had studied aplastic anaemia in culture systems, whether they are studying the laboratory equivalence of morphological changes rather than etiological changes. (It is perhaps significant that we do not have a biochemist here. Culture and immunology is in fashion, biochemistry is out of fashion.) I would like to emphasize one point, which a number of speakers have made, that most of these events take place in the bone marrow. Some of the consequences of the events, including the tissue response to damage and to proliferative changes, produce changes in the peripheral blood. One tries to study them there in the peripheral blood and to extrapolate back to what happens in the bone marrow. It has become clear to me, from Dr. Dexter's discussions, that one has to interpret all data – immunological, cultural and biochemical – with caution. For example, the abnormalities of enzyme systems and surface antigens of the red cells in aplastic anaemia are not thought to be etiological, they are a consequence of this qualitative change in bone marrow. One point we haven't been able to discuss much is the question of recovery. Is it clonal? Some ideas for studying this have been put forward using G6PD clones, though, of course, one needs to have a population with a balanced polymorphism. If recovery is polyclonal, maybe there is removal of an inhibition, of proliferation or differentiation which leads to recovery. The removal of the inhibition may occur with time or as a result of some new event. And then there may be environmental chages which might take place which permit previously damaged stem cells to recover.

Those are the problems – how far have we answered them, I am not sure. One of the difficulties which has been emphasised, even at this meeting, is the difficulty of agreement about diagnosis. As I sit listening to the speakers describe their patients, I sometimes think "Those aren't the patients that I see, they are talking about a different disease. My patients don't respond like this, this doesn't happen". But, of course, some of us work only in referral hospitals where only the most severe aplastic anaemias come. The important contribution of Dr. Böttiger's study is that it includes a whole population. It may well be the same disease in all cases; in diseases of other organs a wide range of severity is accepted as a feature of a common aetiology but one tends to study the severe rather than the mild cases. In aplastic anaemia there is the suspicion that always comes over that we are not diagnosing the same disease. It is a tragic blot on the study of aplastic anaemia. I hope that we do not let it remain so.

Then there is the problem of techniques and how far they are standardised in the study of aplasia. I think that Dr. te Velde's contribution shows the kind of careful consideration of bone marrow material which should be brought to the study of aplasia.

We cannot say what the pathogenesis of aplastic anaemia is but it is evident from this meeting that the prospects, as Dr. Moore said, of unravelling the mysteries are very exciting.

There are new methods, new tools available for the study of normal and abnormal bone marrow function. We should recognise that this interest in aplastic anaemia derives mainly from the work of Don Thomas and his group over the past 10 and more years. It was the feasibility of treating some patients with aplasia successfully that awakened interest in the rare disease itself. The conjunction of clinical and scientific enthusiasm created the right moment to produce a most fascinating and I hope useful symposium at the Reisensburg.

Other Volumes of Interest from this Series

Springer-Verlag
Berlin
Heidelberg
New York

H. Begemann, J. Rastetter

Atlas of Clinical Hematology

Initiated by L. Heilmeyer, H. Begemann.
With contributions on the Ultrastructure
of Blood Cells and Their Precursors by
D. Huhn and on Tropical Diseases by
W. Mohr. Translated from the German
by H. J. Hirsch. 3rd, completely revised
edition. 1979. 228 figures, 194 in color,
12 tables. XVII, 275 pages
ISBN 3-540-09404-0
Distribution rights for Japan: Maruzen
Co. Ltd., Tokyo

E. Kelemen, W. Calvo, T. M. Fliedner

Atlas of Human Hemopoietic Development

Foreword by M. Bessis. 1979. 343 figures
(204 in color), 9 tables. XIV, 266 pages
ISBN 3-540-08741-9

In Vitro Aspects of Erythropoiesis

Editor: M. J. Murphy jr. Co-Editors:
C. Peschle, A. S. Gordon, E. A. Mirand.
1978. 192 figures, 79 tables.
XIX, 280 pages
ISBN 3-540-90320-8

A. Polliack

Normal, Transformed and Leukemic Leukocytes

A Scanning Electron Microscopy Atlas.
1977. 236 figures. IX, 140 pages
ISBN 3-540-08376-6

Red Cell Rheology

Editors: M. Bessis, S. B. Shohet,
N. Mohandas. 1978. 200 figures,
36 tables. 438 pages. (Monograph edition
of "Blood Cells" Vol. 3, Issues 1–2)
ISBN 3-540-09001-0

Springer-Verlag
Berlin
Heidelberg
New York